Advances in Flavivirus Research

Special Issue Editor
Michael R. Holbrook

MDPI • Basel • Beijing • Wuhan • Barcelona • Belgrade

MDPI

Special Issue Editor
Michael R. Holbrook
Battelle Memorial Institute/NIAID Integrated Research Facility
USA

Editorial Office
MDPI AG
St. Alban-Anlage 66
Basel, Switzerland

This edition is a reprint of the Special Issue published online in the open access journal *Viruses* (ISSN 1999-4915) from 2016–2017 (available at: http://www.mdpi.com/journal/viruses/special_issues/flavivirus).

For citation purposes, cite each article independently as indicated on the article page online and as indicated below:

Author 1; Author 2. Article title. *Journal Name* **Year**, *Article number*, page range.

First Edition 2017

ISBN 978-3-03842-486-4 (Pbk)
ISBN 978-3-03842-487-1 (PDF)

Table of Contents

About the Special Issue Editor v

Michael R. Holbrook
Historical Perspectives on Flavivirus Research
Reprinted from: *Viruses* **2017**, *9*(5), 97; doi: 10.3390/v9050097 1

Adam J. Lopez-Denman and Jason M. Mackenzie
The Importance of the Nucleus during Flavivirus Replication
Reprinted from: *Viruses* **2017**, *9*(1), 14; doi: 10.3390/v9010014 20

Alan M. Watson and William B. Klimstra
T Cell-Mediated Immunity towards Yellow Fever Virus and Useful Animal Models
Reprinted from: *Viruses* **2017**, *9*(4), 77; doi: 10.3390/v9040077 31

**Julianna D. Zeidler, Lorena O. Fernandes-Siqueira, Glauce M. Barbosa
and Andrea T. Da Poian**
Non-Canonical Roles of Dengue Virus Non-Structural Proteins
Reprinted from: *Viruses* **2017**, *9*(3), 42; doi: 10.3390/v9030042 45

Monica A. McArthur
Zika Virus: Recent Advances towards the Development of Vaccines and Therapeutics
Reprinted from: *Viruses* **2017**, *9*(6), 143; doi: 10.3390/v9060143 61

Syed Lal Badshah, Abdul Naeem and Yahia Mabkhot
The New High Resolution Crystal Structure of NS2B-NS3 Protease of Zika Virus
Reprinted from: *Viruses* **2017**, *9*(1), 7; doi: 10.3390/v9010007 79

**Rodrigo Delvecchio, Luiza M. Higa, Paula Pezzuto, Ana Luiza Valadão, Patrícia P. Garcez,
Fábio L. Monteiro, Erick C. Loiola, André A. Dias, Fábio J. M. Silva, Matthew T. Aliota,
Elizabeth A. Caine, Jorge E. Osorio, Maria Bellio, David H. O'Connor, Stevens Rehen,
Renato Santana de Aguiar, Andrea Savarino, Loraine Campanati and Amilcar Tanuri**
Chloroquine, an Endocytosis Blocking Agent, Inhibits Zika Virus Infection in
Different Cell Models
Reprinted from: *Viruses* **2016**, *8*(12), 322; doi: 10.3390/v8120322 82

**Jian Yang, Huiqiang Yang, Zhushi Li, Wei Wang, Hua Lin, Lina Liu, Qianzhi Ni, Xinyu Liu,
Xianwu Zeng, Yonglin Wu and Yuhua Li**
Envelope Protein Mutations L107F and E138K Are Important for Neurovirulence Attenuation for
Japanese Encephalitis Virus SA14-14-2 Strain
Reprinted from: *Viruses* **2017**, *9*(1), 20; doi: 10.3390/v9010020 97

**Jing Zhou, Shi-Qi Wang, Jian-Chao Wei, Xiao-Min Zhang, Zhi-Can Gao, Ke Liu, Zhi-Yong Ma,
Pu-Yan Chen and Bin Zhou**
Mx Is Not Responsible for the Antiviral Activity of Interferon-α against Japanese Encephalitis Virus
Reprinted from: *Viruses* **2017**, *9*(1), 5; doi: 10.3390/v9010005 109

Jiannan Cui, Yongkun Zhao, Hualei Wang, Boning Qiu, Zengguo Cao, Qian Li, Yanbo Zhang,
Feihu Yan, Hongli Jin, Tiecheng Wang, Weiyang Sun, Na Feng, Yuwei Gao, Jing Sun,
Yanqun Wang, Stanley Perlman, Jincun Zhao, Songtao Yang and Xianzhu Xia
Equine Immunoglobulin and Equine Neutralizing F(ab')₂ Protect Mice from
West Nile Virus Infection
Reprinted from: *Viruses* **2016**, *8*(12), 332; doi: 10.3390/v8120332..125

Katherine D. Shives, Aaron R. Massey, Nicholas A. May, Thomas E. Morrison and
J. David Beckham
4EBP-Dependent Signaling Supports West Nile Virus Growth and Protein Expression
Reprinted from: *Viruses* **2016**, *8*(10), 287; doi: 10.3390/v8100287..136

Luwanika Mlera, Wessam Melik, Danielle K. Offerdahl, Eric Dahlstrom, Stephen F. Porcella
and Marshall E. Bloom
Analysis of the Langat Virus Genome in Persistent Infection of an *Ixodes scapularis* Cell Line
Reprinted from: *Viruses* **2016**, *8*(9), 252; doi: 10.3390/v8090252..157

Chenxi Li, Junyan Liu, Wulin Shaozhou, Xiaofei Bai, Qingshan Zhang, Ronghong Hua,
Jyung-Hurng Liu, Ming Liu and Yun Zhang
Epitope Identification and Application for Diagnosis of Duck Tembusu Virus Infections in Ducks
Reprinted from: *Viruses* **2016**, *8*(11), 306; doi: 10.3390/v8110306..163

Zhijiang Miao, Li Gao, Yindi Song, Ming Yang, Mi Zhang, Jincheng Lou, Yue Zhao,
Xicheng Wang, Yue Feng, Xingqi Dong and Xueshan Xia
Prevalence and Clinical Impact of Human Pegivirus-1 Infection in HIV-1-Infected Individuals in
Yunnan, China
Reprinted from: *Viruses* **2017**, *9*(2), 28; doi: 10.3390/v9020028..177

About the Special Issue Editor

Michael R. Holbrook is a Research Leader with Battelle Memorial Institute working at the NIAID Integrated Research Facility (IRF) at Ft. Detrick in Frederick, Maryland. Previously, Dr. Holbrook was an Associate Professor at the University of Texas Medical Branch, Director of the Robert E. Shope BSL-4 laboratory, and Director of the Emerging and High-Risk Pathogens Core of the Galveston National Laboratory. Dr. Holbrook has been studying flaviviruses since 1998 with a primary focus on pathogenesis of tick-borne flaviviruses and yellow fever virus. He has also evaluated structural and genetic characteristics of several flaviviruses. Since 2005, Dr. Holbrook has been working with BSL-4 viruses including Ebola, Nipah and Rift Valley fever viruses. His work at the IRF is currently focused primarily on pathogenesis and vaccine development for the Nipah virus.

viruses

MDPI

Review

Historical Perspectives on Flavivirus Research

Michael R. Holbrook

NIAID Integrated Research Facility, 8200 Research Plaza, Ft. Detrick, Frederick, MD 21702, USA;
Michael.holbrook@nih.gov; Tel.: +1-301-631-7265

Academic Editor: Eric O. Freed
Received: 9 February 2017; Accepted: 21 April 2017; Published: 30 April 2017

Abstract: The flaviviruses are small single-stranded RNA viruses that are typically transmitted by mosquito or tick vectors. These "arboviruses" are found around the world and account for a significant number of cases of human disease. The flaviviruses cause diseases ranging from mild or sub-clinical infections to lethal hemorrhagic fever or encephalitis. In many cases, survivors of neurologic flavivirus infections suffer long-term debilitating sequelae. Much like the emergence of West Nile virus in the United States in 1999, the recent emergence of Zika virus in the Americas has significantly increased the awareness of mosquito-borne viruses. The diseases caused by several flaviviruses have been recognized for decades, if not centuries. However, there is still a lot that is unknown about the flaviviruses as the recent experience with Zika virus has taught us. The objective of this review is to provide a general overview and some historical perspective on several flaviviruses that cause significant human disease. In addition, available medical countermeasures and significant gaps in our understanding of flavivirus biology are also discussed.

Keywords: flavivirus; yellow fever; West Nile; Japanese encephalitis; tick-borne encephalitis; Zika; dengue

1. Introduction

The intent of this review is to provide a broad-brush overview of the flaviviruses, some of their historical highlights and to identify significant gaps in our understanding of these very interesting viruses. The diversity in arthropod-vectors, reservoir species and diseases caused in humans is unlike any other virus family. The recent outbreaks of Zika, yellow fever and Usutu viruses should highlight the potential impact of the flaviviruses to human health while the continuing challenges of dengue, Japanese encephalitis and tick-borne encephalitis reminds us of the number of people at risk for infection with one of these viruses.

2. Origins of Flavivirus Research

In the current classification of flaviviruses (Family *Flaviviridae*; genus *Flavivirus*; type species yellow fever virus, Asibi strain) that are associated with infection of mammals, there are two main groupings of viruses: those transmitted by ticks and those transmitted by mosquitoes. The tick-borne flaviviruses are a closely related, monophyletic group consisting of a single "serocomplex", despite distinct differences in the disease caused by representative viruses. The mosquito-borne viruses are far more diverse, consisting of the Japanese encephalitis virus (JEV) serocomplex, yellow fever virus (YFV) and members of the four dengue virus (DENV) serotypes, among many others. Evaluations of phylogenetic divergence times indicate the origins of the flaviviruses go back around 100,000 years to a common ancestor with the split between mosquito- and tick-borne flaviviruses around 40,000 years ago [1]. The flaviviruses were originally grouped among the togaviruses based on early serological assessment, but were separated from the togaviruses into the family *Flaviviridae* in 1984 based on differences in structure, gene sequence and replication strategy [2]. Since that time, delineation of the viral genome, virus structure and viral biology have identified significant differences between the flaviviruses and their historical colleagues in the family *Togaviridae*. In addition, several viruses have been identified as "non-vectored" flaviviruses and a number of insect-specific flaviviruses (ISFs) have also been discovered [3]. Dual-host affiliated ISFs (dISFs) were first identified in 2004 and, to date, have only been identified in mosquitoes [3]. While the ISFs are genetically distinct from other flaviviruses, the dISFs cluster with the vertebrate associated flaviviruses suggesting a certain level of evolutionary congruence.

A number of flaviviruses are considered major human pathogens (Figure 1) and the diseases they cause have been recognized for many years. The mosquito-borne viruses can broadly be grouped as those transmitted by *Culex* spp. mosquitoes (JE serocomplex) and generally associated with neurotropic viruses, and those transmitted by *Aedes* spp. mosquitoes and more closely associated with viscerotropic or hemorrhagic disease in humans. The recent discovery that Zika virus (ZIKV), a virus transmitted by *Aedes aegypti*, can cause severe neurological disease in developing fetuses may stimulate a re-evaluation of the impact of infection by the closely related DENV or YFV on pregnant women. The tick-borne flaviviruses cause of number of significant diseases in humans that are typically associated with neurological symptoms although hemorrhagic manifestations have been documented following infection with some of these viruses. The diversity of arthropod vectors, disease characteristics and the wide geographic distribution of the flaviviruses makes these viruses especially interesting, particularly if one considers that most people throughout the world live in a flavivirus endemic region. The relative ease with which some of these viruses can be introduced into new environments should also raise concerns and highlight the need for extensive additional research on these viruses, both in the lab and in the field.

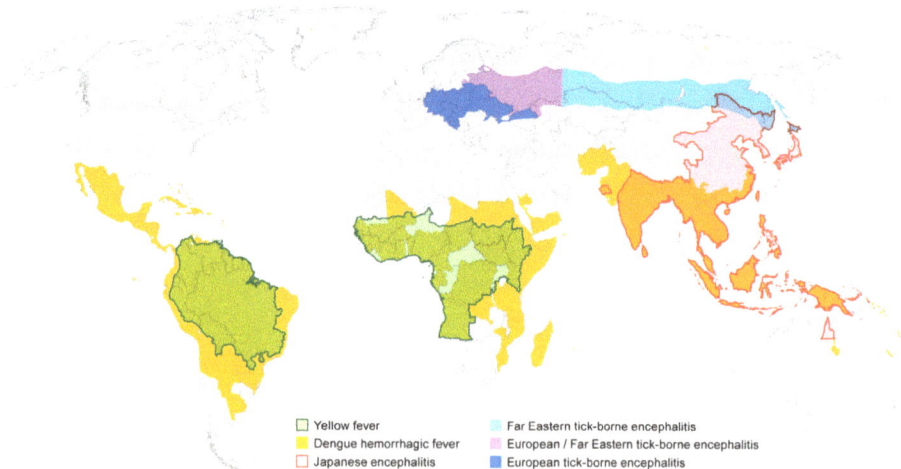

	Yellow fever		Far Eastern tick-borne encephalitis
	Dengue hemorrhagic fever		European / Far Eastern tick-borne encephalitis
	Japanese encephalitis		European tick-borne encephalitis

Figure 1. Distribution of major flaviviruses discussed in this article. Information was adapted from data and figures provided on Centers for Disease Control and Prevention (CDC) and World Health Organization (WHO) websites.

2.1. Yellow Fever

As far back as the mid-1600s the "Black Death" or "Blood Vomit" (Xekik in Mayan) was a known disease that afflicted people primarily in port cities throughout the Caribbean and in the Americas. Outbreaks of this disease were documented in cities in Europe, particularly along the Mediterranean coast where the disease was thought to be imported from Africa. Although evidence of a yellow fever (YF)-like disease was reported in Hispaniola in 1495 [4], the first documented outbreak of yellow fever was either in Barbados [5] or St. Christophe (now St. Kitts) in 1647 [4]. The outbreak in the Caribbean subsequently spread to the Yucatan peninsula in 1648 [4]. After 1648, YF spread throughout the Caribbean, including regular outbreaks in Cuba. In 1793, a significant outbreak of yellow fever hit Philadelphia and killed around 10% of the population [6]. Perhaps the largest outbreak of yellow fever in the Americas hit communities along the southern Mississippi river in 1878 between Memphis and New Orleans. This outbreak killed upwards of 20,000 people with estimates of around 120,000 cases [7]. Outbreaks of yellow fever persisted in the United States through 1905 when the final outbreak was document in New Orleans. A number of clinical descriptions of yellow fever disease have been published, including the extensive description of the Philadelphia outbreak by Benjamin Rush [6]. The cause of yellow fever was unknown and was frequently referred to as a "miasma" transmitted by foul air [7]. A very detailed account of the historical epidemiology of YF was provided by Henry Rose Carter and is available online [4].

In 1897 the Yellow Fever Commission was established to investigate the origins of YF in Cuba. Previous work by J.C. Nott and Carlos Finlay had suggested that mosquitoes may be a means of transmitting the disease YF between people [8,9]. The Yellow Fever Commission established that mosquitoes were, in fact, the vector for the disease [10]. Members of the Commission and other military "volunteers" participated in a series human infection studies to demonstrate that mosquitoes moved the disease from afflicted patients to healthy study participants, in some cases, at the cost of their lives. These studies were pivotal in proving the role of mosquitoes in the transmission of YF and validated the hypotheses of Nott and Finlay from many years earlier.

The concept of "viruses" was not understood at the time of the Yellow Fever Commission. Instead, the prevailing hypothesis was that a bacterium (*Bacillus icteroides* or *Leptospira icteroides*) was the transmissible element causing YF as some studies identified bacteria in cultures taken from YF patients. It was not until 1928 when Stokes and colleagues, in a seminal series of studies, identified a "filterable" agent as the transmissible component of YF infection that the concept of a "yellow fever virus" was understood [11].

Infection with YFV can result in disease ranging from a sub-clinical infection to severe hemorrhagic disease and death [12]. In the more severe forms of the disease, the illness is typically biphasic, progressing from an "infection" phase, through "remission" and into a period of "intoxication". The "infection" phase of disease presents as a "flu-like" illness with fever, malaise, headache and myalgia, but is complicated by hyperemia, conjunctival injection and tenderness in the liver. Many patients recover following the "infection" phase, but others progress though a brief period of remission where symptoms subside, and into the very severe period of "intoxication". This phase of the disease is characterized by hemorrhagic disease and multi-organ dysfunction with symptoms including the characteristic jaundice, nausea, vomiting and frank hemorrhagic manifestations. Terminal patients can develop neurological manifestations, including delirium, convulsions and coma. Neurological symptoms are likely due to generalized inflammatory responses and vascular leakage into the brain rather than a specific neurotropic characteristic of the virus.

The specific mechanisms of YFV induced disease are unclear. Liver dysfunction is evidenced by jaundice and significant changes in liver enzyme profiles and hemorrhagic indications are frequently apparent in extremely ill patients [13]. Unlike some viruses that cause hepatitis by stimulating an inflammatory response, YFV directly infects hepatocytes and Kupffer cells [14–16] leading to a loss of hepatocyte function and acute liver injury. YFV infection can also significantly impact the vascular endothelial cell barrier, but it is not clear whether the onset of vascular leakage is due to changes in liver physiology, inflammatory cytokine response, direct infection by YFV or by an another mechanism [13]. The loss of liver function may also lead to dysregulation of the coagulation cascade, but specific details have not been determined. There have been reports of disseminated intravascular coagulation (DIC) in YF patients and global loss of coagulation factors in YFV infected rhesus macaques has also been reported [17,18]. Much of what is known about YF pathogenesis is the result of a handful of clinical assessments, a few studies in the macaque model and extrapolation from clinical cases of dengue hemorrhagic fever, which is caused by a related virus, but is not the same disease. There are clearly a number of significant questions that need to be addressed to gain a better understanding of YFV pathogenesis.

2.2. Dengue

"Dandy Fever" and break-bone fever were described as early as the late 1700s and pandemics of what was called "dengue" were seen approximately every 50 years from the 1770s through 2005, as described in an intriguing article by Scott Halstead [19]. Halstead suggests that many of the early descriptions of "dengue" were instances of Chikungunya virus (CHIKV) infection as the disease was frequently described as having an "arthritic" component of the disease that persisted once fever had waned. Dengue, caused by DENV infection, was recognized as a disease separate from the "dengue" caused by CHIKV infection with the significant differences being DENV infection was referred to as "break-bone fever" and the presence of headaches, a rash and without the arthritic sequelae [19]. The "official" first description of dengue, or "joint fever" was by David Bylon following an outbreak in Java in 1779 [20]. In the United States, an extensive outbreak occurred in the southern part of the country in 1922 where it was estimated that 1–2 million people were impacted by this disease [21,22].

The hypothesis that YF and dengue (diseases) were transmitted in the same manner was recognized in the 1800s; Dr. William Smart noted " . . . there were those who attributed its (dengue) diffusion to a widely spread so-called 'epidemic constitution of the atmosphere', such as was at the same period maintained to be the sole cause of epidemic yellow fever" [23]. It was also noted that dengue and yellow fever occurred in the same locations and that those having the "milder fever" were not immune against developing yellow fever. Shortly after the discovery that YF was transmitted by *A. aegypti*, dengue was also shown to be transmitted by this mosquito vector [24–27] and that the transmissible agent was a "filterable agent" [28]. In what has complicated the management of dengue in years since, it was also discovered that *A. albopictus* is a vector for transmission of dengue [29].

Using "cross-immunity" and "dermal neutralization" tests in addition to "intracerebral neutralization" tests with sera from convalescent human volunteers, Sabin and Schlesinger demonstrated that there were at least two different "immunological types" of dengue virus [30]. Subsequent serological assessments, including hemagglutination and complement fixation assays, were used to further distinguish DENV serotypes and to demonstrate that the DENV were composed of four distinct virus serotypes [31–34]. While serologically distinct, viruses from each serocomplex cause similar disease in humans. Subsequent genetic analysis, initially by oligonucleotide fingerprint analysis [35] and later by partial and full genome analysis [36–38] have validated serological assessments by identifying four distinct virus genotypes that correlate with virus serotypes and that are divergent by no more than 6% at the nucleotide level [39].

People at risk for dengue disease inhabit tropical and subtropical regions around the world with an estimated 40% of the global population at risk for DENV infection and 390 million cases annually [40]. Dengue is a disease with a range of clinical presentations. In an effort to harmonize the clinical description of dengue by clinicians, in 2009 the WHO developed a classification system for dengue that graded the severity of disease based on clinical observations [41]. DENV infection is predominantly seen as an acute febrile disease that can last up to a week from onset of symptoms and may also follow a biphasic course. This disease, termed dengue fever (DF), is also characterized by headache, myalgia, lumbosacral pain and arthralgia of variable severity. Characteristic in many cases is a macular rash that appears early in the infection that may progress to a secondary rash. Hemorrhagic manifestations including petechiae and other hemorrhagic signs may occur, but are less common [27,42]. Severe dengue takes the form of dengue hemorrhagic fever (DHF) which has four severity grades, with the more severe grades (III and IV) classified as dengue shock syndrome (DSS) [43]. In DHF, patients typically have hemorrhagic manifestations, including petechial hemorrhage, whereas those that progress to DSS have evidence of mild shock with failure of the circulatory system (Grade III) or profound shock with no pulse or blood pressure (Grade IV). The development of DHF/DSS correlates with the onset of thrombocytopenia, prolonged clotting times and other characteristics of DIC. While the occurrence of DHF/DSS can occur in any DENV infection, it appears to be more frequent in secondary DENV infections, particularly in children or in newborns who are partially protected by maternal antibodies [44]. In practical application, the grading system of severe DENV infection is not as clearly defined as the classification scales imply. Subsequently, efforts are being made to improve the classification on severe dengue [45].

There have been a number of studies evaluating the role of antibody dependent enhancement (ADE) and its role in development of severe dengue disease. The prevailing hypothesis is that the presence of low-affinity, cross-reactive, non-neutralizing antibodies from a primary DENV infection (i.e., with one serotype) will enhance DENV infection with a second serotype (heterotypic). The premise is that once the non-neutralizing antibodies bind virus, they then bind a cell presenting an Fc receptor to facilitate virus entry into that cell. The fact that the virus is not "neutralized" in the antibody-virion complex allows the virus to release viral RNA and leads to a productive infection. The occurrence of ADE may be exacerbated by the occurrence of "original antigenic sin" wherein the response of T cells to secondary DENV infection may increase the potential for severe disease [46]. For more complete discussions on the role of ADE in DENV infections, please see reviews by Halstead [47] and Flipse et al. [48]. This topic is particularly pertinent at the current time as there are discussions regarding the possibility that antibodies specific for DENV may be cross-reactive for related viruses (e.g., ZIKV) leading to enhanced disease [49,50].

2.3. Japanese Encephalitis

Epidemics of encephalitis had been noted in Japan as far back as 1871. In 1924, an outbreak of encephalitis that affected 6000 people and killed 60% of those affected, gave rise to a disease called Japanese B summer encephalitis [51]. The agent causing Japanese B summer encephalitis, subsequently termed Japanese encephalitis virus (JEV), was isolated and characterized in non-human primates in 1933 [51,52] and a number of additional isolates were made in mice during an outbreak in 1935 [53]. Japanese B encephalitis was classified among the "B" type togaviruses based on serological studies. Genetic analysis of JEV genomes suggests that the virus originated in the Malay Archipelago several thousand years ago and then spread throughout Asia [54]. There are four distinct genotypes of JEV that have been circulating throughout Asia for the past 50 years. Recently, several isolates have been made of viruses representing the 5th genotype of JEV, which had previously been represented by a single isolate from 1952 [55–57].

There are an estimated three billion people who live within areas of 24 countries impacted by JEV [58]. The annual incidence of JE is around 70,000 cases with a case fatality rate estimated to be 14,000–20,500 per year [59,60]. In countries where JEV is endemic, the incidence rate is 0.6–12.6/100,000 depending upon geographic and climatic factors in addition to vaccination rates in susceptible populations [59,60].

JEV is transmitted by *Culex* spp. mosquitoes in an enzootic cycle that includes pigs and birds [52,61]. Pigs are an important component of the transmission cycle as an amplifying host as they can develop a high titer and long-lasting viremia that does not seem to have a significant health impact on these animals [61,62]. The involvement of birds in transmission of JEV is less relevant to direct transmission to humans than it is to the dissemination of the virus to new geographic areas.

Infection with JEV causes an acute non-specific febrile illness that consists of rapid onset with headache, myalgia, diarrhea and vomiting. In some patients, the disease can be complicated by neurological signs including opisthotonus, acute flaccid paralysis, convulsions, mental confusion, mask-like facies and cogwheel rigidity [63]. Severe disease can progress to severe encephalitis, meningitis, loss of conscious, coma and death. Neurological sequelae occur in about 30% of those who survive severe disease. These sequelae can include seizures, physical disabilities and cognitive deficits [59,64]. An extensive description of the clinical features of JE can be found in a book chapter by Scott Halstead [51].

2.4. Tick-Borne Encephalitis

Tick-borne encephalitis (TBE) was first recognized in 1932 as a severe neurological disease that occurred in forest workers in the far eastern Soviet Union (now Russia). In 1936 the Soviet Union established an exploratory expedition to determine the source and cause of this disease. As a result of this expedition, *Ixodes persulcatus* was identified as the vector for TBE [65]. In 1937, individual groups identified the causative agent of TBE to be a virus that was subsequently called "far-eastern encephalitis virus" [65]. A similar, but less severe, disease that was found in western Russia and Eastern Europe was called Western encephalitis [66]. Western encephalitis was also known as "biphasic milk fever" given its apparent linkage to consumption of unpasteurized milk from infected animals. The causative agent for Western encephalitis was identified during outbreaks in Czechoslovakia as a virus related to the far-eastern TBE virus (TBEV) [67]. Western encephalitis virus (subsequently known as central European encephalitis virus) is transmitted by the *Ix. ricinus* tick. Over the course of outbreak investigations, a third variant of TBEV was identified and termed the Siberian subtype of TBEV. This virus, transmitted by *Ix. persulcatus*, caused a disease that was of intermediate severity between far-eastern TBEV and its European relative. Genetic analysis of far-eastern TBEV, Siberian and central European encephalitis virus demonstrated that these viruses were closely related not only serologically and in the clinical disease they caused, but also genetically [68,69]. The three viruses are now called TBEV-FE (Far-Eastern), TBEV-Sib (Siberian) and TBEV-Eu (European) [70].

The TBEV are maintained in a life cycle that includes their tick hosts and small rodents upon which the ticks feed. While TBEV can be maintained in tick-populations by trans-stadial and trans-ovarial transmission [71], horizontal transmission via co-feeding of ticks on small mammals may also play a significant role in maintaining the virus in ticks [72]. In support of co-feeding transmission, recent studies with Powassan virus (POWV) have shown that a potential mammalian host for POWV, *Peromyscus leucopus*, did not develop disease when infected with POWV and did not develop a sufficient viremia to directly infect feeding ticks [73].

Infection with TBEV-FE can cause a very severe disease following an uneventful prodrome. The disease manifests with a rapid onset, high fever, myalgia and neurological indications including headache, photophobia and clinical evidence of encephalitis or meningitis with complications including flaccid motor neuron paralysis, ascending paralysis or hemiparesis [74,75]. The case fatality rate for TBEV-FE infections is 20–30% with many survivors having long-term neurological sequelae including paresis and atrophy of the neck and brachial plexus muscles, paresis within the lower extremities as well as poliomyelitis-like neurological sequelae [74].

Unlike TBEV-FE infections, the disease caused by TBEV-Eu can be relatively mild with a number of infections resulting in subclinical infections [76]. In those who develop clinical disease, it is typically biphasic with the first phase represented as a "flu-like" illness with fever, myalgia and malaise. In about 65% of symptomatic cases, the clinical course resolves after the first phase. For those who progress to the second phase, high fever and neurological involvement including meningitis and meningoencephalitis are typical symptoms [77]. The case fatality rate for TBEV-Eu infections is 1–2% with long-term sequelae atypical except for in older patients [75].

Infection by TBEV-Sib results in a disease that is described as intermediate between those caused by TBEV-FE and TBEV-Eu. However, a unique characteristic of this virus is that it has been associated with chronic infection in both humans and non-human primates, a complication not typically described for TBEV-FE or TBEV-Eu [76,78–82].

Since the initial discovery of TBEV, a number of related viruses causing human disease have been identified including Omsk hemorrhagic fever virus (OHFV), Kyasanur forest disease virus (KFDV), Alkhumra hemorrhagic fever virus (AHFV) and POWV. OHFV is found in a small region near Novosibirsk in Russia [83], KFDV is found in an ever-expanding range in India [84] and its closely related cousin (AHFV) is found primarily in Saudi Arabia along the coast of the Red Sea [85]. POWV has been suggested as the most ancestral member of the TBEV serocomplex [1,69] and is the only tick-borne flavivirus found in the Americas. Gritsun et al. provide a comprehensive review of TBE [76].

3. Emergence/Re-Emergence

In August 1999, the introduction of West Nile virus (WNV) into the USA [86] significantly heightened awareness of vector-borne viruses. While vector-borne viruses are endemic and common in many parts of the world, the introduction and rapid spread of WNV across the USA [87] was an awakening for many medical and government officials. The expansion of WNV distribution was not limited to the United States. West Nile virus has also spread into Central and South America, parts of Europe and Russia [88]. Since 1999, WNV has become globally distributed and causes severe disease and death worldwide. Recent reviews by Chancey et al. and Kilpatrick discussed the spread of this virus and its impact on the health of humans and wildlife [89,90].

Concurrent with the spread of WNV in the United States was the recognition of an increase in cases of POWV or Deer Tick virus (DTV) infection [91], perhaps due to increased surveillance and testing of clinical cases of encephalitis. In 2005 and 2013, cases of DENV infection were identified in the Brownsville, TX, along the southern border with Mexico [92,93], in 2009–2011, several cases of dengue were identified in south Florida [94] and in 2001 DENV was seen in Hawaii for the first time in nearly 70 years with over 120 cases identified during the outbreak [95]. A subsequent outbreak of dengue fever was reported in Hawaii in the fall of 2015 with over 100 cases confirmed [96]. It has also been reported that DENV was circulating in the area of Houston, TX in 2003–2005 [97]. The Houston metro area is home to over 6.5 million people and potential vectors for DENV are abundant.

In 2001, Usutu virus (USUV) was identified in Austria [98], the first time this virus has been found outside of the African continent. Usutu virus has subsequently spread to a number of different countries throughout Europe and serological testing suggests its presence in other countries [88,99,100]. While the presence of USUV has primarily been detected through ecological surveillance in mosquito and bird populations, symptomatic human cases of USUV infection have been found in Italy and Croatia [101]. The frequency of symptomatic infection of humans by USUV appears to be very low with only a handful of known cases reported since the discovery of this virus in 1959 [102].

The expansion of ZIKV into the Americas has again highlighted the importance of surveillance and of disease reporting and recognition. For many months, clinicians in Brazil were reporting increases in the occurrence of microcephaly in newborns and a potential correlation with ZIKV infection. These reports were met with a healthy dose of skepticism. Now we have come to realize that there is a correlation between microcephaly and ZIKV infection and that the impact of ZIKV infection is much different from other flaviviruses. A more extensive review of ZIKV is presented elsewhere in this Special Issue.

4. Vaccines

There are a number of highly efficacious vaccines for protection against flavivirus infection, including, perhaps, the best vaccine ever produced. In 1937, Max Theiler reported production of the YFV 17D vaccine [103]. This vaccine was generated by serially passing the type strain Asibi in mouse and chicken tissue to produce the attenuated and non-neurovirulent vaccine virus [104]. Since its introduction, over 600 million doses of the 17D vaccine have been delivered [104] with no documented evidence of vaccine reversion to wild-type virus. A single vaccination with the 17D virus provides rapid protection (people are considered immune 10 days post-vaccination) and potentially life-long protection [105]. Serological protection against YFV infection has been defined as having a $\log_{10}$ neutralization index of >0.7 (or a dilution titer of 1:10) [106]. While this threshold has not been empirically proven effective in the case of all flaviviruses, it is generally accepted as the minimum requirement to demonstrate efficacy for all flavivirus vaccines. The 17D vaccine played a critical role in limiting the scale of a 2016 YF outbreak in Angola and the Democratic Republic of Congo where over 18 million doses were given to stop the outbreak. A significant challenge with the 17D vaccine, however, is that it is grown in Specific Pathogen Free eggs, which limits rapid expansion of virus production. A number of cases of vaccine-related viscerotropic or neurotropic disease have been reported following vaccination with YFV 17D [107,108], but these cases appear likely due to co-morbidities impacting immune competence [107]. While the specific cause(s) of vaccine-related disease are unknown, their occurrence has inspired the development of potential vaccine candidates to replace the 17D vaccine. These include adenovirus-based [109], vaccinia-based [110] (ClinicalTrials.gov NCT02743455) DNA-based [111,112], and inactivated vaccines [113] (ClinicalTrials.gov NCT00995865).

The first JE vaccine produced and introduced in Japan in 1954 was a formalin-inactivated vaccine using mouse brain homogenates from JEV-Nakayama infected mice [114,115]. Since the initial introduction, several modifications have been made to the JEV vaccine including efforts to remove brain material, increasing purity and shifting the strain from Nakayama to Beijing-1 in some countries [114]. The mouse-brain derived vaccine was discontinued in 2011. At the current time, there are a number of JE vaccines in use. Use restrictions vary somewhat depending upon the vaccine, but in general, all appear to be effective in those vaccinated who are over one year old [64]. Two vaccines are based on the attenuated SA14-14-2 JEV, an inactivated cell culture vaccine broadly marketed as Ixiaro® or IC51 (Valneva, Vienna, Austria), and a vaccine composed of the live SA-14-14-2 virus itself that is available in China and other Asian countries [64]. An inactivated vaccine based on the Beijing-1 virus is available in Japan under the trade names JEBIK V (BIKEN, Kagawa, Japan) or ENCEVAC (Chemo-Sero Therapeutic Research, Kumamoto, Japan) [64]. A recently developed chimeric vaccine based on the YFV 17D virus back-bone and containing the prM and E protein genes from JEV SA14-14-2 (ChimeriVax-JE) (Sanofi-Pasteur, Lyon, France) is available in Thailand and Australia [64].

The first vaccine for TBEV was an inactivated virus vaccine that was produced shortly after the virus was discovered in 1937 and was used to vaccinate workers during the course of outbreaks [67]. Vaccines using cell culture systems were developed in the 1960s [116] and early 1970s. Early vaccines generated in the Soviet Union were based on a Far-eastern subtype of TBEV. The first vaccines generated in Europe were based on the European strain Neudörfl grown in primary chicken embryo cells [117] and marketed as FSME-IMMUN® (Baxter AG, Vienna, Austria) In the 1980s, a similar European vaccine was produced using the K23 strain of TBEV and is marketed as ENCEPUR® (Novartis/GlaxoSmithKline, Germany) [118]. Both European vaccines have undergone modifications over the years to improve safety and immunogenicity. Currently both the FSME-IMMUN® and ENCEPUR® have similar formulations and both are highly immunogenic and clinically effective [119,120]. Neither FSME-IMMUN®, nor ENCEPUR® are licensed for use outside of Europe despite extensive demonstration of safety and clinical efficacy. In addition to the European vaccines, Russian scientists have generated two vaccines for use against TBEV by utilizing either TBEV-FE strain Sofjin or strain 205 [119].

The development of a vaccine for DENV has been an ongoing effort for several decades. Vaccine development is complicated by the need for the vaccine to be protective against all four DENV serotypes. The potential for enhanced disease during secondary DENV infections could be exacerbated if a vaccine is not fully protective against all four serotypes. At the current time, there are several vaccines that show promise, including CYD-TDV (or Dengvaxia®) from Sanofi-Pasteur (Lyon, France). The CYD-TDV has been through two phase 3 clinical trials and has been approved for use in individuals aged 9–45 years living in endemic areas of Mexico, Brazil, the Philippines, El Salvador and Costa Rica [121,122]. The CYD-TDV vaccine contains four live chimeric viruses that consist of the YFV 17D virus backbone, but has swapped in the membrane and envelope protein genes for the individual DENV serotypes in place of those for YFV [123]. Phase 2 and phase 3 clinical trials with this vaccine have shown this vaccine to induce robust immune responses in children and adults against all four serotypes when following a three-dose vaccination schedule [124,125]. Clinical trials were carried out in DENV endemic countries. There are a number of caveats regarding participant serostatus against flaviviruses at the start of the trials and participant age that may have impacted the study [125]. Analysis of data from trials compiled through early 2016, led the WHO Strategic Advisory Group of Experts on Immunization (SAGE) to provide recommendations for the administration of the CYD-TDV vaccine [126]. These recommendations stipulate that the vaccine should only be given in populations where DENV seroprevalence is >70%, although it may be effective in populations with 50–70% seroprevalence. The vaccine should be given in a three-dose vaccination schedule and it is not recommended for children <9 years old due to an increased risk of hospitalization and severe disease. The risk to young children in endemic areas is due to the presence of existing cross-reactive antibodies of maternal origin in neonates, or virus exposure in older children. The fact that the CYD-TDV vaccine cannot be given to young children to provide complete protection against DENV infection is frustrating, but highlights an additional challenge of producing an effective vaccine for dengue. While this vaccine appears to have some challenges [127], analysis of clinical trial data suggests that routine vaccination programs utilizing the CYD-TDV could have a significant impact on the number of dengue cases in endemic areas [121]. Despite its limitations, the CYD-TDV vaccine does represent a significant step forward in the effort to reduce the impact of DENV infection.

In addition to the CYD-TDV vaccine, several other DENV vaccines are in clinical trials. A tetravalent vaccine (DENVax) developed by Takeda Vaccines (Singapore) is based on work initiated at CDC-Fort Collins a number of years ago. DENVax is similar to the CYD-TDV vaccine, except that it uses an attenuated DENV-2 virus backbone rather than YFV to generate chimeras for DENV-1, -3 and -4 and uses the authentic attenuated virus for DENV-2 [128,129]. The DENVax vaccine has been shown to be safe and immunogenic in phase 1 clinical trials [130] and is currently being evaluated in phase 2 and 3 clinical trials [131].

The TV003 vaccine (NIAID/Johns Hopkins) (Maryland, USA) is an admixture of four recombinant DENV that have mutations in the 3'-non-coding region leading to attenuation of the viruses [132]. Early studies demonstrated development of protective immunity following vaccination [133]. In a human DENV challenge trial, TV003 was shown to be protective when volunteers were challenge with an attenuated DENV-2 virus six months after vaccination [134]. One side effect of the TV003 was the development of a mild rash. More recent trials with a different admixture, TV005, have shown very promising results [135]. This formulation also seems to have resolved the issue of the post-vaccination rash seen in TV003 trials. A phase 2 trial for the TV003 vaccine is currently underway (ClinicalTrials.gov NCT02332733) while phase 1 and efficacy trials for a different formulation of this vaccine are in development (NCT02879266; NCT02873260).

In addition to the above-mentioned vaccines, there are a number of inactivated, subunit or DNA vaccines currently in development or in clinical trials. For more comprehensive reviews on DENV vaccines and the issues related to vaccine development, please see articles by Thomas and Rothman [136], Thomas [137] and McArthur et al. [124].

Major milestones in flavivirus research are highlighted in Figure 2.

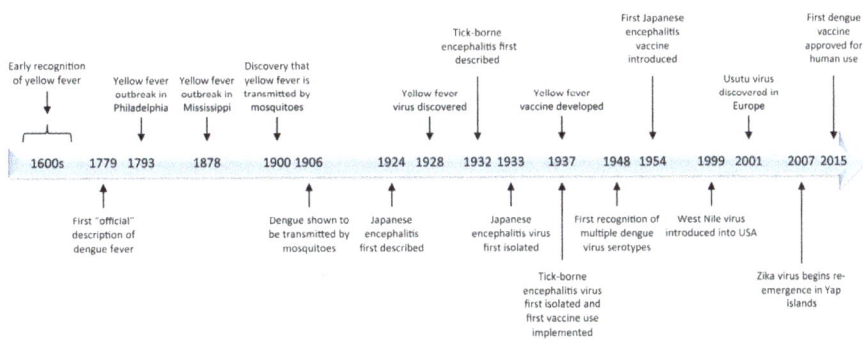

Figure 2. A time-line of historical highlights of flavivirus research.

5. Gaps in Flavivirus Research

There are a number of significant gaps in our understanding of flaviviruses and the diseases they cause. Comparatively, more is known about DENV due to its broad distribution, significant health impact and the lower biocontainment level (BSL-2) required for safely performing research with this virus. The recent global spread of ZIKV has provided impetus and financial support for research with this interesting virus which has allowed for significant advances in our understanding of ZIKV structure, pathogenesis and for vaccine development in a short period of time. However, many questions remain regarding fundamental components of flavivirus biology.

5.1. Identification of the Cognate Receptors for Flaviviruses

Given the divergent diseases caused by the different flaviviruses, it is probable that there is not a single receptor for all of the viruses, but rather a family or groups of structurally similar cell surface proteins that function as receptors for these viruses. Receptor proteins probably have different functions and cell expression patterns, given that there appears to be variable target cell types between different flaviviruses. For example, most of the mosquito-borne flaviviruses will infect both Vero and C6/36 cells, while tick-borne flaviviruses do not easily infect C6/36 cells and many replicate poorly in Vero cells. Previous work has suggested that highly glycosylated molecules, such as DC-SIGN (Dendritic Cell-Specific Intercellular adhesion molecule-3-Grabbing Non-integrin), may be receptors for flaviviruses. However, interactions with molecules such as DC-SIGN appear to be low affinity interactions rather than high affinity and are not the sole component required for viral entry [138,139]. Furthermore, published work suggests that the flaviviruses typically enter cells via receptor-mediated endocytosis, suggesting a specific receptor-ligand interaction. Interestingly, a recent study with YFV demonstrated that the 17D vaccine strain and wild-type Asibi virus utilized different mechanisms for cell entry, potentially suggesting alternative receptors for two nearly identical viruses [140]. In addition, an important point to consider is that the receptor to which a virus binds may be a critical component of the response to viral infection as many cell surface proteins have signaling functions that could impact permissiveness to viral entry replication.

5.2. An Understanding of the Role of T Cell in Mediated Immunity in Flavivirus Infection

With the exception of dengue, research efforts toward understanding of the role of T cell mediated immunity during flavivirus infection in humans has only recently become of significant interest. In addition to extensive work with dengue virus, a number of studies with WNV and TBEV have evaluated T cell responses to wild-type virus infection. The majority of work with WNV has focused primarily in mouse models [141,142]. While these studies provide considerable detail regarding mechanisms of the host response in mice, they are of limited utility as it is not clear how findings in these mice apply to the human condition. Work focused on stimulation of T cell immunity following TBEV infection has been very limited [143–145]. Studies with YFV have evaluated the importance of T cells in response to vaccination [146–148]. The broad distribution and number of DENV infections has allowed evaluation of T cell mediated immunity in patients with dengue, dengue hemorrhagic fever or dengue shock syndrome since the early 1990s. These studies have found that the role of T cell immunity during DENV infections is complicated. Some studies have suggested a limited role for T cells in primary DENV infection while others suggest that the T cell response correlates with the severity of disease [46]. As mentioned previously, the role of T cells during secondary DENV infection may contribute to enhanced disease through expansion of low avidity cross-reactive T cells. For a more thorough review of the role of T cells in DENV infection, see a review by Screaton et al. [46].

5.3. Animal Models that Faithfully Recapitulate Human Disease

The animal models routinely used for evaluation of flavivirus pathogenesis and the testing of medical countermeasures are not generally representative of disease as it manifests in humans. Mouse models for neurotropic flaviviruses develop neurological disease, but the disease is typically monophasic and lethal, depending upon virus inoculum and mouse strain, within 7–14 days. Primate models for neurotropic flaviviruses are also limited as most do not develop disease when the virus is delivered peripherally [149]. In the case of TBEV, non-human primates (NHP) are susceptible to infection, develop a disease similar to a mild case of human infection and have been shown to develop chronic infection [150].

Our understanding of YFV-induced pathogenesis is limited by a lack of useful animal models. As established by Stokes et al. [11], the rhesus macaque develops disease similar to that seen in humans, but the disease course is monophasic and of shorter duration [18]. Experimental work with non-human primates is also limited due to the cost associated with working with these animals. A hamster model for YFV was identified in the early 2000s by adapting either the Jimenez or Asibi strains of YFV by serial passage in hamsters [151,152]. Animals infected with the adapted viruses develop disease similar to that seen in humans, but this model is limited by the availability of reagents that would help to understand the immunopathology of the disease. More recently, the AG129 mouse, an interferon-deficient strain, was found to develop disease similar to humans when infected with a non-adapted virus [153,154]. Given the limited innate immune response engendered by the lack of interferon, the applicability of the AG129 mouse to understanding human disease is up for debate.

Similar to the case for YFV, there are a limited number of animal models for DENV that are useful for studying human disease. The rhesus macaque may be a faithful model for mild cases of dengue in that they have a productive infection with reasonably high viremia, but do not have the clinical picture that is typically seen in humans. The laboratory mouse is also a poor model for DENV pathogenesis. Since the early 2000s, several modified mouse models have been developed to address specific questions of DENV pathogenesis. These include animals that have some component of their innate immune response, specifically interferon-related, removed or are humanized mouse models. While not necessarily accurately representing human disease, each of these models provides information regarding aspects of DENV-induced pathophysiology that might be applicable to humans. For comprehensive reviews of the existing models for DENV infection see Sarathy et al. [155] and Chan et al. [156].

5.4. The Role of Sexual Transmission in Flavivirus Infection

The discovery that ZIKV can be transmitted sexually and that virus appears to persist in semen for some time [157,158] was novel, as flaviviruses had not previously been shown to be transmitted sexually. However, the question to be asked is whether anyone looked. Presumably, if sexual transmission had played a significant role in flavivirus dissemination, it would have been noticed. The discovery of sexual transmission in the case of ZIKV, suggests the ability of other flaviviruses to persist in immune-privileged sites and to be transmitted sexually. This phenomenon clearly needs to be investigated further.

5.5. Mother to Child Transmission

The discovery that ZIKV could be transmitted from mother to fetus and cause neurological disease in the fetus was an alarming discovery. The mechanism the virus uses to cross the placental barrier is still being evaluated, but this virus infects a number of cells in the placenta [159]. Previous studies with WNV have suggested that this virus might also be able to cross the placental barrier [160,161], demonstrating that further evaluation of this mechanism of transmission is warranted with all of the flaviviruses. Viable ZIKV has also been identified in breast milk and may be a potential source of transmission from mother to child [162]. The transmission of other flaviviruses from mother to child via breast-feeding has been suspected or documented for several other flaviviruses [163–165].

5.6. No Therapeutic Options

To date, there are no therapeutic options for the treatment of any flavivirus infection. Supportive care is the norm and has shown some success for the treatment of flavivirus infection, particularly yellow fever and DHF/DSS where fluid loss is a critical concern due to vascular leakage. The lack of treatment options is a significant problem as is evidenced by the continuing dilemma with DENV infections, the recent outbreaks of YFV in Angola and Brazil, and the continuing spread of ZIKV. The typical approach to drug screening focuses on direct antiviral effects in cultured or primary cells. In the case of most flavivirus infections, once symptoms are apparent the virus has been cleared from the blood (no viremia), is seeded in tissues and the host immune response is a significant contributor to the disease. In order to develop successful therapeutic approaches for treatment of flavivirus infections, a combination approach utilizing antivirals and host response directed countermeasures might be required.

6. Summary

The objective of this review was to provide a broad overview of flaviviruses and the diseases they cause. A secondary objective was to heighten awareness of the risks of flavivirus induced disease, the potential for continued spread of these viruses and the relative lack of understanding regarding the mechanisms these viruses use to cause disease. Diseases caused by the flaviviruses have been recognized for more than 200 years, but there is a lot of work yet to do before we have a proper understanding of these very interesting viruses.

Conflicts of Interest: The author declares no conflict of interest.

References

1. Pettersson, J.H.; Fiz-Palacios, O. Dating the origin of the genus *Flavivirus* in the light of Beringian biogeography. *J. Gen. Virol.* **2014**, *95*, 1969–1982. [CrossRef] [PubMed]
2. Westaway, E.G.; Brinton, M.A.; Gaidamovich, S.; Horzinek, M.C.; Igarashi, A.; Kaariainen, L.; Lvov, D.K.; Porterfield, J.S.; Russell, P.K.; Trent, D.W. *Flaviviridae* . *Intervirolgy* **1985**, *24*, 183–192. [CrossRef]
3. Blitvich, B.J.; Firth, A.E. Insect-specific flaviviruses: A systematic review of their discovery, host range, mode of transmission, superinfection exclusion potential and genomic organization. *Viruses* **2015**, *7*, 1927–1959. [CrossRef] [PubMed]
4. Carter, H.R. *Yellow Fever, an Epidemiologicaland Historical Study of Its Place of Origin*; William and Wilkins Company: Baltimore, MD, USA, 1931.
5. Garrison, F.H. *An Introduction to the History of Medicine*, 3rd ed.; W.B. Saunders: Philadelphia, PA, USA, 1921.
6. Rush, B. *Medical Inquiries and Observations*; Thomas Dobson: Philadelphia, PA, USA, 1796; Volume 4.
7. Bloom, K.J. *The Mississippi Valley's Great Yellow Fever Epidemic*; Louisiana State University Press: Baton Rouge LA, USA, 1993; p. 296.
8. Finlay, C.J. The mosquito hypotheitically considered as the transmission agent of yellow fever (In Spanish). *Anales de la Real Academia de Ciencias Medicas Fisicas y Naturales de la Habana* **1881**, *18*, 147–169.
9. Finlay, C.J. The mosquito hypothetically considered as an agent in the transmission of yellow fever poison. *New Orleans Med. Surg. J.* **1881**, *9*, 601–616.
10. Reed, W.; Carroll, J.; Agramonte, A.; Lazear, J.W. The etiology of yellow fever-A preliminary note. *Public Health Pap. Rep.* **1900**, *26*, 37–53. [PubMed]
11. Stokes, A.; Bauer, J.H.; Hudson, N.P. Experimental transmission of yellow fever to laboratory animals. *Am. J. Trop. Med.* **1928**, *8*, 103–164. [CrossRef]
12. Monath, T.P.; Barrett, A.D. Pathogenesis and pathophysiology of yellow fever. *Adv. Virus Res.* **2003**, *60*, 343–395. [PubMed]
13. Quaresma, J.A.; Pagliari, C.; Medeiros, D.B.; Duarte, M.I.; Vasconcelos, P.F. Immunity and immune response, pathology and pathologic changes: Progress and challenges in the immunopathology of yellow fever. *Rev. Med. Virol.* **2013**, *23*, 305–318. [CrossRef] [PubMed]
14. Woodson, S.E.; Freiberg, A.N.; Holbrook, M.R. Differential cytokine responses from primary human Kupffer cells following infection with wild-type or vaccine strain yellow fever virus. *Virology* **2011**, *412*, 188–195. [CrossRef] [PubMed]
15. Woodson, S.E.; Freiberg, A.N.; Holbrook, M.R. Coagulation factors, fibrinogen and plasminogen activator inhibitor-1, are differentially regulated by yellow fever virus infection of hepatocytes. *Virus Res.* **2013**, *175*, 155–159. [CrossRef] [PubMed]
16. Woodson, S.E.; Holbrook, M.R. Infection of hepatocytes with 17-D vaccine-strain yellow fever virus induces a strong pro-inflammatory host response. *J. Gen. Virol.* **2011**, *92*, 2262–2271. [CrossRef] [PubMed]

17. Dennis, L.H.; Reisberg, B.E.; Crosbie, J.; Crozier, D.; Conrad, M.E. The original haemorrhagic fever: Yellow fever. *Br. J. Haematol.* **1969**, *17*, 455–462. [CrossRef] [PubMed]

18. Monath, T.P.; Brinker, K.R.; Chandler, F.W.; Kemp, G.E.; Cropp, C.B. Pathophysiologic correlations in a rhesus monkey model of yellow fever with special observations on the acute necrosis of B cell areas of lymphoid tissues. *Am. J. Trop. Med. Hyg.* **1981**, *30*, 431–443. [PubMed]

19. Halstead, S.B. Reappearance of chikungunya, formerly called dengue, in the Americas. *Emerg. Infect. Dis.* **2015**, *21*, 557–561. [CrossRef] [PubMed]

20. Pepper, O.H.P. A note on David Bylon and dengue. *Ann. Med. Hist.* **1941**, *3*, 363–368.

21. Chandler, A.C.; Rice, L.M. Observations on the etiology of dengue fever. *Am. J. Trop. Med. Hyg.* **1923**, *3*, 233–262. [CrossRef]

22. Siler, J.F. Dengue Fever. In *The Georgraphy of Disease*; McKinley, E.B., Ed.; George Washington University Press: Washington, DC, USA, 1935; pp. 402–408.

23. Smart, W.R. On Dengue or Dandy Fever. *Br. Med. J.* **1877**, *1*, 382–383. [CrossRef] [PubMed]

24. Bancroft, T.L. On the etiology of dengue fever. *Australas. Med. Gaz.* **1906**, *25*, 17–18.

25. Cleland, J.B.; Bradley, B.; Macdonald, W. Further Experiments in the Etiology of Dengue Fever. *J. Hyg.* **1919**, *18*, 217–254. [CrossRef] [PubMed]

26. Cleland, J.B.; Bradley, B.; McDonald, W. On the transmission of Australian dengue by the mosquito *Stegomyia faciata*. *Med. J. Aust.* **1916**, *11*, 179–184.

27. Siler, J.F.; Hall, M.W.; Hitchens, A.P. Dengue. *Philipp. J. Sci.* **1926**, *29*, 1–304.

28. Ashburn, P.M.; Craig, C.F. Experimental investigations regarding the etiology of dengue fever. *J. Infect. Dis.* **1907**, *4*, 440–475. [CrossRef]

29. Simmons, J.S.; St. John, J.H.; Reynolds, F.H.K. Experimental studies of dengue. *Philipp. J. Sci.* **1931**, *44*, 1–247.

30. Sabin, A.B. Dengue. In *Viral and Rickettsial Infections of Man*, 1st ed.; Rivers, T.M., Ed.; J.B. Lippincott: Philadelphia, PA, USA, 1948; pp. 445–453.

31. Carey, D.E. Use of a Combined Complement-Fixing Antigen to Detect Arthropod-Borne Viral Infection. *Nature* **1963**, *200*, 1024–1025. [CrossRef] [PubMed]

32. Casals, J.; Brown, L.V. Hemagglutination with arthropod-borne viruses. *J. Exp. Med.* **1954**, *99*, 429–449. [CrossRef] [PubMed]

33. Sabin, A.B.; Young, I. A complement fixation test for dengue. *Proc. Soc. Exp. Biol. Med.* **1948**, *69*, 478–480. [CrossRef] [PubMed]

34. Sweet, B.H.; Sabin, A.B. Properties and antigenic relationships of hemagglutinins associated with the dengue viruses. *J. Immunol.* **1954**, *73*, 363–373. [PubMed]

35. Vezza, A.C.; Rosen, L.; Repik, P.; Dalrymple, J.; Bishop, D.H. Characterization of the viral RNA species of prototype dengue viruses. *Am. J. Trop. Med. Hyg.* **1980**, *29*, 643–652. [CrossRef] [PubMed]

36. Dunham, E.J.; Holmes, E.C. Inferring the timescale of dengue virus evolution under realistic models of DNA substitution. *J. Mol. Evol.* **2007**, *64*, 656–661. [CrossRef] [PubMed]

37. Tolou, H.; Couissinier-Paris, P.; Mercier, V.; Pisano, M.R.; de Lamballerie, X.; de Micco, P.; Durand, J.P. Complete genomic sequence of a dengue type 2 virus from the French West Indies. *Biochem. Biophys. Res. Commun.* **2000**, *277*, 89–92. [CrossRef] [PubMed]

38. Zanotto, P.M.; Gould, E.A.; Gao, G.F.; Harvey, P.H.; Holmes, E.C. Population dynamics of flaviviruses revealed by molecular phylogenies. *Proc. Natl. Acad. Sci. USA* **1996**, *93*, 548–553. [CrossRef] [PubMed]

39. Rico-Hesse, R. Molecular evolution and distribution of dengue viruses type 1 and 2 in nature. *Virology* **1990**, *174*, 479–493. [CrossRef]

40. WHO, Dengue and severe dengue. Available online: http://www.who.int/mediacentre/factsheets/fs117/en/ (accessed on 9 February 2017).

41. World Health Organization (WHO). *Dengue: Guidelines for Diagnosis, Treatment, Prevention and Control*; World Health Organization: Geneva, Switzerland, 2009.

42. Sabin, A.B. Research on dengue during World War II. *Am. J. Trop. Med. Hyg.* **1952**, *1*, 30–50. [PubMed]

43. World Health Organization (WHO). *Dengue Haemorrhagic Fever: Diagnosis, Prevention, Treatment and Control*; World Health Organization: Geneva, Switzerland, 1997; pp. 1–84.

44. Jain, A.; Chaturvedi, U.C. Dengue in infants: An overview. *FEMS Immunol. Med. Microbiol.* **2010**, *59*, 119–130. [CrossRef] [PubMed]

45. Horstick, O.; Martinez, E.; Guzman, M.G.; Martin, J.L.; Ranzinger, S.R. WHO dengue case classification 2009 and its usefulness in practice: An expert consensus in the Americas. *Pathog. Glob. Health* **2015**, *109*, 19–25. [CrossRef] [PubMed]

46. Screaton, G.; Mongkolsapaya, J.; Yacoub, S.; Roberts, C. New insights into the immunopathology and control of dengue virus infection. *Nat. Rev. Immunol.* **2015**, *15*, 745–759. [CrossRef] [PubMed]

47. Halstead, S.B. Dengue Antibody-Dependent Enhancement: Knowns and Unknowns. *Microbiol. Spectr.* **2014**, *2*. [CrossRef] [PubMed]

48. Flipse, J.; Wilschut, J.; Smit, J.M. Molecular mechanisms involved in antibody-dependent enhancement of dengue virus infection in humans. *Traffic* **2013**, *14*, 25–35. [CrossRef] [PubMed]

49. Castanha, P.M.; Nascimento, E.J.; Cynthia, B.; Cordeiro, M.T.; de Carvalho, O.V.; de Mendonca, L.R.; Azevedo, E.A.; Franca, R.F.; Rafael, D.; Marques, E.T., Jr. Dengue virus (DENV)-specific antibodies enhance Brazilian Zika virus (ZIKV) infection. *J. Infect. Dis.* **2017**, *215*, 781–785. [CrossRef] [PubMed]

50. Paul, L.M.; Carlin, E.R.; Jenkins, M.M.; Tan, A.L.; Barcellona, C.M.; Nicholson, C.O.; Michael, S.F.; Isern, S. Dengue virus antibodies enhance Zika virus infection. *Clin. Transl. Immunol.* **2016**, *5*, e117. [CrossRef] [PubMed]

51. Halstead, S.B.; Jacobson, J. Japanese encephalitis. *Adv. Virus Res.* **2003**, *61*, 103–138. [PubMed]

52. Rosen, L. The natural history of Japanese encephalitis virus. *Annu. Rev. Microbiol.* **1986**, *40*, 395–414. [CrossRef] [PubMed]

53. Mitamura, T.; Kitaoka, M.; Miura, T. On the geographical distribution of Japanese B encephalitis in the Far East Asia. *Jpn. Med. J.* **1950**, *3*, 257–264. [CrossRef]

54. Solomon, T.; Ni, H.; Beasley, D.W.; Ekkelenkamp, M.; Cardosa, M.J.; Barrett, A.D. Origin and evolution of Japanese encephalitis virus in southeast Asia. *J. Virol.* **2003**, *77*, 3091–3098. [CrossRef] [PubMed]

55. Li, M.H.; Fu, S.H.; Chen, W.X.; Wang, H.Y.; Guo, Y.H.; Liu, Q.Y.; Li, Y.X.; Luo, H.M.; Da, W.; Duo Ji, D.Z.; et al. Genotype V Japanese encephalitis virus is emerging. *PLoS Negl. Trop. Dis.* **2011**, *5*, e1231. [CrossRef] [PubMed]

56. Mohammed, M.A.; Galbraith, S.E.; Radford, A.D.; Dove, W.; Takasaki, T.; Kurane, I.; Solomon, T. Molecular phylogenetic and evolutionary analyses of Muar strain of Japanese encephalitis virus reveal it is the missing fifth genotype. *Infect. Genet. Evol.* **2011**, *11*, 855–862. [CrossRef] [PubMed]

57. Takhampunya, R.; Kim, H.C.; Tippayachai, B.; Kengluecha, A.; Klein, T.A.; Lee, W.J.; Grieco, J.; Evans, B.P. Emergence of Japanese encephalitis virus genotype V in the Republic of Korea. *Virol. J.* **2011**, *8*, 449. [CrossRef] [PubMed]

58. Erlanger, T.E.; Weiss, S.; Keiser, J.; Utzinger, J.; Wiedenmayer, K. Past, present, and future of Japanese encephalitis. *Emerg. Infect. Dis.* **2009**, *15*, 1–7. [CrossRef] [PubMed]

59. World Health Organization (WHO). Japanese Encephalitis Vaccines: WHO position paper, February 2015—Recommendations. *Vaccine* **2016**, *34*, 302–303.

60. Campbell, G.L.; Hills, S.L.; Fischer, M.; Jacobson, J.A.; Hoke, C.H.; Hombach, J.M.; Marfin, A.A.; Solomon, T.; Tsai, T.F.; Tsu, V.D.; et al. Estimated global incidence of Japanese encephalitis: A systematic review. *Bull. World Health Organ.* **2011**, *89*, 766–774. [CrossRef] [PubMed]

61. Weaver, S.C.; Barrett, A.D. Transmission cycles, host range, evolution and emergence of arboviral disease. *Nat. Rev. Microbiol.* **2004**, *2*, 789–801. [CrossRef] [PubMed]

62. Le Flohic, G.; Porphyre, V.; Barbazan, P.; Gonzalez, J.P. Review of climate, landscape, and viral genetics as drivers of the Japanese encephalitis virus ecology. *PLoS Negl. Trop. Dis.* **2013**, *7*, e2208. [CrossRef] [PubMed]

63. Griffiths, M.J.; Turtle, L.; Solomon, T. Japanese encephalitis virus infection. *Handb. Clin. Neurol.* **2014**, *123*, 561–576. [PubMed]

64. Yun, S.I.; Lee, Y.M. Japanese encephalitis: The virus and vaccines. *Hum. Vaccine Immunother.* **2014**, *10*, 263–279. [CrossRef] [PubMed]

65. Zilber, L.A.; Soloviev, V.D. Far eastern tick-borne spring-summer (spring) encephalitis. *Am. Rev. Sov. Med.* **1946**, *3*, 1–75.

66. Smorodintsev, A.A. Tick-borne spring-summer encephalitis. *Prog. Med. Virol.* **1958**, *1*, 210–248. [CrossRef] [PubMed]

67. Smorodintsev, A.A.; Kagan, N.W.; Levkovitsch, E.N.; Dankovskij, N.L. Experimenteller und epidemiologischer Beitrag zur activen Immunisierung gegen die Fruhling-Sommer-zecken-encephalitis. *Arch. Ges. Virusforsch.* **1941**, *2*, 1–25. [CrossRef]

68. Calisher, C.H. Antigenic classification and taxonomy of flaviviruses (family *Flaviviridae*) emphasizing a universal system for the taxonomy of viruses causing tick-borne encephalitis. *Acta Virol.* **1988**, *32*, 469–478. [PubMed]
69. Zanotto, P.M.; Gao, G.F.; Gritsun, T.; Marin, M.S.; Jiang, W.R.; Venugopal, K.; Reid, H.W.; Gould, E.A. An arbovirus cline across the northern hemisphere. *Virology* **1995**, *210*, 152–159. [CrossRef] [PubMed]
70. Fauquet, C.M.; Mayo, M.A.; Maniloff, J.; Desselberger, U.; Ball, L.A. *Virus Taxonomy: VIIIth Report of the International Committee on Taxonomy of Viruses*; Elsevier Academic Press: San Diego, CA, USA, 2005.
71. Slovak, M.; Kazimirova, M.; Siebenstichova, M.; Ustanikova, K.; Klempa, B.; Gritsun, T.; Gould, E.A.; Nuttall, P.A. Survival dynamics of tick-borne encephalitis virus in Ixodes ricinus ticks. *Ticks Tick-Borne Dis.* **2014**, *5*, 962–969. [CrossRef] [PubMed]
72. Randolph, S.E. Transmission of tick-borne pathogens between co-feeding ticks: Milan Labuda's enduring paradigm. *Ticks Tick-Borne Dis.* **2011**, *2*, 179–182. [CrossRef] [PubMed]
73. Mlera, L.; Meade-White, K.; Saturday, G.; Scott, D.; Bloom, M.E. Modeling Powassan virus infection in *Peromyscus leucopus*, a natural host. *PLoS Negl. Trop. Dis.* **2017**, *11*, e0005346. [CrossRef] [PubMed]
74. Clarke, D.H.; Casals, J. Arboviruses: Group B. In *Viral and Rickettsial Infections of Man*, 4th ed.; Horsfall, F.L., Jr., Tamm, I., Eds.; J.B. Lippincott: Philadelphia, PA, USA, 1965.
75. Ruzek, D.; Dobler, G.; Donoso Mantke, O. Tick-borne encephalitis: Pathogenesis and clinical implications. *Travel Med. Infect. Dis.* **2010**, *8*, 223–232. [CrossRef] [PubMed]
76. Gritsun, T.S.; Lashkevich, V.A.; Gould, E.A. Tick-borne encephalitis. *Antivir. Res.* **2003**, *57*, 129–146. [CrossRef]
77. Bodemann, H.H.; Pausch, J.; Schmitz, H.; Hoppe-Seyler, G. Tick-born encephalitis (ESME) as laboratory infection. *Die Med. Welt* **1977**, *28*, 1779–1781.
78. Gritsun, T.S.; Nuttall, P.A.; Gould, E.A. Tick-borne flaviviruses. *Adv. Virus Res.* **2003**, *61*, 317–371. [PubMed]
79. Pogodina, V.V.; Frolova, M.P.; Malenko, G.V.; Fokina, G.I.; Levina, L.S.; Mamonenko, L.L.; Koreshkova, G.V.; Ralf, N.M. Persistence of tick-borne encephalitis virus in monkeys. I. Features of experimental infection. *Acta Virol.* **1981**, *25*, 337–343. [PubMed]
80. Pogodina, V.V.; Levina, L.S.; Fokina, G.I.; Koreshkova, G.V.; Malenko, G.V.; Bochkova, N.G.; Rzhakhova, O.E. Persistence of tic-borne encephalitis virus in monkeys. III. Phenotypes of the persisting virus. *Acta Virol.* **1981**, *25*, 352–360. [PubMed]
81. Pogodina, V.V.; Malenko, G.V.; Fokina, G.I.; Levina, L.S.; Koreshkova, G.V.; Rzhakhova, O.E.; Bochkova, N.G.; Mamonenko, L.L. Persistence of tick-borne encephalitis virus in monkeys. II. Effectiveness of methods used for virus detection. *Acta Virol.* **1981**, *25*, 344–351. [PubMed]
82. Poponnikova, T.V. Specific clinical and epidemiological features of tick-borne encephalitis in Western Siberia. *Int. J. Med. Microbiol. IJMM* **2006**, *296* (Suppl. 40), 59–62. [CrossRef] [PubMed]
83. Ruzek, D.; Yakimenko, V.V.; Karan, L.S.; Tkachev, S.E. Omsk haemorrhagic fever. *Lancet* **2010**, *376*, 2104–2113. [CrossRef]
84. Sadanandane, C.; Elango, A.; Marja, N.; Sasidharan, P.V.; Raju, K.H.; Jambulingam, P. An outbreak of Kyasanur forest disease in the Wayanad and Malappuram districts of Kerala, India. *Ticks Tick-Borne Dis.* **2017**, *8*, 25–30. [CrossRef] [PubMed]
85. Memish, Z.A.; Fagbo, S.F.; Osman Ali, A.; AlHakeem, R.; Elnagi, F.M.; Bamgboye, E.A. Is the epidemiology of alkhurma hemorrhagic fever changing?: A three-year overview in Saudi Arabia. *PLoS ONE* **2014**, *9*, e85564. [CrossRef] [PubMed]
86. Briese, T.; Jia, X.Y.; Huang, C.; Grady, L.J.; Lipkin, W.I. Identification of a Kunjin/West Nile-like *Flavivirus* in brains of patients with New York encephalitis. *Lancet* **1999**, *354*, 1261–1262. [CrossRef]
87. Murray, K.O.; Mertens, E.; Despres, P. West Nile virus and its emergence in the United States of America. *Vet. Res.* **2010**, *41*, 67. [CrossRef] [PubMed]
88. Nikolay, B. A review of West Nile and Usutu virus co-circulation in Europe: How much do transmission cycles overlap? *Trans. R. Soc. Trop. Med. Hyg.* **2015**, *109*, 609–618. [CrossRef] [PubMed]
89. Chancey, C.; Grinev, A.; Volkova, E.; Rios, M. The global ecology and epidemiology of West Nile virus. *Biomed. Res. Int.* **2015**, *2015*, 376230. [CrossRef] [PubMed]
90. Kilpatrick, A.M. Globalization, land use, and the invasion of West Nile virus. *Science* **2011**, *334*, 323–327. [CrossRef] [PubMed]

91. Hinten, S.R.; Beckett, G.A.; Gensheimer, K.F.; Pritchard, E.; Courtney, T.M.; Sears, S.D.; Woytowicz, J.M.; Preston, D.G.; Smith, R.P., Jr.; Rand, P.W.; et al. Increased recognition of Powassan encephalitis in the United States, 1999–2005. *Vector-Borne Zoonotic Dis.* **2008**, *8*, 733–740. [CrossRef] [PubMed]

92. Ramos, M.M.; Mohammed, H.; Zielinski-Gutierrez, E.; Hayden, M.H.; Lopez, J.L.; Fournier, M.; Trujillo, A.R.; Burton, R.; Brunkard, J.M.; Anaya-Lopez, L.; et al. Epidemic dengue and dengue hemorrhagic fever at the Texas-Mexico border: Results of a household-based seroepidemiologic survey, December 2005. *Am. J. Trop. Med. Hyg.* **2008**, *78*, 364–369. [PubMed]

93. Thomas, D.L.; Santiago, G.A.; Abeyta, R.; Hinojosa, S.; Torres-Velasquez, B.; Adam, J.K.; Evert, N.; Caraballo, E.; Hunsperger, E.; Munoz-Jordan, J.L.; et al. Reemergence of dengue in Southern Texas, 2013. *Emerg. Infect. Dis.* **2016**, *22*, 1002–1007. [CrossRef] [PubMed]

94. Adalja, A.A.; Sell, T.K.; Bouri, N.; Franco, C. Lessons learned during dengue outbreaks in the United States, 2001–2011. *Emerg. Infect. Dis.* **2012**, *18*, 608–614. [CrossRef] [PubMed]

95. Effler, P.V.; Pang, L.; Kitsutani, P.; Vorndam, V.; Nakata, M.; Ayers, T.; Elm, J.; Tom, T.; Reiter, P.; Rigau-Perez, J.G.; et al. Dengue fever, Hawaii, 2001–2002. *Emerg. Infect. Dis.* **2005**, *11*, 742–749. [CrossRef] [PubMed]

96. Johnston, D.; Viray, M.; Ushiroda, J.; Whelen, A.C.; Sciulli, R.; Gose, R.; Lee, R.; Honda, E.; Park, S.Y.; Hawaii dengue response, T. Notes from the field: Outbreak of locally acquired cases of dengue fever—Hawaii, 2015. *MMWR Morb. Mortal. Wkly. Rep.* **2016**, *65*, 34–35. [CrossRef] [PubMed]

97. Murray, K.O.; Rodriguez, L.F.; Herrington, E.; Kharat, V.; Vasilakis, N.; Walker, C.; Turner, C.; Khuwaja, S.; Arafat, R.; Weaver, S.C.; et al. Identification of dengue fever cases in Houston, Texas, with evidence of autochthonous transmission between 2003 and 2005. *Vector-Borne Zoonotic Dis.* **2013**, *13*, 835–845. [CrossRef] [PubMed]

98. Weissenbock, H.; Kolodziejek, J.; Url, A.; Lussy, H.; Rebel-Bauder, B.; Nowotny, N. Emergence of Usutu virus, an African mosquito-borne Flavivirus of the Japanese encephalitis virus group, central Europe. *Emerg. Infect. Dis.* **2002**, *8*, 652–656. [CrossRef] [PubMed]

99. Ashraf, U.; Ye, J.; Ruan, X.; Wan, S.; Zhu, B.; Cao, S. Usutu virus: An emerging Flavivirus in Europe. *Viruses* **2015**, *7*, 219–238. [CrossRef] [PubMed]

100. Cadar, D.; Luhken, R.; van der Jeugd, H.; Garigliany, M.; Ziegler, U.; Keller, M.; Lahoreau, J.; Lachmann, L.; Becker, N.; Kik, M.; et al. Widespread activity of multiple lineages of Usutu virus, western Europe, 2016. *Euro Surveill* **2017**, *22*. [CrossRef] [PubMed]

101. Grottola, A.; Marcacci, M.; Tagliazucchi, S.; Gennari, W.; Di Gennaro, A.; Orsini, M.; Monaco, F.; Marchegiano, P.; Marini, V.; Meacci, M.; et al. Usutu virus infections in humans: A retrospective analysis in the municipality of Modena, Italy. *Clin. Microbiol. Infect.* **2017**, *23*, 33–37. [CrossRef] [PubMed]

102. Woodall, J.P. The viruses isolated from arthropods at the East African virus research institute in the 26 years ending December 1963. *Proc. E Afr. Acad.* **1964**, *2*, 141–146.

103. Theiler, M.; Smith, H.H. The use of yellow fever virus modified by in vitro cultivation for human immunization. *J. Exp. Med.* **1937**, *65*, 787–800. [CrossRef] [PubMed]

104. Beck, A.S.; Barrett, A.D. Current status and future prospects of yellow fever vaccines. *Expert Rev. Vaccines* **2015**, *14*, 1479–1492. [CrossRef] [PubMed]

105. Staples, J.E.; Bocchini, J.A., Jr.; Rubin, L.; Fischer, M.; Centers for Disease Control and Prevention (CDC). Yellow fever vaccine booster doses: Recommendations of the Advisory Committee on Immunization Practices, 2015. *MMWR Morb. Mortal. Wkly. Rep.* **2015**, *64*, 647–650. [PubMed]

106. Mason, R.A.; Tauraso, N.M.; Spertzel, R.O.; Ginn, R.K. Yellow fever vaccine: Direct challenge of monkeys given graded doses of 17D vaccine. *Appl. Microbiol.* **1973**, *25*, 539–544. [PubMed]

107. Seligman, S.J. Risk groups for yellow fever vaccine-associated viscerotropic disease (YEL-AVD). *Vaccine* **2014**, *32*, 5769–5775. [CrossRef] [PubMed]

108. Thomas, R.E. Yellow fever vaccine-associated viscerotropic disease: Current perspectives. *Drug Des. Dev. Ther.* **2016**, *10*, 3345–3353. [CrossRef] [PubMed]

109. Bassi, M.R.; Larsen, M.A.; Kongsgaard, M.; Rasmussen, M.; Buus, S.; Stryhn, A.; Thomsen, A.R.; Christensen, J.P. Vaccination with Replication Deficient Adenovectors Encoding YF-17D Antigens Induces Long-Lasting Protection from Severe Yellow Fever Virus Infection in Mice. *PLoS Negl. Trop. Dis.* **2016**, *10*, e0004464. [CrossRef] [PubMed]

110. Schafer, B.; Holzer, G.W.; Joachimsthaler, A.; Coulibaly, S.; Schwendinger, M.; Crowe, B.A.; Kreil, T.R.; Barrett, P.N.; Falkner, F.G. Pre-clinical efficacy and safety of experimental vaccines based on non-replicating vaccinia vectors against yellow fever. *PLoS ONE* **2011**, *6*, e24505. [CrossRef] [PubMed]

111. Maciel, M., Jr.; Cruz Fda, S.; Cordeiro, M.T.; da Motta, M.A.; Cassemiro, K.M.; Maia Rde, C.; de Figueiredo, R.C.; Galler, R.; Freire Mda, S.; August, J.T.; et al. A DNA vaccine against yellow fever virus: Development and evaluation. *PLoS Negl. Trop. Dis.* **2015**, *9*, e0003693. [CrossRef] [PubMed]

112. Tretyakova, I.; Nickols, B.; Hidajat, R.; Jokinen, J.; Lukashevich, I.S.; Pushko, P. Plasmid DNA initiates replication of yellow fever vaccine in vitro and elicits virus-specific immune response in mice. *Virology* **2014**, *468–470*, 28–35. [CrossRef] [PubMed]

113. Pereira, R.C.; Silva, A.N.; Souza, M.C.; Silva, M.V.; Neves, P.P.; Silva, A.A.; Matos, D.D.; Herrera, M.A.; Yamamura, A.M.; Freire, M.S.; et al. An inactivated yellow fever 17DD vaccine cultivated in Vero cell cultures. *Vaccine* **2015**, *33*, 4261–4268. [CrossRef] [PubMed]

114. Hoke, C.H.; Nisalak, A.; Sangawhipa, N.; Jatanasen, S.; Laorakapongse, T.; Innis, B.L.; Kotchasenee, S.; Gingrich, J.B.; Latendresse, J.; Fukai, K.; et al. Protection against Japanese encephalitis by inactivated vaccines. *N. Engl. J. Med.* **1988**, *319*, 608–614. [CrossRef] [PubMed]

115. Igarashi, H. Control of Japanese encephalitis in Japan: Immunization of humans and animals, and vector control. In *Japanese Encephalitis and West Nile Viruses*; Mackenzie, J., Barrett, A.D.T., Deubel, V., Eds.; Springer: Berlin, Germany, 2002; pp. 139–152.

116. Chumakov, M.P.; Gagarina, A.V.; Vilner, L.M.; Khanina, M.K.; Rodin, I.M.; Vasenovich, M.I.; Lakina, V.I.; Finogenova, E.V. Experience in the Experimental Production and Control of Tissue Culture Vaccine against Tick Encephalitis. *Vopr. Virusol.* **1963**, *29*, 415–420. [PubMed]

117. Kunz, C. TBE vaccination and the Austrian experience. *Vaccine* **2003**, *21*, S50–S55. [CrossRef]

118. Girgsdies, O.E.; Rosenkranz, G. Tick-borne encephalitis: Development of a paediatric vaccine. A controlled, randomized, double-blind and multicentre study. *Vaccine* **1996**, *14*, 1421–1428. [CrossRef]

119. Lehrer, A.T.; Holbrook, M.R. Tick-borne encephalitis vaccines. In *Vaccines for Biodefense and Emerging and Neglected Diseases*; Barrett, A.D.T., Stanberry, L.R., Eds.; Academic Press: London, UK, 2009; pp. 713–718.

120. Zent, O.; Broker, M. Tick-borne encephalitis vaccines: Past and present. *Expert Rev. Vaccines* **2005**, *4*, 747–755. [CrossRef] [PubMed]

121. Coudeville, L.; Baurin, N.; L'Azou, M.; Guy, B. Potential impact of dengue vaccination: Insights from two large-scale phase III trials with a tetravalent dengue vaccine. *Vaccine* **2016**, *34*, 6426–6435. [CrossRef] [PubMed]

122. Coudeville, L.; Baurin, N.; Vergu, E. Estimation of parameters related to vaccine efficacy and dengue transmission from two large phase III studies. *Vaccine* **2016**, *34*, 6417–6425. [CrossRef] [PubMed]

123. Guy, B.; Barrere, B.; Malinowski, C.; Saville, M.; Teyssou, R.; Lang, J. From research to phase III: Preclinical, industrial and clinical development of the Sanofi Pasteur tetravalent dengue vaccine. *Vaccine* **2011**, *29*, 7229–7241. [CrossRef] [PubMed]

124. McArthur, M.A.; Sztein, M.B.; Edelman, R. Dengue vaccines: Recent developments, ongoing challenges and current candidates. *Expert Rev. Vaccines* **2013**, *12*, 933–953. [CrossRef] [PubMed]

125. Scott, L.J. Tetravalent Dengue Vaccine: A Review in the Prevention of Dengue Disease. *Drugs* **2016**, *76*, 1301–1312. [CrossRef] [PubMed]

126. Anonymous. Dengue vaccine: WHO position paper—July 2016. *Wkly. Epidemiol. Rec.* **2016**, *91*, 349–364.

127. Halstead, S.B.; Russell, P.K. Protective and immunological behavior of chimeric yellow fever dengue vaccine. *Vaccine* **2016**, *34*, 1643–1647. [CrossRef] [PubMed]

128. Huang, C.Y.; Butrapet, S.; Tsuchiya, K.R.; Bhamarapravati, N.; Gubler, D.J.; Kinney, R.M. Dengue 2 PDK-53 virus as a chimeric carrier for tetravalent dengue vaccine development. *J. Virol.* **2003**, *77*, 11436–11447. [CrossRef] [PubMed]

129. Osorio, J.E.; Huang, C.Y.; Kinney, R.M.; Stinchcomb, D.T. Development of DENVax: A chimeric dengue-2 PDK-53-based tetravalent vaccine for protection against dengue fever. *Vaccine* **2011**, *29*, 7251–7260. [CrossRef] [PubMed]

130. Osorio, J.E.; Velez, I.D.; Thomson, C.; Lopez, L.; Jimenez, A.; Haller, A.A.; Silengo, S.; Scott, J.; Boroughs, K.L.; Stovall, J.L.; et al. Safety and immunogenicity of a recombinant live attenuated tetravalent dengue vaccine (DENVax) in flavivirus-naive healthy adults in Colombia: A randomised, placebo-controlled, phase 1 study. *Lancet Infect. Dis.* **2014**, *14*, 830–838. [CrossRef]

131. Osorio, J.E.; Partidos, C.D.; Wallace, D.; Stinchcomb, D.T. Development of a recombinant, chimeric tetravalent dengue vaccine candidate. *Vaccine* **2015**, *33*, 7112–7120. [CrossRef] [PubMed]

132. Whitehead, S.S. Development of TV003/TV005, a single dose, highly immunogenic live attenuated dengue vaccine; what makes this vaccine different from the Sanofi-Pasteur CYD vaccine? *Expert Rev. Vaccines* **2016**, *15*, 509–517. [CrossRef] [PubMed]

133. Lindow, J.C.; Durbin, A.P.; Whitehead, S.S.; Pierce, K.K.; Carmolli, M.P.; Kirkpatrick, B.D. Vaccination of volunteers with low-dose, live-attenuated, dengue viruses leads to serotype-specific immunologic and virologic profiles. *Vaccine* **2013**, *31*, 3347–3352. [CrossRef] [PubMed]

134. Kirkpatrick, B.D.; Whitehead, S.S.; Pierce, K.K.; Tibery, C.M.; Grier, P.L.; Hynes, N.A.; Larsson, C.J.; Sabundayo, B.P.; Talaat, K.R.; Janiak, A.; et al. The live attenuated dengue vaccine TV003 elicits complete protection against dengue in a human challenge model. *Sci. Transl. Med.* **2016**, *8*, 330ra36. [CrossRef] [PubMed]

135. Kirkpatrick, B.D.; Durbin, A.P.; Pierce, K.K.; Carmolli, M.P.; Tibery, C.M.; Grier, P.L.; Hynes, N.; Diehl, S.A.; Elwood, D.; Jarvis, A.P.; et al. Robust and Balanced Immune Responses to All 4 Dengue Virus Serotypes Following Administration of a Single Dose of a Live Attenuated Tetravalent Dengue Vaccine to Healthy, Flavivirus-Naive Adults. *J. Infect. Dis.* **2015**, *212*, 702–710. [CrossRef] [PubMed]

136. Thomas, S.J.; Rothman, A.L. Trials and Tribulations on the Path to Developing a Dengue Vaccine. *Am. J. Prev. Med.* **2015**, *49* (Suppl. 4), S334–S344. [CrossRef] [PubMed]

137. Thomas, S.J. Developing a dengue vaccine: Progress and future challenges. *Ann. N. Y. Acad. Sci.* **2014**, *1323*, 140–159. [CrossRef] [PubMed]

138. Kroschewski, H.; Allison, S.L.; Heinz, F.X.; Mandl, C.W. Role of heparan sulfate for attachment and entry of tick-borne encephalitis virus. *Virology* **2003**, *308*, 92–100. [CrossRef]

139. Liu, P.; Ridilla, M.; Patel, P.; Betts, L.; Gallichotte, E.; Shahidi, L.; Thompson, N.L.; Jacobson, K. Beyond attachment: Roles of DC-SIGN in dengue virus infection. *Traffic* **2017**, *18*, 218–231. [CrossRef] [PubMed]

140. Fernandez-Garcia, M.D.; Meertens, L.; Chazal, M.; Hafirassou, M.L.; Dejarnac, O.; Zamborlini, A.; Despres, P.; Sauvonnet, N.; Arenzana-Seisdedos, F.; Jouvenet, N.; et al. Vaccine and wild-type strains of yellow fever virus engage distinct entry mechanisms and differentially stimulate antiviral immune responses. *mBio* **2016**, *7*, e01956-15. [CrossRef] [PubMed]

141. Netland, J.; Bevan, M.J. CD8 and CD4 T cells in west nile virus immunity and pathogenesis. *Viruses* **2013**, *5*, 2573–2584. [CrossRef] [PubMed]

142. Wang, T.; Welte, T. Role of natural killer and Gamma-delta T cells in West Nile virus infection. *Viruses* **2013**, *5*, 2298–2310. [CrossRef] [PubMed]

143. Aberle, J.H.; Schwaiger, J.; Aberle, S.W.; Stiasny, K.; Scheinost, O.; Kundi, M.; Chmelik, V.; Heinz, F.X. Human CD4+ T Helper Cell Responses after Tick-Borne Encephalitis Vaccination and Infection. *PLoS ONE* **2015**, *10*, e0140545. [CrossRef] [PubMed]

144. Blom, K.; Braun, M.; Pakalniene, J.; Dailidyte, L.; Beziat, V.; Lampen, M.H.; Klingstrom, J.; Lagerqvist, N.; Kjerstadius, T.; Michaelsson, J.; et al. Specificity and dynamics of effector and memory CD8 T cell responses in human tick-borne encephalitis virus infection. *PLoS Pathog.* **2015**, *11*, e1004622. [CrossRef] [PubMed]

145. Ruzek, D.; Salat, J.; Palus, M.; Gritsun, T.S.; Gould, E.A.; Dykova, I.; Skallova, A.; Jelinek, J.; Kopecky, J.; Grubhoffer, L. CD8+ T-cells mediate immunopathology in tick-borne encephalitis. *Virology* **2009**, *384*, 1–6. [CrossRef] [PubMed]

146. Ahmed, R.; Akondy, R.S. Insights into human CD8(+) T-cell memory using the yellow fever and smallpox vaccines. *Immunol. Cell Biol.* **2011**, *89*, 340–345. [CrossRef] [PubMed]

147. Wieten, R.W.; Goorhuis, A.; Jonker, E.F.; de Bree, G.J.; de Visser, A.W.; van Genderen, P.J.; Remmerswaal, E.B.; Ten Berge, I.J.; Visser, L.G.; Grobusch, M.P.; et al. 17D yellow fever vaccine elicits comparable long-term immune responses in healthy individuals and immune-compromised patients. *J. Infect.* **2016**, *72*, 713–722. [CrossRef] [PubMed]

148. Wieten, R.W.; Jonker, E.F.; van Leeuwen, E.M.; Remmerswaal, E.B.; Ten Berge, I.J.; de Visser, A.W.; van Genderen, P.J.; Goorhuis, A.; Visser, L.G.; Grobusch, M.P.; et al. A single 17D yellow fever vaccination provides lifelong immunity; characterization of yellow-fever-specific neutralizing antibody and T-cell responses after vaccination. *PLoS ONE* **2016**, *11*, e0149871. [CrossRef] [PubMed]

149. Holbrook, M.R.; Gowen, B.B. Animal models of highly pathogenic RNA viral infections: Encephalitis viruses. *Antivir. Res.* **2008**, *78*, 69–78. [CrossRef] [PubMed]

150. Kenyon, R.H.; Rippy, M.K.; McKee, K.T., Jr.; Zack, P.M.; Peters, C.J. Infection of Macaca radiata with viruses of the tick-borne encephalitis group. *Microb. Pathog.* **1992**, *13*, 399–409. [CrossRef]

151. McArthur, M.A.; Suderman, M.T.; Mutebi, J.P.; Xiao, S.Y.; Barrett, A.D. Molecular characterization of a hamster viscerotropic strain of yellow fever virus. *J. Virol.* **2003**, *77*, 1462–1468. [CrossRef] [PubMed]

152. Tesh, R.B.; Guzman, H.; da Rosa, A.P.; Vasconcelos, P.F.; Dias, L.B.; Bunnell, J.E.; Zhang, H.; Xiao, S.Y. Experimental yellow fever virus infection in the Golden Hamster (Mesocricetus auratus). I. Virologic, biochemical, and immunologic studies. *J. Infect. Dis.* **2001**, *183*, 1431–1436. [CrossRef] [PubMed]

153. Meier, K.C.; Gardner, C.L.; Khoretonenko, M.V.; Klimstra, W.B.; Ryman, K.D. A mouse model for studying viscerotropic disease caused by yellow fever virus infection. *PLoS Pathog.* **2009**, *5*, e1000614. [CrossRef] [PubMed]

154. Watson, A.M.; Lam, L.K.; Klimstra, W.B.; Ryman, K.D. The 17D-204 vaccine strain-induced protection against virulent yellow fever virus is mediated by humoral immunity and CD4+ but not CD8+ T cells. *PLoS Pathog.* **2016**, *12*, e1005786. [CrossRef] [PubMed]

155. Sarathy, V.V.; Milligan, G.N.; Bourne, N.; Barrett, A.D. Mouse models of dengue virus infection for vaccine testing. *Vaccine* **2015**, *33*, 7051–7060. [CrossRef] [PubMed]

156. Chan, K.W.; Watanabe, S.; Kavishna, R.; Alonso, S.; Vasudevan, S.G. Animal models for studying dengue pathogenesis and therapy. *Antivir. Res.* **2015**, *123*, 5–14. [CrossRef] [PubMed]

157. Foy, B.D.; Kobylinski, K.C.; Chilson Foy, J.L.; Blitvich, B.J.; Travassos da Rosa, A.; Haddow, A.D.; Lanciotti, R.S.; Tesh, R.B. Probable non-vector-borne transmission of Zika virus, Colorado, USA. *Emerg. Infect. Dis.* **2011**, *17*, 880–882. [CrossRef] [PubMed]

158. Moreira, J.; Peixoto, T.M.; Machado de Siqueira, A.; Lamas, C.C. Sexually acquired Zika virus: A systematic review. *Clin. Microbiol. Infect.* **2017**. [CrossRef] [PubMed]

159. Tabata, T.; Petitt, M.; Puerta-Guardo, H.; Michlmayr, D.; Wang, C.; Fang-Hoover, J.; Harris, E.; Pereira, L. Zika virus targets different primary human placental cells, suggesting two routes for vertical transmission. *Cell Host Microbe* **2016**, *20*, 155–166. [CrossRef] [PubMed]

160. Alpert, S.G.; Fergerson, J.; Noel, L.P. Intrauterine West Nile virus: Ocular and systemic findings. *Am. J. Ophthalmol.* **2003**, *136*, 733–735. [CrossRef]

161. O'Leary, D.R.; Kuhn, S.; Kniss, K.L.; Hinckley, A.F.; Rasmussen, S.A.; Pape, W.J.; Kightlinger, L.K.; Beecham, B.D.; Miller, T.K.; Neitzel, D.F.; et al. Birth outcomes following West Nile Virus infection of pregnant women in the United States: 2003–2004. *Pediatrics* **2006**, *117*, e537–e545. [CrossRef] [PubMed]

162. Colt, S.; Garcia-Casal, M.N.; Pena-Rosas, J.P.; Finkelstein, J.L.; Rayco-Solon, P.; Weise Prinzo, Z.C.; Mehta, S. Transmission of Zika virus through breast milk and other breastfeeding-related bodily-fluids: A systematic review. *PLoS Negl. Trop. Dis.* **2017**, *11*, e0005528. [CrossRef] [PubMed]

163. Anonymous. From the Centers for Disease Control and Prevention. Possible West Nile virus transmission to an infant through breast-feeding—Michigan, 2002. *JAMA* **2002**, *288*, 1976–1977.

164. Barthel, A.; Gourinat, A.-C.; Cazorla, C.; Joubert, C.; Dupont-Rouzeyrol, M.; Descloux, E. Breast milk as a possible route of vertical transmission of dengue virus? *Clin. Infect. Dis.* **2013**, *57*, 415–417. [CrossRef] [PubMed]

165. Kuhn, S.; Twele-Montecinos, L.; MacDonald, J.; Webster, P.; Law, B. Case report: Probable transmission of vaccine strain of yellow fever virus to an infant via breast milk. *CMAJ* **2011**, *183*, E243–E245. [CrossRef] [PubMed]

viruses

MDPI

Review

The IMPORTance of the Nucleus during Flavivirus Replication

Adam J. Lopez-Denman [1,2] and Jason M. Mackenzie [1,*]

[1] Department of Microbiology & Immunology, Peter Doherty Institute for Infection and Immunity,
     University of Melbourne, Victoria 3010, Australia; adam.lopez@unimelb.edu.au
[2] Department of Physiology, Anatomy & Microbiology, La Trobe University, Melbourne 3086, Australia
*   Correspondence: jason.mackenzie@unimelb.edu.au; Tel.: +61-3-9035-8376

Academic Editor: Michael Holbrook
Received: 16 December 2016; Accepted: 12 January 2017; Published: 19 January 2017

Abstract: Flaviviruses are a large group of arboviruses of significant medical concern worldwide. With outbreaks a common occurrence, the need for efficient viral control is required more than ever. It is well understood that flaviviruses modulate the composition and structure of membranes in the cytoplasm that are crucial for efficient replication and evading immune detection. As the flavivirus genome consists of positive sense RNA, replication can occur wholly within the cytoplasm. What is becoming more evident is that some viral proteins also have the ability to translocate to the nucleus, with potential roles in replication and immune system perturbation. In this review, we discuss the current understanding of flavivirus nuclear localisation, and the function it has during flavivirus infection. We also describe—while closely related—the functional differences between similar viral proteins in their nuclear translocation.

Keywords: flavivirus; nucleus; transport; karyopherins; nuclear pore complex; nuclear localisation sequence

1. Introduction

Flaviviruses are a large group of diverse arboviruses within the *Flaviviridae* family of viruses that include human pathogens of great medical and social burden worldwide, with billions of US Dollars being spent annually on patient care and vector control. Key members of this group include dengue virus (DENV), Japanese encephalitis virus (JEV), yellow fever virus (YFV), West Nile virus (WNV), and Zika virus (ZIKV). Flaviviruses are seemingly simple viruses, consisting of a positive sense single stranded RNA genome of ~11 kb (with 5′ and 3′ UTRs) that encodes for a single polyprotein that is co- and post-translationally modified into mature proteins. The genes encode for three structural proteins: capsid (C), pre-membrane (prM) and envelope (E); and seven non-structural (NS) proteins, many of which are multifunctional, responsible for viral replication and immune system perturbation [1] (Figure 1).

Flavivirus replication has been extensively studied, and shown to involve the considerable redistribution of host membranes to facilitate efficient replication within the cytoplasm [2–5]. This membrane redistribution allows for the creation of virally induced structures known as replication complexes (RCs) which act as a replicative niche for the virus in the cytoplasm, and help in shielding from immune detection [5–8]. The membranous origin of the RC is complex and differs between viruses, but accumulation within the perinuclear region seems common between them [2]. Throughout their replication, Flaviviruses are associated with three distinct structures; the rough endoplasmic reticulum (rER), convoluted membranes/paracrystalline arrays (CM/PC), and vesicle packets (VPs). Due to the nature of the flavivirus positive sense RNA genome, replication can fully occur within the cytoplasm, with no nuclear intermediaries seemingly necessary. Although considered cytoplasmic viruses, in recent years it has come to the attention of many groups that some viral proteins localise to the nucleus during infection or overexpression. When nuclear transport has been inhibited, either by use of mutagenesis or chemical inhibition, there have been many observable effects (with varying levels of significance) to viral replication and modulating host transcription. This has led us, and many groups, to believe that there may be a previously unrealised significant role of viral protein nuclear localisation amongst the flaviviruses.

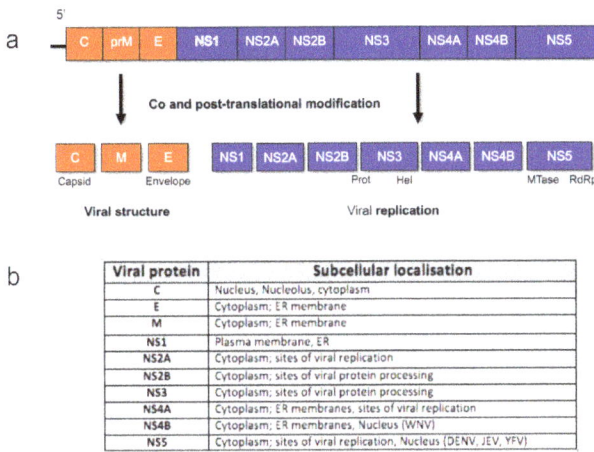

Figure 1. Representation of Flavivirus genome organisation. (**a**) Ten viral genes encode for three structural and seven non-structural (NS) viral proteins that are co- and post-translationally modified into mature viral proteins by host factors and the viral NS2B-3 protease complex [2]; (**b**) Viral protein subcellular localisation is generally cytoplasmic or associated with the endoplasmic reticulum (ER) membrane, due to the localisation of replication, but in specific viruses (listed), proteins such as NS5 can also localise to the nucleus.

2. Nuclear Trafficking

The nuclear envelope is the physical barrier between the nucleoplasm and the cytoplasm, consisting of a double membrane that is studded with large proteinaceous channels known as nuclear pore complexes (NPCs). Both sides of the nuclear membrane exhibit different protein composition; the outer membrane is contiguous with the ER and, like the ER, is associated with ribosomes undergoing active protein synthesis. The inner membrane appears to have more structural functions and is involved with anchoring chromatin and attaching the nuclear lamina [9–11]. Bidirectional nuclear translocation is a tightly regulated selective process occurring through the NPC that segregates the contents of the cytoplasm and the nucleoplasm [9] (Figure 2). This is an essential function that facilitates the transport of an enormous range of proteins, ribonucleoproteins, and various macromolecules, while simultaneously preventing non-specific molecules from passing through.

The NPC itself is a very large structure (>120 mDa) made up of ~30 different proteins (known as nucleoporins) in multiple copies, leading to its huge structure of ~400 polypeptides [9]. Free diffusion of small proteins/macromolecules can occur if they are less than 50 kDa in size, anything larger than this needs to undergo active transport to gain access to the nucleus. To facilitate nuclear transport, a specific targeting sequencing and a transport molecular (that recognises the receptor) is required. Nuclear bound trafficking is guided through specific amino acid sequences known as a nuclear localisation sequence (NLS) (typically Arg and Lys), whilst cytoplasmic trafficking is guided by a nuclear export sequence (NES). There are variations of composition of these sequences, with the classical being simply five contiguous basic amino acids (KKKRK), but often they can also be bi-partite sequences containing two clusters of basic amino acids with a spacer ~10 amino acids in between [12–14]. Although these two examples have been defined, there is likely a myriad of others that work in similar ways that are currently unknown.

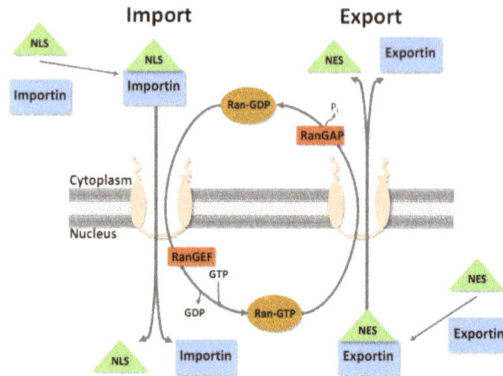

Figure 2. Simplified nucleocytoplasmic protein transport. The nuclear envelope is the physical barrier between the nucleus and the cytoplasm, and is studded with channels known as nuclear pore complexes. For proteins to undergo nucleocytoplasmic transport, they require a nuclear localisation sequence (NLS) or nuclear export sequence (NES), and its corresponding karyopherin. Transport then occurs through the nuclear pore, with directionality determined through interactions with the Ran cycle. When in the cytoplasm, Ran is in a GDP bound state and can facilitate nuclear import. Once in the nucleus, the GDP gets converted to GTP and now can facilitate export [9].

As mentioned, specific molecules are required for nucleocytoplasmic transport to occur, typically by a member of the large superfamily of proteins known as karyopherins (also known as importins, exportins and transportins) [15,16]. Karyopherins mediate transport by binding to an NLS/NES and interact directly with components of the NPC. To further regulate this process, karyopherins can bind to members of the Ras-like GTPase Ran [17]. This helps provide the directionality of nuclear transport by creating a Ran gradient across the NPC. In the cytoplasm, Ran is in a GDP bound state, whilst in the nucleus it is bound to GTP. This is determined by the Ran GTPase activating protein (RanGAP) in the cytoplasm and the Ran guanine nucleotide exchange factor (RanGEF) in the nucleus [17]. In the cytoplasm, a karyopherin binds to its associated NLS/Cargo and forms a protein-complex to facilitate transport to the NPC. As it traverses, it interacts with the RanGTP in the nucleus, releasing the protein and is recycled back into the cytoplasm. Export occurs in an analogous way, with RanGTP affinity being essential within the nucleus, losing binding as it travels through the NPC towards the cytoplasm (as shown in Figure 2).

3. Flavivirus Protein Nuclear Localisation

3.1. Core/Capsid

The flavivirus structural protein C has been observed to localise to the nucleus across many flavivirus species (DENV, Kunjin virus (KUNV), JEV), and in some cases C nuclear translocation has been shown to enhance viral replication [18–20]. C is a small (~12 kb) basic protein with charged N- and C-termini, separated by a hydrophobic region. Although the hydrophobicity remains similar amongst the flaviviruses, the amino acid homology can vary greatly between them. The structure of C has eluded to the RNA binding and membrane interaction functions that have been described during infection [21,22]. During the flavivirus lifecycle, multiple C subunits bind together to form the viral nucleocapsid, encasing the RNA genome, allowing for further pathogenesis. It has also been shown in WNV and tick-borne encephalitis virus (TBEV) that C is permissive to major deletions, with large segments able to be deleted without greatly affecting the ability to encapsulate the viral genome, although they are less infective [23–25]. Even so, the underlying role for C nuclear localisation remains poorly understood, however there have been some insights into its function. The introduction of mutations within a putative NLS within JEV C protein were shown to inhibit its nuclear localisation, leading to a reduced amount of infectious virus produced in mammalian cells [20,24,26]. The consequence of this was explained in two potential ways: (i) the introduction of these mutations had an effect on the efficiency of virus particle production; or (ii) a potential role for C in the nucleus not being completed. This is particularly interesting, as it was also observed that similar mutations in TBEV C protein, which has not been shown to go to the nucleus, increased the production of subviral particles, however reduced their infectivity. Interactions with C and the host cell nuclear components are varied between the flaviviruses but include DENV—with Death Domain Associated Protein (DAXX) interaction and Fas-mediated apoptosis [27,28]; JEV-host protein B23, allowing for increased replication [29]; WNV-binding of DDX56, for efficient assembly of infectious particles [30]; and sequestration of HDM2, influencing apoptosis [31].

Even though the nuclear function of flavivirus C is not fully elucidated, groups have mapped predicted NLSs within the protein. There are three proposed NLSs located within DENV at amino acids 6–9, 73–76 and a bipartite sequence found at 85–100 [19,32,33]. Additionally, it has also been identified that in WNV, nuclear import is mediated by importin-alpha/beta, specifically through interactions with amino acids 42–43 and 85–101 [34]. JEV, whilst similar again, showed different NLS functionality compared to the others, with an active NLS being found with amino acids 42 and 43 [20]. It is also speculated that as nuclear transport is often regulated by phosphorylation [35], there may be a phosphorylated C that gains nuclear access and it was shown that specific phosphorylation on residues 83, 99 and 100 influences nuclear localisation (by altering binding efficacy to importin-alpha) [36]. Whilst C protein performs a similar role in viral capsid assembly across the flaviviruses, it is also clear that there are differences in alternate functionality between them.

3.2. Non-Structural (NS) Protein 5 (NS5)

Of the NS proteins, NS5 has garnered much interest regarding its nuclear localisation over the past few years. NS5 is a large (~103 kd), highly conserved enzymatic protein with both RNA-Dependent RNA-polymerase (RdRp) and methyltransferase abilities, and has been shown to be intimately involved with viral RNA replication within the RCs. Thus, it is especially interesting that such a large viral protein, particularly one with RdRp activity, may also have a role within the nucleus. During YFV, JEV and DENV infections, there have been demonstrated cases of NS5's ability to localise to the nucleus [37–39]. As NS5 is too large to passively diffuse, the only way it can access the nucleus is through active transport through the NPC. The two enzymatic regions of NS5 are separated by a 37-amino acid linker region (NS5 residue 369–405) in DENV. This region has been thoroughly interrogated, and found to contain two functional NLSs, which have been shown to function through both importin alpha-beta, and importin beta transport [38,40–42]. Unsurprisingly, as NS5 is highly conserved amongst the flaviviruses, the NLSs also share this conservation [41]. When nuclear localiation of NS5 was inhibited through mutagenesis of these NLSs, a significant decrease in the amount of virus was observed [41]. Why this is occuring is not fully understood, but may be due to DENV NS5's ability to modulate IL-8 [43,44], but could also be due to (i) potential structural conformational changes in NS5 that occur from introducing these mutations; or (ii) altering NS5 binding affinity with viral or host proteins during replication [45]. Interestingly, when alanine mutations were introduced to the C terminal of the bipartite NLS, NS5 nuclear translocation was immensely impaired but either did not effect replication, or had only a minimal effect [45]; this further suggests that perhaps nuclear NS5 is not needed, or that only a small portion of cellular NS5 is required within the nucleus. Through the use of the CRM1-inhibitor leptomycin B (LMB), it has also been observed that DENV NS5 can accumulate within the nucleus. This leads to the belief that NS5 also has a NES for nuclear export, and may lead to futher understanding of the function of NS5 upon exit from the nucleus. Rawlinson et al. were able to identify a functional NES on residues 327–343, and confirmed its interaction with CRM1 [43]. While the two aformentioned NLSs have been the most highly characterised, recently a motif within the C-terminal of NS5 has also been shown to be a determinant of subscullular localsiation and RNA binding for DENV 1–4 [46].

Whilst NS5 sub-nuclear localisation has not been fully elucidated, it has recently been shown in DENV that it can localise to the nucleolus in a pH dependent manner [47]. This gives further insight into NS5 nuclear functionality in manipulating cellular responses, and shows that it could be using a multifunctional approach within the nucleus. The majority of this research was performed in DENV-2, however the nuclear localisation of NS5 in all strains of DENV has also been observed, with notable differences in localisation and nuclear accumulation between them [48,49], with serotypes 2 and 3 accummulating more within the nucleus, whereas 1 and 4 do not. Although there are no direct links between NS5 localicalisation and pathogenesis, the exact role that NS5 plays in the nucleus may impact on disease outcome and thus would imply that a targeted inhibition of nuclear transport may only be appropriate for particular strains.

However, it should be noted that a recent paper has revealed that the nuclear localisation of DENV NS5 is neither critical for replication nor the ability of DENV NS5 to impede immune evasion via the degradation of STAT-1 [45]. Via mutation, the investigators showed that mutation of the putative NLS actually impaired the RdRp activity of the mutants' NS5 species, thus impacting virus replication. Interestingly, the NLS mutants that impaired virus replication could not be rescued in trans, indicating that the links between nuclear NS5 with replication and intracellular survival are not that straight forward and that interactions with nuclear and/or cytoplasmic host factors may be quite complex, influencing the different functions of the flavivirus NS5 protein. These observations are additionally confirmed by the Vasudevan laboratory [46] where they show that the nuclear localisation of some DENV serotype NS5 proteins is mediated via a C-terminal motif and that removal or mutation of this motif in DENV1 results in a completely cytoplasmically localised virus that replicates efficiently. Interestingly, our own reassessment of KUNV NS5 indicates that the protein shuttles very quickly between the nucleus and the cytoplasm, yet does not accummulate within the nucleus as observed for some DENV NS5 proteins (unpublished results). Thus, the exact role for nuclear NS5 in the replication cycle or for immune evasion is still unknown, particuarly if these roles are conserved amongst the different flaviviruses.

In light of these observations, the realisation and significance of nuclear localised NS5 has now been seen to be important in the flavivirus lifecycle, and as such its import pathway has become a target for drug therapy. Through the use of highthroughput screening, thousands of compounds have been able to be analysed for their effectiveness in inhibiting nuclear transport [50]. This has led to the discovery of several compounds that have some effectiveness against blocking DENV infection [51,52]. Ivermectin was one such drug that has been shown to inhibit importin alpha/beta1 nuclear import, the main pathway for NS5 nuclear localisation. When treated with Ivermectin, the inhibition of NS5 and Importin interactions, (therefore stopping NS5 nuclear import) leads to a significant decrease in the amount of DENV present, leading to its potential use as an antivial compound [48,51]. The drawback of chemical inhibitors is that they often have unspecific effects, and could inadvertantly target normal host functions required for cell survival [53]. The benefit of these compounds is that they could have broad antiviral activity against the actions of the flavivirus core, NS5 and potentially NS4B, in the nucleus of infected cells.

3.3. NS Proteins NS3 and NS4B and RNA Synthesis

The C and NS5 proteins have received the most attention with respect to nuclear localisation during virus replication and transient expression, however, some reports have also suggested that the flavivirus NS3 and NS4B proteins and RNA synthesis also appear within the nucleus [18,54]. These observations have not been so well defined and are somewhat difficult to interpret, especially the possibility of viral RNA synthesis in the nucleus, as both more sophisticated and specific labelling methods have not consistently detected either viral single-stranded RNA (ssRNA) or double-stranded RNA (dsRNA) within the nucleus [55,56]. In addition, higher resolution confocal microscopy and immunogold labelling of cryopreserved material also reveals a confinement of viral RNA replication in the cytoplasm [8,55–59]. A recent analysis of DENV NS4B has revealed a new role for the protein in mitochondrial localisation but limited protein was observed in the nucleus [60,61]. Overall, the localisation of the NS3, NS4B and viral RNA within the nucleus has not been consistently observed across the flavivirus genus, and thus requires further investigation.

4. Sequestration of Host Nuclear Components

While this review is focused on flavivirus protein nuclear trafficking, it is also important to note that proteins encoded within the flavivirus genome also have the ability to sequester cellular nuclear proteins, and inhibit host protein nuclear import. Type I interferon (IFN) is produced in response to viral infections, and plays an important role in innate immunity towards viral infection, with the ability to activate sevaral antiviral pathways [62–64]. It is a common theme for flavivirus NS5 to inhibit the IFN mediated Janus kinase/signal transducers and activators of transcription (JAK-STAT) signalling [65–68], aiding in virus replication. DEAD-box RNA helicases are a large group of proteins, with varied function, that reside in both the nucleus and cytoplasm that have also been shown to be redistributed to sites of flavivirus replication. These have varied functions, but many have been identified as cofactors for replication, or assisting with the antivial response [30,69–73], demonstraing that their sequestration could be essential for flavivirus replication. DDX3 cellular localisaiton has been shown to be both cytoplasmic and nuclear, but is redistributed to sites of JEV replication in the cytoplasm with interactions in binding 5′ and 3′ UTRs, and with NS5 and NS3—proving to be necessary for JEV infection [72]. DDX21 is nuclear localised in uninfected cells, but is redistributed to the cytoplasm during DENV infection, where it has a role in directly inhibiting viruses, and through other intermediaries activating the innate immune response [70,74]. Once in the cytoplasm, DENV NS2B/3 degrades DDX21, aiding viral replication. DDX56 has been shown to redistribute from its normal nucleolar localisation, to sites of WNV replication in the ER [30,69]. This relocalisation interacts with the WNV capsid, with the helicase activity of DDX56 having a role in correct infectious particle assembly [73].

Synthesis of both positive and negative stranded RNA in flaviviruses is a complex process that requires the interaction of viral RNA secondary structures with host nuclear components to facilitate efficient replication (Table 1). These interactions occur at varying stages and have roles in RNA synthesis, genome circularization, and sequestering cytoplasmic proteins [75–77]. During DENV infections, a clear distribution of polypyrimidine-tract-binding (PTB) protein from the nucleus to the cytoplasm has been suggested to play a positive role in viral replication, with siRNA knockdown of PTB inhibiting replication, and overexpression stimulating replication [78]. La autoantigen (La) is a ribonucleoprotein found in the nucleus of healthy cells [79]; it has been shown to bind to both the 5′ and 3′ flavivirus UTRs, with the potential role of stabilising RNA loop structure, and/or recruiting NS5 and NS3 [76]. Flaviviruses have also been shown to interact with TIAR/TIA1, another set of proteins that are shuttled from the nucleus to the cytoplasm during cell stress. They have both been shown to interact with WNV 3′ UTR, aiding with viral replication [80]. The additional advantage of WNV sequestration is that TIAR/TIA1 are also associated with the formation of stress granules (SG) and, by inhibiting SG formation, may prevent priming of the innate immune response [80,81].

Although it is yet to be elucidated exactly which viral proteins mediate the sequestration of nuclear proteins within the cytoplasm, it is clear that many flaviviruses have evolved to specifically restrict the intra-nuclear shuttling of many host proteins that aid in the replication of the viral genome but equally restrict host cell immune sensing and communication. It is also yet to be determined if the different flaviviruses selectively bind cellular factors aiding in intracellular replication or globally restrict nuclear trafficking. One tends to favour the former as some flavivirus proteins shuttle in and out of the nucleus and import, rather than export, appears to be a critical requirement for flavivirus replication in cells.

Table 1. Sequestered host nuclear factors (adapted from [82]).

Virus	Host Factor	Role in Sequestration
Flavi	La	IRES conformation, translation-replication switching, genome circularization
Flavi	PTB	IRES conformation, eIF4G1 recruitment to IRES, genome circularization
DENV	RNA Helicase A	Genome circularization protein bridge, promote RNA replication
Flavi/WNV	Tia1, TIAR	Promotes + strand RNA synthesis
Flavi/DENV	Tudor-DN/p-100	Promotes RNA synthesis

IRES, internal ribosomal entry site; La, La autoantigen; PTB, polypyrimidine-tract-binding; DENV, dengue virus; WNV, West Nile virus.

5. Future Considerations

From the current literature, we now have some basic insight into the role that the nucleus, and viral protein nuclear localisation has during flavivirus replication. As the flavivirus lifecycle has been mainly observed within the cytoplasm, nuclear transport of flavivirus components has been an underrepresented field of study. Of the flaviviruses, DENV has been the main focus of research, but it is evident that nuclear localisation is occurring amongst other flaviviruses. It is also interesting to note that while the corresponding proteins (i.e., C and NS5) have been observed trafficking into the nucleus across flaviviruses, they may have different functions/roles between related viruses. The targeted inhibition of nuclear transport is also a potential therapeutic target for positive stranded RNA virus infection that has not been robustly interrogated. With the knowledge of nuclear transport being more significant than previously thought, therapies could be developed to target viral nuclear transport mechanisms to combat disease. Finally, most of these studies have been completed in mammalian cells. As a key part of the viral life cycle occurs within arthropod vectors, it would be advantageous to pursue the role, if any, of nuclear localisation within them.

Acknowledgments: Our research is supported by a Project Grant (No. 1004619) to J.M.M. from the National Health and Medical Research Council of Australia.

Author Contributions: A.J.L.-D. and J.M.M. wrote the paper.

Conflicts of Interest: The authors declare no conflict of interest. The founding sponsors had no role in the design or in the writing of the manuscript, and in the decision to publish the manuscript.

References

1. Chambers, T.J.; Hahn, C.S.; Galler, R.; Rice, C.M. Flavivirus genome organization, expression, and replication. *Annu. Rev. Microbiol.* **1990**, *44*, 649–688. [CrossRef] [PubMed]
2. Mackenzie, J. Wrapping Things up about Virus RNA Replication. *Traffic* **2005**, *6*, 967–977. [CrossRef] [PubMed]
3. Westaway, E.G.; Mackenzie, J.M.; Khromykh, A.A. Replication and gene function in Kunjin virus. *Curr. Top. Microbiol. Immunol.* **2002**, *267*, 323–351. [PubMed]
4. Leary, K.; Blair, C.D. Sequential events in the morphogenesis of Japanese encephalitis virus. *J. Ultrastruct. Res.* **1980**, *72*, 123–129. [CrossRef]
5. Gillespie, L.K.; Hoenen, A.; Morgan, G.; Mackenzie, J.M. The endoplasmic reticulum provides the membrane platform for biogenesis of the flavivirus replication complex. *J. Virol.* **2010**, *84*, 10438–10447. [CrossRef] [PubMed]

6. Uchil, P.D.; Satchidanandam, V. Architecture of the flaviviral replication complex: Protease, nuclease, and detergents reveal encasement within double-layered membrane compartments. *J. Biol. Chem.* **2003**, *278*, 24388–24398. [CrossRef] [PubMed]
7. Mackenzie, J.M.; Jones, M.K.; Westaway, E.G. Markers for trans-Golgi membranes and the intermediate compartment localize to induced membranes with distinct replication functions in flavivirus-infected cells. *J. Virol.* **1999**, *73*, 9555–9567. [PubMed]
8. Welsch, S.; Miller, S.; Romero-Brey, I.; Merz, A.; Bleck, C.K.E.; Walther, P.; Fuller, S.D.; Antony, C.; Krijnse-Locker, J.; Bartenschlager, R. Composition and Three-Dimensional Architecture of the Dengue Virus Replication and Assembly Sites. *Cell Host Microbe* **2009**, *5*, 365–375. [CrossRef] [PubMed]
9. Wente, S.R.; Rout, M.P. The nuclear pore complex and nuclear transport. *Cold Spring Harb. Perspect. Biol.* **2010**, *10*, a000562. [CrossRef] [PubMed]
10. Grossman, E.; Medalia, O.; Zwerger, M. Functional architecture of the nuclear pore complex. *Annu. Rev. Biophys.* **2012**, *41*, 557–584. [CrossRef] [PubMed]
11. Hinshaw, J.E.; Carragher, B.O.; Milligan, R.A. Architecture and design of the nuclear pore complex. *Cell* **1992**, *69*, 1133–1141. [CrossRef]
12. Dingwall, C.; Robbins, J.; Dilworth, S.M.; Roberts, B.; Richardson, W.D. The nucleoplasmin nuclear location sequence is larger and more complex than that of SV-40 large T antigen. *J. Cell Biol.* **1988**, *107*, 841–849. [CrossRef] [PubMed]
13. Goldfarb, D.S.; Gariepy, J.; Schoolnik, G.; Kornberg, R.D. Synthetic peptides as nuclear localization signals. *Nature* **1986**, *322*, 641–644. [CrossRef] [PubMed]
14. Dingwall, C.; Laskey, R.A. Nuclear import: A tale of two sites. *Curr. Biol.* **1998**, *8*, R922–R924. [CrossRef]
15. Pemberton, L.F.; Paschal, B.M. Mechanisms of receptor-mediated nuclear import and nuclear export. *Traffic* **2005**, *6*, 187–198.
16. Yuh, M.C.; Blobel, G. Karyopherins and nuclear import. *Curr. Opin. Struct. Biol.* **2001**, *11*, 703–715.
17. Macara, I.G. Transport into and out of the nucleus. *Microbiol. Mol. Biol. Rev.* **2001**, *65*, 570–594. [CrossRef] [PubMed]
18. Westaway, E.G.; Khromykh, A.A.; Kenney, M.T.; Mackenzie, J.M.; Jones, M.K. Proteins C and NS4B of the flavivirus Kunjin translocate independently into the nucleus. *Virology* **1997**, *234*, 31–41. [CrossRef] [PubMed]
19. Bulich, R.; Aaskov, J.G. Nuclear localization of dengue 2 virus core protein detected with monoclonal antibodies. *J. Gen. Virol.* **1992**, *73*, 2999–3003. [CrossRef] [PubMed]
20. Mori, Y.; Okabayashi, T.; Yamashita, T.; Zhao, Z.; Wakita, T.; Yasui, K.; Hasebe, F.; Tadano, M.; Konishi, E.; Moriishi, K.; et al. Nuclear Localization of Japanese Encephalitis Virus Core Protein Enhances Viral Replication. *J. Virol.* **2005**, *79*, 3448–3458. [CrossRef] [PubMed]
21. Dokland, T.; Walsh, M.; Mackenzie, J.M.; Khromykh, A.A.; Ee, K.H.; Wang, S. West Nile Virus Core Protein. *Structure* **2004**, *12*, 1157–1163. [CrossRef] [PubMed]
22. Jones, C.T.; Ma, L.; Burgner, J.W.; Groesch, T.D.; Post, C.B.; Kuhn, R.J. Flavivirus Capsid Is a Dimeric Alpha-Helical Protein. *J. Virol.* **2003**, *77*, 7143–7149. [CrossRef] [PubMed]
23. Schlick, P.; Kofler, R.M.; Schittl, B.; Taucher, C.; Nagy, E.; Meinke, A.; Mandl, C.W. Characterization of West Nile virus live vaccine candidates attenuated by capsid deletion mutations. *Vaccine* **2010**, *28*, 5903–5909. [CrossRef] [PubMed]
24. Kofler, R.M.; Heinz, F.X.; Mandl, C.W. Capsid protein C of tick-borne encephalitis virus tolerates large internal deletions and is a favorable target for attenuation of virulence. *J. Virol.* **2002**, *76*, 3534–3543. [CrossRef] [PubMed]
25. Khromykh, A.A.; Westaway, E.G. Subgenomic replicons of the flavivirus Kunjin: Construction and applications. *J. Virol.* **1997**, *71*, 1497–1505. [PubMed]
26. Kofler, R.M.; Leitner, A.; O'Riordain, G.; Heinz, F.X.; Mandl, C.W. Spontaneous mutations restore the viability of tick-borne encephalitis virus mutants with large deletions in protein C. *J. Virol.* **2003**, *77*, 443–451. [CrossRef] [PubMed]
27. Netsawang, J.; Noisakran, S.; Puttikhunt, C.; Kasinrerk, W.; Wongwiwat, W.; Malasit, P.; Yenchitsomanus, P.; Limjindaporn, T. Nuclear localization of dengue virus capsid protein is required for DAXX interaction and apoptosis. *Virus Res.* **2010**, *147*, 275–283. [CrossRef] [PubMed]

28. Limjindaporn, T.; Netsawang, J.; Noisakran, S.; Thiemmeca, S.; Wongwiwat, W.; Sudsaward, S.; Avirutnan, P.; Puttikhunt, C.; Kasinrerk, W.; Sriburi, R.; et al. Sensitization to Fas-mediated apoptosis by dengue virus capsid protein. *Biochem. Biophys. Res. Commun.* **2007**, *362*, 334–339. [CrossRef] [PubMed]
29. Tsuda, Y.; Mori, Y.; Abe, T.; Yamashita, T.; Okamoto, T.; Ichimura, T.; Moriishi, K.; Matsuura, Y. Nucleolar protein B23 interacts with Japanese encephalitis virus core protein and participates in viral replication. *Microbiol. Immunol.* **2006**, *50*, 225–234. [CrossRef] [PubMed]
30. Xu, Z.; Anderson, R.; Hobman, T.C. The capsid-binding nucleolar helicase DDX56 is important for infectivity of West Nile virus. *J. Virol.* **2011**, *85*, 5571–5580. [CrossRef] [PubMed]
31. Yang, M.R.; Lee, S.R.; Oh, W.; Lee, E.W.; Yeh, J.Y.; Nah, J.J.; Joo, Y.S.; Shin, J.; Lee, H.W.; Pyo, S.; et al. West Nile virus capsid protein induces p53-mediated apoptosis via the sequestration of HDM2 to the nucleolus. *Cell. Microbiol.* **2008**, *10*, 165–176. [CrossRef] [PubMed]
32. Wang, S.H.; Syu, W.J.; Huang, K.J.; Lei, H.Y.; Yao, C.W.; King, C.C.; Hu, S.T. Intracellular localization and determination of a nuclear localization signal of the core protein of dengue virus. *J. Gen. Virol.* **2002**, *83*, 3093–3102. [CrossRef] [PubMed]
33. Sangiambut, S.; Keelapang, P.; Aaskov, J.; Puttikhunt, C.; Kasinrerk, W.; Malasit, P.; Sittisombut, N. Multiple regions in dengue virus capsid protein contribute to nuclear localization during virus infection. *J. Gen. Virol.* **2008**, *89*, 1254–1264. [CrossRef] [PubMed]
34. Bhuvanakantham, R.; Chong, M.K.; Ng, M.L. Specific interaction of capsid protein and importin-a/b influences West Nile virus production. *Biochem. Biophys. Res. Commun.* **2009**, *389*, 63–69. [CrossRef] [PubMed]
35. Jans, D.A.; Hübner, S. Regulation of protein transport to the nucleus: Central role of phosphorylation. *Physiol. Rev.* **1996**, *76*, 651–685. [PubMed]
36. Bhuvanakantham, R.; Cheong, Y.K.; Ng, M.L. West Nile virus capsid protein interaction with importin and HDM2 protein is regulated by protein kinase C-mediated phosphorylation. *Microbes Infect.* **2010**, *12*, 615–625. [CrossRef] [PubMed]
37. Edward, Z.; Takegami, T. Localization and functions of Japanese encephalitis virus nonstructural proteins NS3 and NS5 for viral RNA synthesis in the infected cells. *Microbiol. Immunol.* **1993**, *37*, 239–243. [CrossRef] [PubMed]
38. Forwood, J.K.; Brooks, A.; Briggs, L.J.; Xiao, C.Y.; Jans, D.A.; Vasudevan, S.G. The 37-amino-acid interdomain of dengue virus NS5 protein contains a functional NLS and inhibitory CK2 site. *Biochem. Biophys. Res. Commun.* **1999**, *257*, 731–737. [CrossRef] [PubMed]
39. Buckley, A.; Gaidamovich, S.; Turchinskaya, A.; Gould, E.A. Monoclonal antibodies identify the NS5 yellow fever virus non-structural protein in the nuclei of infected cells. *J. Gen. Virol.* **1992**, *73*, 1125–1130. [CrossRef] [PubMed]
40. Johansson, M.; Brooks, A.J.; Jans, D.A.; Vasudevan, S.G. A small region of the dengue virus-encoded RNA-dependent RNA polymerase, NS5, confers interaction with both the nuclear transport receptor importin-β and the viral helicase, NS3. *J. Gen. Virol.* **2001**, *82*, 735–745. [CrossRef] [PubMed]
41. Pryor, M.J.; Rawlinson, S.M.; Butcher, R.E.; Barton, C.L.; Waterhouse, T.A.; Vasudevan, S.G.; Bardin, P.G.; Wright, P.J.; Jans, D.A.; Davidson, A.D. Nuclear Localization of Dengue Virus Nonstructural Protein 5 Through Its Importin α/β-Recognized Nuclear Localization Sequences is Integral to Viral Infection. *Traffic* **2007**, *8*, 795–807. [CrossRef] [PubMed]
42. Brooks, A.J.; Johansson, M.; John, A.V.; Xu, Y.; Jans, D.A.; Vasudevan, S.G. The interdomain region of dengue NS5 protein that binds to the viral helicase NS3 contains independently functional importin-β1 and importin α/β-recognized nuclear localization signals. *J. Biol. Chem.* **2002**, *277*, 36399–36407. [CrossRef] [PubMed]
43. Rawlinson, S.M.; Pryor, M.J.; Wright, P.J.; Jans, D.A. CRM1-mediated nuclear export of dengue virus RNA polymerase NS5 modulates interleukin-8 induction and virus production. *J. Biol. Chem.* **2009**, *284*, 15589–15597. [CrossRef] [PubMed]
44. Medin, C.L.; Fitzgerald, K.A.; Alan, L.; Rothman, A.L. Dengue Virus Nonstructural Protein NS5 Induces Interleukin-8 Transcription and Secretion. *J. Virol.* **2005**, *79*, 11053–11061. [CrossRef] [PubMed]
45. Kumar, A.; Bühler, S.; Selisko, B.; Davidson, A.; Mulder, K.; Canard, B.; Miller, S.; Bartenschlager, R. Nuclear localization of dengue virus nonstructural protein 5 does not strictly correlate with efficient viral RNA replication and inhibition of type I interferon signaling. *J. Virol.* **2013**, *87*, 4545–4557. [CrossRef] [PubMed]

46. Tay, M.Y.F.; Smith, K.; Ng, I.H.W.; Chan, K.W.K.; Zhao, Y.; Ooi, E.E.; Lescar, J.; Luo, D.; Jans, D.A.; Forwood, J.K.; et al. The C-terminal 18 Amino Acid Region of Dengue Virus NS5 Regulates its Subcellular Localization and Contains a Conserved Arginine Residue Essential for Infectious Virus Production. *PLoS Pathog.* **2016**, *12*, e1005886. [CrossRef] [PubMed]

47. Fraser, J.E.; Rawlinson, S.M.; Heaton, S.M.; Jans, D.A. Dynamic nucleolar targeting of dengue virus polymerase NS5 in response to extracellular pH. *J. Virol.* **2016**, *90*, 5797–5807. [CrossRef] [PubMed]

48. Tay, M.Y.F.; Fraser, J.E.; Chan, W.K.K.; Moreland, N.J.; Rathore, A.P.; Wang, C.; Vasudevan, S.G.; Jans, D.A. Nuclear localization of dengue virus (DENV) 1–4 non-structural protein 5; protection against all 4 DENV serotypes by the inhibitor Ivermectin. *Antivir. Res.* **2013**, *99*, 301–306. [CrossRef] [PubMed]

49. Hannemann, H.; Sung, P.Y.; Chiu, H.C.; Yousuf, A.; Bird, J.; Lim, S.P.; Davidson, A.D. Serotype-specific differences in dengue virus non-structural protein 5 nuclear localization. *J. Biol. Chem.* **2013**, *288*, 22621–22635. [CrossRef] [PubMed]

50. Wagstaff, K.M.; Rawlinson, S.M.; Hearps, A.C.; Jans, D.A. An AlphaScreen®-based assay for high-throughput screening for specific inhibitors of nuclear import. *J. Biomol. Screen.* **2011**, *16*, 192–200. [CrossRef] [PubMed]

51. Wagstaff, K.M.; Sivakumaran, H.; Heaton, S.M.; Harrich, D.; Jans, D.A. Ivermectin is a specific inhibitor of importin α/β-mediated nuclear import able to inhibit replication of HIV-1 and dengue virus. *Biochem. J.* **2012**, *443*, 851–856. [CrossRef] [PubMed]

52. Fraser, J.E.; Watanabe, S.; Wang, C.; Chan, W.K.K.; Maher, B.; Lopez-Denman, A.; Hick, C.; Wagstaff, K.M.; Mackenzie, J.M.; Sexton, P.M.; et al. A nuclear transport inhibitor that modulates the unfolded protein response and provides in vivo protection against lethal dengue virus infection. *J. Infect. Dis.* **2014**, *210*, 1780–1791. [CrossRef] [PubMed]

53. Faustino, R.S.; Nelson, T.J.; Terzic, A.; Perez-Terzic, C. Nuclear transport: Target for therapy. *Clin. Pharmacol. Ther.* **2007**, *81*, 880–886. [CrossRef] [PubMed]

54. Uchil, P.D.; Kumar, A.V.A.; Satchidanandam, V. Nuclear Localization of Flavivirus RNA Synthesis in Infected Cells. *J. Virol.* **2006**, *80*, 5451–5464. [CrossRef] [PubMed]

55. Westaway, E.G.; Mackenzie, J.M.; Kenney, M.T.; Jones, M.K.; Khromykh, A.A. Ultrastructure of Kunjin virus-infected cells: Colocalization of NS1 and NS3 with double-stranded RNA, and of NS2B with NS3, in virus-induced membrane structures. *J. Virol.* **1997**, *71*, 6650–6661. [PubMed]

56. Westaway, E.G.; Khromykh, A.A.; Mackenzie, J.M. Nascent flavivirus RNA colocalized in situ with double-stranded RNA in stable replication complexes. *Virology* **1999**, *258*, 108–117. [CrossRef] [PubMed]

57. Mackenzie, J.M.; Jones, M.K.; Young, P.R. Immunolocalization of the dengue virus nonstructural glycoprotein NS1 suggests a role in viral RNA replication. *Virology* **1996**, *220*, 232–240. [CrossRef] [PubMed]

58. Kopek, B.G.; Perkins, G.; Miller, D.J.; Ellisman, M.H.; Ahlquist, P. Three-dimensional analysis of a viral RNA replication complex reveals a virus-induced mini-organelle. *PLoS Biol.* **2007**, *5*, e220. [CrossRef] [PubMed]

59. Overby, A.K.; Popov, V.L.; Niedrig, M.; Weber, F. Tick-borne encephalitis virus delays interferon induction and hides its double-stranded RNA in intracellular membrane vesicles. *J. Virol.* **2010**, *84*, 8470–8483. [CrossRef] [PubMed]

60. Barbier, V.; Lang, D.; Valois, S.; Rothman, A.L.; Medin, C.L. Dengue virus induces mitochondrial elongation through impairment of Drp1-triggered mitochondrial fission. *Virology* **2016**, *500*, 149–160. [CrossRef] [PubMed]

61. Chatel-Chaix, L.; Cortese, M.; Romero-Brey, I.; Bender, S.; Neufeldt, C.J.; Fischl, W.; Scaturro, P.; Schieber, N.; Schwab, Y.; Fischer, B.; et al. Dengue Virus Perturbs Mitochondrial Morphodynamics to Dampen Innate Immune Responses. *Cell Host Microbe* **2016**, *20*, 342–356. [CrossRef] [PubMed]

62. Randall, R.E.; Goodbourn, S. Interferons and viruses: An interplay between induction, signalling, antiviral responses and virus countermeasures. *J. Gen. Virol.* **2008**, *89*, 1–47. [CrossRef] [PubMed]

63. Levy, D.E.; García-Sastre, A. The virus battles: IFN induction of the antiviral state and mechanisms of viral evasion. *Cytokine Growth Factor Rev.* **2001**, *12*, 143–156. [CrossRef]

64. Fensterl, V.; Sen, G.C. Interferons and viral infections. *BioFactors* **2009**, *35*, 14–20. [CrossRef] [PubMed]

65. Laurent-Rolle, M.; Boer, E.F.; Lubick, K.J.; Wolfinbarger, J.B.; Carmody, A.B.; Rockx, B.; Liu, W.; Ashour, J.; Shupert, W.L.; Holbrook, M.R.; et al. The NS5 protein of the virulent West Nile virus NY99 strain is a potent antagonist of type I interferon-mediated JAK-STAT signaling. *J. Virol.* **2010**, *84*, 3503–3515. [CrossRef] [PubMed]

66. Ashour, J.; Laurent-Rolle, M.; Shi, P.Y.; García-Sastre, A. NS5 of dengue virus mediates STAT2 binding and degradation. *J. Virol.* **2009**, *83*, 5408–5418. [CrossRef] [PubMed]
67. Lin, R.J.; Chang, B.L.; Yu, H.P.; Liao, C.L.; Lin, Y.L. Blocking of interferon-induced Jak-Stat signaling by Japanese encephalitis virus NS5 through a protein tyrosine phosphatase-mediated mechanism. *J. Virol.* **2006**, *80*, 5908–5918. [CrossRef] [PubMed]
68. Mazzon, M.; Jones, M.; Davidson, A.; Chain, B.; Jacobs, M. Dengue virus NS5 inhibits interferon-alpha signaling by blocking signal transducer and activator of transcription 2 phosphorylation. *J. Infect. Dis.* **2009**, *200*, 1261–1270. [CrossRef] [PubMed]
69. Reid, C.R.; Hobman, T.C. The nucleolar helicase DDX56 redistributes to West Nile virus assembly sites. *Virology* **2017**, *500*, 169–177. [CrossRef] [PubMed]
70. Dong, Y.; Ye, W.; Yang, J.; Han, P.; Wang, Y.; Ye, C.; Weng, D.; Zhang, F.; Xu, Z.; Lei, Y. DDX21 translocates from nucleus to cytoplasm and stimulates the innate immune response due to dengue virus infection. *Biochem. Biophys. Res. Commun.* **2016**, *473*, 648–653. [CrossRef] [PubMed]
71. Lin, C.W.; Cheng, C.W.; Yang, T.C.; Li, S.W.; Cheng, M.H.; Wan, L.; Lin, Y.J.; Lai, C.H.; Lin, W.Y.; Kao, M.C. Interferon antagonist function of Japanese encephalitis virus NS4A and its interaction with DEAD-box RNA helicase DDX42. *Virus Res.* **2008**, *137*, 49–55. [CrossRef] [PubMed]
72. Li, C.; Ge, L.L.; Li, P.P.; Wang, Y.; Dai, J.J.; Sun, M.X.; Huang, L.; Shen, Z.Q.; Hu, X.C.; Ishag, H.; et al. Cellular DDX3 regulates Japanese encephalitis virus replication by interacting with viral un-translated regions. *Virology* **2014**, *449*, 70–81. [CrossRef] [PubMed]
73. Xu, Z.; Hobman, T.C. The helicase activity of DDX56 is required for its role in assembly of infectious West Nile virus particles. *Virology* **2012**, *433*, 226–235. [CrossRef] [PubMed]
74. Zhang, Z.; Kim, T.; Bao, M.; Facchinetti, V.; Jung, S.Y.; Ghaffari, A.A.; Qin, J.; Cheng, G.; Liu, Y.J. DDX1, DDX21, and DHX36 Helicases Form a Complex with the Adaptor Molecule TRIF to Sense dsRNA in Dendritic Cells. *Immunity* **2011**, *34*, 866–878. [CrossRef] [PubMed]
75. De Nova-Ocampo, M.; Villegas-Sepúlveda, N.; del Angel, R.M. Translation elongation factor-1alpha, La, and PTB interact with the 3′ untranslated region of dengue 4 virus RNA. *Virology* **2002**, *295*, 337–347. [CrossRef] [PubMed]
76. García-Montalvo, B.M.; Medina, F.; Del Angel, R.M. La protein binds to NS5 and NS3 and to the 5′ and 3′ ends of Dengue 4 virus RNA. *Virus Res.* **2004**, *102*, 141–150. [CrossRef] [PubMed]
77. Paranjape, S.M.; Harris, E. Y box-binding protein-1 binds to the dengue virus 3′-untranslated region and mediates antiviral effects. *J. Biol. Chem.* **2007**, *282*, 30497–30508. [CrossRef] [PubMed]
78. Agis-Juárez, R.A.; Galván, I.; Medina, F.; Daikoku, T.; Padmanabhan, R.; Ludert, J.E.; del Angel, R.M. Polypyrimidine tract-binding protein is relocated to the cytoplasm and is required during dengue virus infection in Vero cells. *J. Gen. Virol.* **2009**, *90*, 2893–2901. [CrossRef] [PubMed]
79. Van Venrooij, W.J.; Pruijn, G.J. Ribonucleoprotein complexes as autoantigens. *Curr. Opin. Immunol.* **1995**, *7*, 819–824. [CrossRef]
80. Li, W.; Kedersha, N.; Anderson, P.; Emara, M.; Swiderek, K.M.; Moreno, G.T.; Brinton, M.A. Cell proteins TIA-1 and TIAR interact with the 3′stem-loop of the West Nile virus complementary minus-strand RNA and facilitate virus replication. *J. Virol.* **2002**, *76*, 11989. [CrossRef] [PubMed]
81. Emara, M.M.; Liu, H.; Davis, W.G.; Brinton, M.A. Mutation of mapped TIA-1/TIAR binding sites in the 3′ terminal stem-loop of West Nile virus minus-strand RNA in an infectious clone negatively affects genomic RNA amplification. *J. Virol.* **2008**, *82*, 10657–10670. [CrossRef] [PubMed]
82. Lloyd, R.E. Nuclear proteins hijacked by mammalian cytoplasmic plus strand RNA viruses. *Virology* **2015**, *479*, 457–474. [CrossRef] [PubMed]

viruses

MDPI

Review

T Cell-Mediated Immunity towards Yellow Fever Virus and Useful Animal Models

Alan M. Watson * and William B. Klimstra

Center for Vaccine Research, Departments of Microbiology and Molecular Genetics, and Immunology, University of Pittsburgh, 3501 Fifth Avenue, Pittsburgh, PA 15261, USA; klimstra@pitt.edu
* Correspondence: alan.watson@pitt.edu

Academic Editor: Michael R. Holbrook
Received: 6 March 2017; Accepted: 6 April 2017; Published: 11 April 2017

Abstract: The 17D line of yellow fever virus vaccines is among the most effective vaccines ever created. The humoral and cellular immunity elicited by 17D has been well characterized in humans. Neutralizing antibodies have long been known to provide protection against challenge with a wild-type virus. However, a well characterized T cell immune response that is robust, long-lived and polyfunctional is also elicited by 17D. It remains unclear whether this arm of immunity is protective following challenge with a wild-type virus. Here we introduce the 17D line of yellow fever virus vaccines, describe the current state of knowledge regarding the immunity directed towards the vaccines in humans and conclude with a discussion of animal models that are useful for evaluating T cell-mediated immune protection to yellow fever virus.

Keywords: yellow fever virus; YFV; flavivirus; 17D; 17DD; 17D-204; T cell; vaccine; animal models; dengue; West Nile; Zika; live attenuated vaccine; chimerivax; vaccine development

1. Yellow Fever Virus: History, Legacy and Future

1.1. Introduction to the Yellow Fever Virus

Yellow fever virus (YFV) simultaneously shares an important historical legacy and a modern-day urgency. Once one of the most feared diseases in the world, this virus terrorized Africa, Europe and the Americas. In the Americas, over 250 years of documented epidemics of YFV killed hundreds of thousands and influenced the outcomes of wars and the stability of economies. YFV is the prototypic member of the genus *Flavivirus*, family *Flaviviridae*, and contains a single-stranded, positive sense RNA genome. In the urban YFV cycle, in which most human cases occur, the virus is transmitted by the bite of an infected *Aedes aegypti* mosquito. A subset of infected people develops a severe hemorrhagic yellow fever (YF) disease, presenting with fever, nausea, vomiting, hepatitis, and jaundice. This severe disease leads to death in 20%–60% of cases (reviewed in [1]).

Originating in Africa, YFV was trafficked to the Americas as a consequence of the slave trade [2]. Eventually, better sanitation led to a precipitous decline in outbreaks of YF. Even as local outbreaks decreased, YFV remained a threat to the United States because of foreign conflicts and foreign economic development. Two prominent examples of this include Cuba during the Spanish-American war where YF killed more American soldiers than battle, and the construction of the Panama Canal which was devastated by ongoing outbreaks of YF. Following the end of the Spanish-American war, YF remained a concern to the United States regarding both the protection of soldiers during foreign conflicts and the possibility of domestic outbreaks. The U.S. Army's Yellow Fever Commission, led famously by Walter Reed, traveled to Cuba and established that mosquitoes were responsible for transmission [3]. Subsequently, mosquito control efforts were used to reduce the impact of the last major U.S. epidemic in New Orleans in 1905, and bring an end to the outbreaks at the Panama Canal in 1906.

1.2. A Brief History of the Yellow Fever Virus Vaccine

In the four decades following the yellow fever commission, an international effort developed to isolate, propagate and create a vaccine against YFV. Integral to this effort was the development of animal models that were required to produce a vaccine. During the fall of 1925 Adrian Stokes led an expedition to study yellow fever in West Africa. In the course of their studies they isolated a virulent virus from a Ghanaian man named Asibi with a mild case of YF [4,5]. The Asibi virus was passaged through rhesus macaques by direct blood/serum transfer and then through infected mosquitoes. Except for two monkeys, the Asibi virus proved lethal causing symptoms that were reportedly similar to human cases of yellow fever. The studies carried out by Stoke's expedition were ground-breaking on various levels as they were the first to establish experimental animal models of YF and show that serum from convalescent humans could protect experimentally infected animals.

The Asibi virus was transported to the Rockefeller Institute where Max Theiler and colleagues discovered that the virus, which was refractory to growth in small laboratory animals through most routes of injection, would grow in the brains of mice following intracranial injection [6], the first record of mice being used as an animal model. Passage in mouse brains reduced the viscerotropic virulence of the virus in monkeys but enhanced the neurotropic properties, causing lethal disease when injected into the brain [7]. Concerns over neurotropism led Theiler's group to passage the virus over 200 times in tissue culture medium made with chicken embryos from which the neurologic tissue was removed. They designated one subculture of the Asibi virus, 17D. Although the 17D culture remained virulent when injected into mouse brains, the virus had lost its neurovirulence in monkeys, causing only a 'moderate febrile reaction' when injected intracerebrally [8]. Moreover, the virus no longer caused viscertropic disease in monkeys when injected subcutaneously but only a very mild infection.

Simultaneously with the above findings, Theiler published a report showing that when the 17D subculture was inoculated into monkeys, immune serum could be detected within one month of infection. Within seven days of infection, the monkeys were completely protected against challenge with the virulent Asibi virus. At seven days and beyond, no circulating Asibi virus was detected in the blood of vaccinated monkeys. In humans injected with 17D, anti-yellow fever immune serum was detected as early as two days following immunization. The eight test subjects experienced only a slight 'fever' (maximum temperature 37.4 °C), a mild headache and a backache that reportedly did not prevent normal daily activities [9]. The 17D subculture of the Asibi virus [8] became the seed strain for the modern day yellow fever virus vaccines, 17DD (passage 195) and 17D-204 (passage 204). Since then, over 500 million people have been administered the 17D-based vaccines (hereafter referred to collectively as 17D). Remarkably, only 32 cases of vaccine failure to protect against a virulent strain of YFV have ever been documented [1]. This is likely due to the high seroconversion rate of human vaccinees, between 80% and 100%, and the 30-year persistence of detectable immunity in some individuals. Max Theiler was awarded the Nobel prize in 1951 [10].

For a more through appreciation of the history of the yellow fever vaccine, the authors suggest reading the excellent discussion by J. Gordon Frierson [11] and the thorough review by Monath, Gershman and Staples [1].

1.3. Status of the Yellow Fever Virus and 17D

Today, YFV remains endemic in South American and African countries where monkeys provide sylvatic reservoirs of virus that spur regular outbreaks (jungle yellow fever). Estimates are as high as 200,000 infections with YFV per year and up to 30,000 deaths [12]. In these countries, vaccination against the virus is a primary means of disease control. However, the virus remains of major concern in Sub-Saharan Africa where the prevalence of vaccination is low and millions are at risk for infection. Indeed, in 2016 Angola experienced its biggest outbreak of YFV in 30 years and the virus has spread to neighboring countries and resulted in numerous traveler-associated cases throughout the globe. Most notably, YFV had never been recorded in Asia; however, China experienced its first traveler-associated cases in 2016 [13]. Due to high population densities and sufficient mosquito populations, yellow fever could be particularly devastating in Asia if it were to become established [14]. The reintroduction of YFV into North America is also becoming a greater threat as populations of *A. aegypti* mosquitoes are repopulating areas that have ceased aggressive control efforts [15]. For similar reasons, local transmission of the *Flaviviruses* Dengue and Zika and the related alphavirus Chikungunya have already been documented in the contiguous U.S. The risk to the residents of the U.S. is particularly high with these viruses as Food and Drug Administration (FDA)-approved vaccines are currently unavailable.

The yellow fever vaccine should not simply be viewed as an empirical success that warrants no further study. 17D, arguably, sits at the pinnacle of vaccine success, alongside those for small pox and polio. The importance of 17D remains high as a reverse genetics system [16] has made the vaccine a vector for other antigens [17–21] and a backbone for the development of vaccines against other flaviviruses, specifically [22]. The highly efficacious and long-lasting immunity elicited by the 17D makes it an important target of future research for the development of vaccines against related viruses and the understanding attenuation and immune inductive processes for highly efficacious vaccines in general [23]. Despite the ongoing use and development of 17D, the relatively recent recognition that the vaccine causes rare but serious adverse events (SAE) has led to concerns over its continued use [24,25]. The causes of SAE are unknown, but patient-specific immune irregularities may be involved [26,27].

While the humoral immune response elicited by 17D has historically been considered protective against challenge with a virulent strain of YFV, it is unknown whether the T cell arm of the immune response plays a role. A thorough understanding of these immune responses remains the most glaring knowledge gap regarding 17D. Bridging this divide is likely to illuminate mechanisms that contribute to SAE as well as facilitate the rational design of vaccines, particularly those based off the 17D platform. The remainder of this review will briefly describe the human immune response to 17D with a focus on T cell mediated immunity. Attention will then be turned to animal models of immunity that are proving useful for characterizing the role of T cell immunity for protection against challenge with virulent YFV and the control of 17D during vaccination.

2. The Immune Response to Yellow Fever Virus in Humans

The abundance of vaccinated individuals and the safety of 17D has ensured ample availability of subjects for the study of human immune responses following vaccination. For example, such subjects have been derived from mass vaccination of U.S. troops during the World War II and recent targeted studies that have recruited volunteers for immunization. Subjects like the former, studied many years after vaccination, have demonstrated the striking longevity of immunity to 17D whereas the latter are opening windows into the earliest events following vaccination. These studies have been limited to evaluating what can be obtained from blood sampling. Thus, since its development, the most studied aspect of human immunity to 17D was the humoral response. However, studies of T cell-mediated immunity to 17D have become prolific in recent years.

2.1. Neutralizing Antibodies as Protective Immunity

Since wild-type virus cannot be used for directed experimentation on humans, empirical evaluation of specific immune components involved in protecting humans is practically non-existent. However, serum from naturally infected convalescent humans exposed to wild-type virus protects rhesus macaques against infection [4], and mice are protected against intracerebral challenge with 17D when mixed with serum from 17D-vaccinated humans [9]. Furthermore, convalescent serum collected from humans located on another continent could protect monkeys infected with local isolates of virus [28]. These studies suggested that (1) humoral immunity was a primary protective element in previously exposed individuals and (2) a single vaccine could offer protection to global strains of YFV. Thus, neutralizing antibodies remain the accepted correlate of protection against YFV.

2.2. The Neutralizing Antibody Response to 17D

Consistently, 90% or greater of the individuals immunized with 17D develop neutralizing antibodies (reviewed in [29]). Early studies quantified neutralizing antibody titers as dilutions of immune serum that protected 50% of intracranially infected mice against 17D [8,9]. By this method, neutralizing antibodies have been detected as early as two weeks following vaccination [9]. Plaque reduction neutralization tests (PRNT) demonstrated robust neutralizing antibody responses beginning as early as six days following vaccination [30], with most individuals peaking after thirty days [30]. Neutralizing responses have been recovered for up to 60 years after vaccination [31–34]. The longevity of the 17D-specific neutralizing antibody responses suggests that 17D elicits a long-lived and functional memory response, supported by a boost in titer after revaccination [30,35]. This and other evidence for the potency of neutralizing antibody responses has been presented in detail (reviewed in [29]), and suggests that a single vaccination offers life-long protection against YFV.

2.3. Innate Immune Responses to Vaccination with 17D

17D elicits a complex modulation of innate immune cytokines in humans. Elevated levels of plasma interferon (IFN)-γ are seen on day 15 post vaccination [36]. Although plasma levels of other cytokines have not been assessed, re-stimulation of innate immune cell cultures of natural killer (NK) cells, neutrophils, and monocytes from 17D vaccinated humans with YF antigen results in the increased production of IFN-γ [36], interleukin (IL)-1β [37], IL-12 [38], tumor necrosis factor (TNF)-α [38,39] and IL-10 [39] in some subsets and a concomitant decrease of TNF-α [38], IL-10, and IL-4 [39] in other subsets. A wide array of increased innate immune gene expression signatures is detected as early as one day following vaccination. The cytokines include inducers and mediators of antiviral and interferon activity such as Toll-like receptor (TLR)-7, interferon-induced protein with tetratricopeptide repeats (IFIT) 1, IFIT2, IFIT3, interferon regulatory factor (IRF) 7, and signal transducer and activator of transcription (STAT) 1 [40]. Gene signatures of directly antiviral cytokines IFN-α and IFN-γ are highly variable among vaccine recipients, with a near even distribution of vaccinees showing an increase or decrease of IFN-α and IFN-γ transcripts between days three and seven post vaccination [40].

The earliest induction of cytokines by 17D is most likely the result of direct interactions with innate immune cells like dendritic cells (DC). Stimulation of human monocyte-derived DCs and plasmacytoid DCs with the 17D vaccine preparation induces the production of IL-6, TNF-α, monocyte chemoattractant protein (MCP)-1, interferon-γ-induced protein 10 kDa (IP-10), IL-12p40, IL-12p70 and IFN-α. Murine studies suggested that the activation and production of cytokines by DCs is mediated through interactions with TLR-2, -7, -8 and -9 [41]. Engagement of multiple TLRs results in a complex relationship with the downstream T cell response, resulting in the production of both T_H1 and T_H2 cytokines. Although TLR-2 suppresses cytotoxic T cell responses to 17D in mice, TLR signaling more broadly promote the very same responses [41]. 17D infects and replicates only minimally in human DCs [41,42], likely due to triggering of maturation [43]. However, 17D antigens are processed and presented by DCs [43], which are most likely involved in eliciting downstream T cell responses.

2.4. T Cell-Mediated Immune Responses to 17D

Prior to 1998 [30] the T cell response to 17D in humans had not been studied. While a T cell response to vaccination had been expected, the role of T cells was unclear. Since neutralizing antibodies were the accepted means of protection against YFV, it was strongly expected that cluster of differentiation (CD)4+ T cells would act primarily as a form of 'help' during the formation of neutralizing antibody responses. CD8+ T cells recognize and lyse virus-infected cells and thus it was expected that CD8+ T cells could play a role in clearance of 17D during vaccination or for protection following challenge with wild-type YFV. The specific role that T cells play in humans during immunization, or for protection following challenge remains uncertain, and is likely to remain so due to the limitations of working with human subjects. Since the first T cell studies, a great deal of information has been gathered about the dynamics of T cell activation, proliferation, and the phenotypes that result from 17D immunization.

Both CD4+ and CD8+ T cells respond strongly to 17D. Activated cells are detected as early as three days post-immunization (CD8+) [44]. CD4+ T cells peak between 7 and 14 days after immunization [45–48], before CD8+ T cells which peak between 14 and 30 days [44–46,49,50]. CD8+ T cell proliferation appears to be largely antigen-driven since the size of the T cell response correlates directly with the virus load. The peak of the CD8+ T cell response comes three days after the highest levels of virus genomes are detected in the plasma, and CD8+ T cells peak only after virus is no longer detected [44]. Five to six percent of the approximately 2000 CD8+ T cell clones that respond to 17D [51] differentiate into various memory populations displaying conventional markers including central memory (T_{CM}) and effector memory (T_{EM}/T_{EMRA}). These cells slowly decrease over time, but remain detectable for over 25 years [33,52]. In contrast, a population of self-renewing and highly responsive 17D-specific memory cells similar to the stem cell-like memory subset (T_{scm}) maintains stable numbers for the same 25-year period (discussed below) [52].

2.4.1. CD4+ T Cells

17D-specific CD4+ T cells produce a mix of T_H1 and T_H2 cytokines, including IL-2, IFN-γ, TNF-α, IL-12 (T_H1), IL-4, IL-5, IL-10, and IL-13 (T_H2) [39,45,47,53]. T_H1 T cells characteristically promote CD8+ responses whereas T_H2 cytokines promote B-cell and antibody responses. T follicular helper (Tfh) cells are a subset of CD4+ T cells characterized in part by the expression of C–X–C chemokine receptor type 5 (CXC-R5) and the production of IL-4. Tfh cells promote healthy B cell germinal centers (reviewed in [54]). It is unclear whether the CD4+ T cells provide a 'helper' function that promotes the production of neutralizing antibodies or CD8+ T cell responses or if these responses are independent of T cells. Whether CD8+ T cells are dependent on CD4+ T cell help, in the context of 17D immunization, has not been studied.

An antibody 'helper' scenario is supported by one study where a greater proportion of CD4+ IL-2 and IFN-γ producing T cells on days 1 and 2 were associated with higher neutralizing antibody titers and greater numbers of plasmablasts on day 14 [45]. Murine studies demonstrate that both T_H1 and T_H2 cytokines promote the generation of neutralizing antibody responses to viruses. IFN-γ favors the production of immunoglobulin G2a (IgG2a) and IL-4 favors immunoglobulin G1 (IgG1) while both cytokines promote class switching and affinity maturation (reviewed in [55]). It remains unknown whether Tfh cells play a central role in antibody development towards 17D; although in one murine model, Tfh cells may have a limited role in eliciting protective neutralizing antibodies [56].

Human CD4+ forkhead box P3 (FOXP3)+ T regulatory cells are transiently activated during immunization with 17D [46] although their relationship to the immune outcomes following 17D vaccination has not been investigated.

2.4.2. CD8+ T Cells

17D-specific CD8+ T cells respond to epitopes within every mature product of the 17D polyprotein [46,50]. Upon peptide re-stimulation, 17D-specific CD8+ T cells produce multiple cytokines including proinflammatory cytokines IFN-γ, TNF-α, macrophage inflammatory protein (MIP1)-β [46,50] and the proliferation promoting cytokine IL-2 [46]. Nearly all 17D-specific CD8+ T cells express granzyme B, and the majority of stimulated cells stain positive for surface CD107a [50], suggesting that they are functionally capable of degranulating [57] and are likely to be cytolytic. Following the peak of CD8+ T cell expansion, the cells begin to differentiate in to long-lived memory which retain a polyfunctional phenotype for at least two years [50].

Memory CD8+ T cells are composed of conventional T_{CM} (CCR7+, CD45RA−), T_{EM} (CCR7−, CD45RA−) and T_{EMRA} (CCR7−, CD45RA+), all of which are induced strongly by vaccination and slowly decrease over a 25+ year period. However, a recently identified population of the 'naïve-like' T_{scm} was shown to persist at nearly unchanged levels for the same 25-year period. These cells appear naïve according to CCR7+, CD45RA+ expression but, unlike truly naïve cells, they also express CD58 and CD95. The 17D-specific T_{scm} cells demonstrate a self-renewing behavior and rapid proliferation in response to re-stimulation [52]. These data suggest that 17D-specific memory T cells can respond quite rapidly to re-exposure of YFV antigen in vitro; however, it is unclear whether this happens in humans. One study demonstrated that biomarkers of T cell activity, neopterin and β2-microglobulin, do not increase after a 17D boost, whereas a strong increase is seen during the primary vaccination [33]. Another study indicated that, compared to individuals that had only one vaccination, multiple vaccinations do not result in higher levels of memory T cells as is typical following a secondary memory T cell response [30].

2.5. Naïve Host Infection with Wild-Type YFV: A Limited Understanding of Human Immunity

Very few studies have evaluated the immune response to a primary infection with virulent YFV. Most infections are only identified when patients are in fulminant hemorrhagic disease, if not after death. Thus, much of the information regarding immunity towards wild-type infection is from severe cases of disease and not from the majority, for which the immune response clears the infection. In one study, 18 patients with non-fatal non-hemorrhagic YF were more likely to have measurable neutralizing antibody titers and undetectable virus by reverse transcription polymerase chain reaction (RT-PCR) compared to fatal or non-fatal hemorrhagic cases. Cases of severe disease have elevated chemoattractant cytokines MCP-1 and IP-10 and proinflammatory cytokines TNF-a, IL-6, and IL-1a [58]. These data suggest that size of the neutralizing antibody response to wild-type YFV infection may be a factor determining the severity of disease. Therefore, elevated levels of cytokines seen in severe cases of disease may be the result of virus burden in individuals that mount an insufficient neutralizing antibody response.

Livers from patients succumbing to fatal yellow fever show apoptosis, steatosis and lytic necrosis of hepatocytes. This pattern is associated with a strong infiltration of T cells, primarily CD4+, but also CD8+ T cells [59,60] and B cells into the portal region [59]. Macrophages, NK cells and antigen presenting cells also infiltrate the liver, but in fewer numbers [60]. Expression of TNF-α, IFN-γ, and transforming growth factor (TGF)-β were elevated most strongly in the lobular regions that showed strong macrophage infiltration [59]. The heavy T cell infiltration associated with cell death and elevated cytokine expression suggests that severe disease may have a prominent immunopathologic component [60,61]. These data suggest that patients presenting with severe disease mount a T cell response that may be either (1) ineffective at controlling virus replication in the liver or (2) is strongly cytolytic, causing damage and exacerbating disease.

3. Animals Models of T Cell-Mediated Immunity towards YFV

Human studies are ideal for recording the phenotypes and reactivity of developing or developed immunity to 17D. Retrospective studies have provided an insight into the statistical effectiveness of vaccination and the resulting immune phenotypes. However, questions about mechanisms of protection and anti-YFV immune development cannot currently be asked in humans. Human volunteers can neither be experimentally infected with wild-type YFV nor have their visceral or neurological tissues probed for the presence of virus and immune mediators. Thus, a more thorough understanding of the virus–host interactions must be addressed using animal models.

3.1. Non-Human Primates

Non-human primates serve as the most relevant models of YF due to their close relationship to humans and their natural susceptibility to infection. Rhesus macaques develop severe hemorrhagic yellow fever, presenting with symptoms that resemble human disease [4]. Vaccination with 17D induces neutralizing antibodies that correlate with protection against a virulent YFV [62]. However, only a few studies have addressed T cell responses to 17D in these animals. Activated T cells peak between 12 and 14 days following immunization [63]. CD8+ T cells respond to peptides throughout the genome of 17D. Low levels of CD4+ T cells can be detected as well [64]. 17D-specific T cells produce IFN-γ, IL-4 and TNF-α in response to peptide stimulation. A majority of IFN-γ producing CD8+ T cells may be $\gamma\delta$ T cell receptor (TCR)+ [65].

3.2. Murine Models

Murine models are particularly advantageous for studies of cellular immunity due to known major histocompatibility complex (MHC) haplotypes, well characterized biology and the ubiquitous availability of reagents. However, few studies have addressed the development or role of T cell-mediated responses to YFV in murine model systems. This is primarily because YFV is refractory to replication in adult mice due to restriction by type-I interferon [66] (with the exception of the central nervous system when introduced by intracranial inoculation [6]). Thus, when virus is introduced into the mouse through a peripheral route that mimics vaccination—e.g., subcutaneous (SC)—, little to no replication is detected [67]. Adaptive immune responses play no known role in eliminating virus from mice infected via this route, as infection of recombination activation gene (RAG)$^{-/-}$ animals lacking functional B cells and T cells does not result in the development of disease [56]. Mice do mount an immune response to 17D when it is administered SC that includes neutralizing antibodies and CD8+ and CD4+ T cells [56,68,69]. However, 17D-specific T cells proliferate modestly in this model, compared to humans where responses to specific epitopes can account for 10–20% of total CD8+ T cells [46,49], most likely due to the lack of virus replication which limits viral antigen. In 17D-immunized mice, neutralizing antibodies and CD8+ T cells protect mice against intracranial challenge with 17D through an IFN-γ and perforin mediated mechanism [56]. A similar mechanism is observed when 17D-derived antigens are delivered by an unrelated viral vector [70]. These studies suggest that CD8+ T cells and neutralizing antibodies may be important for the clearance of 17D from the central nervous system (CNS) during vaccination. No studies have been published that examine clearance of the wild-type virus in this model. Although, wild-type virus and 17D are equally virulent by intracranial (IC) infection [71], making it difficult to differentiate the two strains experimentally. The virulence of 17D following IC infection suggests that mice may be a useful model of yellow fever vaccine-associated neurotropic disease (YEL-AND), a rare SAE seen in human vaccinees. These studies suggest that an inadequate neutralizing antibody or CD8+ T cell immune response may increase susceptibility to YEL-AND.

YFV replication in mice is primarily restricted by type-I interferon and to a lesser extent type-II interferon [66]. In type-I interferon receptor knockout mice (IFNAR$^{-/-}$), which lack all type-I interferon responses, YFV replicates when introduced by the SC route. Notably, 17D and wild-type viruses can be differentiated by both replication and clinical disease in the IFNAR$^{-/-}$ mouse. 17D remains attenuated with replication restricted to the lymphoid compartments including the lymph nodes, spleen, bone marrow and blood, whereas the wild-type virus produces a viscerotropic disease resembling that seen in humans, including virus replication, immune infiltrates and steatosis in the liver. Like 17D, wild-type virus also replicates in lymphoid tissues but additionally in the kidneys, heart, and brain [66,67]. Importantly, 17D is cleared from all IFNAR$^{-/-}$ animals whereas wild-type virus is lethal. Furthermore, 17D immunization protects mice against challenge with a wild-type virus. The mechanisms of 17D clearance in IFNAR$^{-/-}$ mice is not known, but unpublished observations [72] indicate that CD8+ T cells are required, consistent with studies of IC inoculation in normal mice.

The IFNAR$^{-/-}$ model is not ideal for the study of T cell responses. Depending on the pathogen, type-I interferon can significantly influence the activation and proliferation of T cells [73,74]. Type-I interferon is produced by human and rhesus macaque cells infected with 17D in vitro [41,42,75–78]; however, a role for type-I interferon in YFV-specific adaptive immune responses in vivo has not been defined. Despite these concerns, 17D-specific T cells induced in IFNAR$^{-/-}$ mice appear to be normal, and their phenotypes align well with those seen in human studies. The T cells are polyfunctional, producing combinations of IFN-γ, TNF-α (CD8+, CD4+) and IL-2 (CD8+) and both CD4+ and CD8+ T cells present CD107a on their surface during peptide stimulation and functionally lyse 17D-specific targets in vivo. The T cells contract into long-lived memory that persists for at least two and a half years, essentially for the life of the mouse. In this model, adoptive transfer experiments determined that both neutralizing antibodies and CD4+ T cells contribute to protection. Surprisingly, no protective effect was seen with CD8+ T cells [67]. This study, when considered with the IC model of YEL-AND, suggests that CD8+ and CD4+ T cells may be differentially important in regard to protection of naïve animals against YEL-AND (CD8+ T cell-mediated), and protection following challenge of vaccinated animals with a virulent strain of YFV (CD4+ T cell-mediated).

When both type-I and type-II interferon receptors are absent (IFNAGR$^{-/-}$), mice become susceptible to disease from both 17D [66,79] and the wild-type virus [66]. Thus, 17D is attenuated by IFN-γ. However, the disease associated with these viruses is dramatically different. Wild-type YFV infected IFNAGR$^{-/-}$ mice develop an accelerated viscerotropic disease, like that seen in IFNAR$^{-/-}$ mice, with an average survival time that is one day less (6–7 days post infection) than in the presence of type-II interferon signaling. 17D infected IFNAGR$^{-/-}$ mice develop a protracted infection that resembles a mild viscerotropic-like disease at the onset but mice then succumb to a disease associated with neurological signs 12 days post infection [66]. The development of neurologic disease suggests that IFNAGR$^{-/-}$ mice may be a useful model of YEL-AND. Although adaptive immune responses to 17D in IFNAGR$^{-/-}$ mice have not been studied, one possible mechanism resulting in neurologic disease is the ineffective clearance of 17D from the brain since infected cells in the brain cannot respond to IFN-γ. This would be consistent with the role that IFN-γ producing CD8+ T cells play in clearance of 17D from IC infection of normal mice [56].

The IFNAGR$^{-/-}$ model can be partially reproduced using a conditional knockout of STAT1, a primary signaling adapter of type I and type II interferon responses [80,81]. This model (Stat1$^{loxP/loxP}$/Vav-cre) [82] can be thought of as a chimeric model where visceral organs respond competently to type I and type II interferons but the response in cells of the hematopoietic lineage is blunted. In this model, 17D replicated in every hematopoietic cell type tested in the blood and spleen, indicating that 17D is tropic for these cell types in mice, absent interferon responses. Unlike human vaccination, 17D was introduced intravenously and proved lethal to approximately 75% of mice over a time-frame similar to that seen in IFNAGR$^{-/-}$ mice. Virus genomes were identified in the liver, kidney and brain at the same levels as those seen in wild-type mice. Consistent with STAT1 deficiency among hematopoietic cells, the spleen showed significantly higher levels of 17D genomes than the other organs. These data suggest that lymphocytic infiltrates and inflammation that was identified in the liver, may be due to non-specific immunopathologic mechanisms, perhaps cytokine-induced. Signs of viscerotropic disease (e.g., liver pathology) may make this a useful model for understanding the contribution of immune pathology to YFV-AVD. In this model, both CD4+ and CD8+ T cells proliferated in response to infection, with CD4+ T cells becoming most enriched, and both cell types displaying markers of memory differentiation. Among the minority of animals that survived infection, it is unclear if T cells contributed to survival or what role T cells played in the clearance of virus. Although, virus was mostly cleared from the peripheral blood at day eleven post-infection (the latest timepoint studied), replication in the spleen continued. An analogous model where the murine hematopoietic system was reconstituted with (type I and type II interferon competent) human cells, 17D replicated in all human cell types tested and caused minimal non-lethal disease. T cells became activated with minor proliferation, but their role in the clearance of virus was not assessed.

4. Conclusions

The 17D-induced adaptive immune response in humans has been well characterized in recent years. There is a robust induction of both neutralizing antibodies and T cells, and both responses have been detected up to three decades following a single immunization. Neutralizing antibodies are the primary correlate of protections in vaccines and animal studies suggest that neutralizing antibodies can be sufficient for protection against challenge. Yet it remains unclear whether T cells also serve a protective function following challenge. Human 17D-specific CD4+ and CD8+ T cell compartments are polyfunctional, suggesting that they may play a role in protection following challenge. However, no studies have addressed this question in humans, and it is unlikely that it can ever be addressed without the use of experimental animal models.

Non-human primates (NHP) are by far the best experimental models for of YFV due to being a natural host and closely related to humans. NHP models have demonstrated the importance of neutralizing antibodies for protection following challenge with virulent YFV. T cells become activated and expand in response to vaccination; although, it is unclear whether they are functionally protective. Only a few studies have addressed the T cell response in these animals. This is not surprising since the outbred nature of NHPs—resulting in variety of MHC haplotypes—makes the study of specific T cell responses difficult. Since two individual NHPs are unlikely to be MHC compatible, complex experiments like adoptive transfers are not feasible. This shortcoming could be addressed with innovative strategies like partial MHC haplotype adoptive transfer approaches [83], the use of novel models like marmosets which give birth to MHC chimeric twins [84] or future mastery of genetic engineering in NHPs.

With regards to T cell immune studies, murine models hold the most promise due to the availability of highly characterized inbred animals and available reagents. IFNAR$^{-/-}$ mice differentiate between 17D and wild-type virus regarding the extent of attenuation and disease. This model has been used to demonstrate that the immunity elicited by 17D is proportional to that found in humans regarding magnitude, function, and long-lived memory. The flexibility of the inbred mouse enables complex immune studies, and therefore this is the only model where 17D-specific T cells have been shown to exhibit protective efficacy during challenge with a wild-type virus. The IC infection model of YEL-AND has demonstrated the importance of CD8+ T cells for clearance of virus from the brain. An alternative model of YEL-AND is the IFNAGR$^{-/-}$ mouse in which 17D can be delivered SC, to simulate human vaccination. However, ultimately, murine immunodeficient models are not without substantial concern. In normal mice, IC inoculation of 17D is not a physiological route of exposure but is required to induce disease. Additionally, the lack of proliferation when 17D is administered by routes other than IC is not consistent with human vaccination and may result in T cell phenotypes that do not fully resemble human T cells. The IFNAR$^{-/-}$ model demonstrates YFV replication outside of the CNS, but may have deficiencies in the magnitude and the quality of the adaptive immune response.

17D continues to be administered to travelers and in YFV endemic countries as the only prevention for YF. 17D is also being used as a vector for the delivery of foreign antigens against the related dengue virus, and this approach has been licensed in Mexico. 17D continues to be developed for the delivery of novel antigens aimed at protection against other pathogens like human immunodeficiency virus (HIV), malaria, trypanosomes and the Lassa virus [17–21]. The risk of SAE in these chimeric 17D viruses remains unknown. Moreover, the mechanisms by which the 17D vaccine is cleared and mediates protection against a virulent virus is not well understood. Establishing an understanding of these mechanisms serves both to prevent cases of SAE and improve upon 17D-chimeric vaccines. 17D has many secrets yet to be revealed regarding factors that contribute to vaccine efficacy, and most of these questions cannot be asked using human studies. Animal models provide the only means to assess the primary immune response to virulent YFV, the protective immunity elicited by 17D, map mutations that contribute to the attenuation and immunogenicity of 17D, and, perhaps most importantly, facilitate the improvement of chimeric vaccines regarding both safety and immunogenicity towards delivered antigens. The authors acknowledge that existing animal models for YFV are less than ideal; however, for the sake of progress, it remains important that the scientific community embrace these types of models and continually work towards their improvement.

Conflicts of Interest: The authors declare no conflicts of interest.

References

1.	Monath, T.P.; Gershman, M.; Erin Staples, J.; Barrett, A.D.T. Yellow fever vaccine. In *Vaccines*; Plotkin, S.A., Orenstein, W.A., Offit, P.A., Eds.; Saunders Elsevier: London, UK, 2012; pp. 870–968.

2. Bryant, J.E.; Holmes, E.C.; Barrett, A.D.T. Out of Africa: A molecular perspective on the introduction of yellow fever virus into the Americas. *PLoS Pathog.* **2007**, *3*, e75. [CrossRef] [PubMed]
3. Reed, W.; Carroll, J.; Agramonte, A.; Lazear, J.W. The etiology of yellow fever—A preliminary note. *Public Health Pap. Rep.* **1900**, *26*, 37–53. [PubMed]
4. Stokes, A.; Bauert, J.H.; Hudson, N.P. Experimental transmission of yellow fever virus to laboratory animals. *Int. J. Infect. Dis.* **1997**, *2*, 54–59. [CrossRef]
5. Stokes, A.; Bauer, J.H.; Hudson, N.P. The Transmission of yellow fever to *Macacus rhesus*: Preliminary note. *J. Am. Med. Assoc.* **1928**, *90*, 253. [CrossRef]
6. Theiler, M. Studies on the action of yellow fever virus in mice. *Ann. Trop. Med. Parasitol.* **1930**, *24*, 249–272. [CrossRef]
7. Sellards, A.W. The behavior of the virus of yellow fever in monkeys and mice. *PNAS* **1931**, *17*, 339–343. [CrossRef] [PubMed]
8. Theiler, M.; Smith, H.H. The effect of prolonged cultivation in vitro upon the pathogenicity of yellow fever virus. *J. Exp. Med.* **1937**, *65*, 767–786. [CrossRef] [PubMed]
9. Theiler, M.; Smith, H.H. The Use of yellow fever virus modified by in vitro cultivation for human immunization. *J. Exp. Med.* **1937**, *65*, 787–800. [CrossRef] [PubMed]
10. Norrby, E. Yellow fever and Max Theiler: The only Nobel Prize for a virus vaccine. *J. Exp. Med.* **2007**, *204*, 2779–2784. [CrossRef] [PubMed]
11. Frierson, J.G. The yellow fever vaccine: A history. *Yale J. Biol. Med.* **2010**, *83*, 77–85. [PubMed]
12. World Health Organization, Division of Epidemiological Surveillance and Health Situation Trend Assessment. In *Global Health Situation and Projections—Estimates*; World Health Organization: Geneva, Switzerland, 1992.
13. Li, R.; Ding, J.; Li, H. Six cases of imported yellow fever in China: 12 March 2016–24 March 2016. *Radiol. Infect. Dis.* **2016**, *3*, 143–144. [CrossRef]
14. Wasserman, S.; Tambyah, P.A.; Lim, P.L. Yellow fever cases in Asia: Primed for an epidemic. *Int. J. Infect. Dis.* **2016**, *48*, 98–103. [CrossRef] [PubMed]
15. Monaghan, A.J.; Morin, C.W.; Steinhoff, D.F.; Wilhelmi, O.; Hayden, M.; Quattrochi, D.A.; Reiskind, M.; Lloyd, A.L.; Smith, K.; Schmidt, C.A.; et al. On the seasonal occurrence and abundance of the Zika virus vector mosquito *Aedes aegypti* in the contiguous United States. *PLoS Curr.* **2016**. [CrossRef] [PubMed]
16. Bredenbeek, P.J.; Kooi, E.A.; Lindenbach, B.; Huijkman, N.; Rice, C.M.; Spaan, W.J.M. A stable full-length yellow fever virus cDNA clone and the role of conserved RNA elements in *Flavivirus* replication. *J. Gen. Virol.* **2003**, *84*, 1261–1268. [CrossRef] [PubMed]
17. Tao, D.; Barba-Spaeth, G.; Rai, U.; Nussenzweig, V.; Rice, C.M.; Nussenzweig, R.S. Yellow fever 17D as a vaccine vector for microbial CTL epitopes: Protection in a rodent malaria model. *J. Exp. Med.* **2005**, *201*, 201–209. [CrossRef] [PubMed]
18. Bredenbeek, P.J.; Molenkamp, R.; Spaan, W.J.M.; Deubel, V.; Marianneau, P.; Salvato, M.S.; Moshkoff, D.; Zapata, J.; Tikhonov, I.; Patterson, J.; et al. A recombinant yellow fever 17D vaccine expressing Lassa virus glycoproteins. *Virology* **2006**, *345*, 299–304. [CrossRef] [PubMed]
19. Franco, D.; Li, W.; Qing, F.; Stoyanov, C.T.; Moran, T.; Rice, C.M.; Ho, D.D. Evaluation of yellow fever virus 17D strain as a new vector for HIV-1 vaccine development. *Vaccine* **2010**, *28*, 5676–5685. [CrossRef] [PubMed]
20. Stoyanov, C.T.; Boscardin, S.B.; Deroubaix, S.; Barba-Spaeth, G.; Franco, D.; Nussenzweig, R.S.; Nussenzweig, M.; Rice, C.M. Immunogenicity and protective efficacy of a recombinant yellow fever vaccine against the murine malarial parasite *Plasmodium yoelii*. *Vaccine* **2010**, *28*, 4644–4652. [CrossRef] [PubMed]
21. Nogueira, R.T.; Nogueira, A.R.; Pereira, M.C.S.; Rodrigues, M.M.; Neves, P.C.D.C.; Galler, R.; Bonaldo, M.C. Recombinant yellow fever viruses elicit CD8+ T cell responses and protective immunity against *Trypanosoma cruzi*. *PLoS ONE* **2013**, *8*, e59347. [CrossRef]
22. Monath, T.P.; Soike, K.; Levenbook, I.; Zhang, Z.-X.; Arroyo, J.; Delagrave, S.; Myers, G.; Barrett, A.D.T.; Shope, R.E.; Ratterree, M.; et al. Recombinant, chimaeric live, attenuated vaccine (ChimeriVax™) incorporating the envelope genes of Japanese encephalitis (SA14-14-2) virus and the capsid and nonstructural genes of yellow fever (17D) virus is safe, immunogenic and protective in non-human primates. *Vaccine* **1999**, *17*, 1869–1882. [PubMed]
23. Pulendran, B. Learning immunology from the yellow fever vaccine: Innate immunity to systems vaccinology. *Nat. Rev. Immunol.* **2009**, *9*, 741–747. [CrossRef] [PubMed]

24. Thomas, R.E.; Lorenzetti, D.L.; Spragins, W.; Jackson, D.; Williamson, T. Active and passive surveillance of yellow fever vaccine 17D or 17DD-associated serious adverse events: Systematic review. *Vaccine* **2011**, *29*, 4544–4555. [CrossRef] [PubMed]

25. Hayes, E.B. Is it time for a new yellow fever vaccine? *Vaccine* **2010**, *28*, 8073–8076. [CrossRef] [PubMed]

26. Belsher, J.L.; Gay, P.; Brinton, M.; DellaValla, J.; Ridenour, R.; Lanciotti, R.; Perelygin, A.; Zaki, S.; Paddock, C.; Querec, T.; et al. Fatal multiorgan failure due to yellow fever vaccine-associated viscerotropic disease. *Vaccine* **2007**, *25*, 8480–8485. [CrossRef] [PubMed]

27. Pulendran, B.; Miller, J.; Querec, T.D.; Akondy, R.; Moseley, N.; Laur, O.; Glidewell, J.; Monson, N.; Zhu, T.; Zhu, H.; et al. Case of yellow fever vaccine-associated viscerotropic disease with prolonged viremia, robust adaptive immune responses, and polymorphisms in CCR5 and RANTES genes. *J. Infect. Dis.* **2008**, *198*, 500–507. [CrossRef] [PubMed]

28. Theiler, M.; Sellards, A.W.; Sellards, A.W. The immunological relationship of yellow fever as it occurs in West Africa and in South America. *Ann. Trop. Med. Parasitol.* **1928**, *22*, 449–460. [CrossRef]

29. Gotuzzo, E.; Yactayo, S.; Cordova, E. Efficacy and duration of immunity after yellow fever vaccination: Systematic review on the need for a booster every 10 years. *Am. J. Trop. Med. Hyg.* **2013**, *89*, 434–444. [CrossRef] [PubMed]

30. Reinhardt, B.; Jaspert, R.; Niedrig, M.; Kostner, C.; L'age-Stehr, J. Development of viremia and humoral and cellular parameters of immune activation after vaccination with yellow fever virus strain 17D: A model of human *Flavivirus* infection. *J. Med. Virol.* **1998**, *56*, 159–167. [CrossRef]

31. Niedrig, M.; Lademann, M.; Emmerich, P.; Lafrenz, M. Assessment of IgG antibodies against yellow fever virus after vaccination with 17D by different assays: Neutralization test, haemagglutination inhibition test, immunofluorescence assay and ELISA. *Trop. Med. Int. Health* **1999**, *4*, 867–871. [CrossRef] [PubMed]

32. Poland, J.D.; Calisher, C.H.; Monath, T.P.; Downs, W.G.; Murphy, K. Persistence of neutralizing antibody 30–35 years after immunization with 17D yellow fever vaccine. *Bull. World Health Organ.* **1981**, *59*, 895–900. [PubMed]

33. Wieten, R.W.; Jonker, E.F.F.; Leeuwen, E.M.M.V.; Remmerswaal, E.B.M.; Berge, I.J.M.T.; Visser, A.W.D.; Genderen, P.J.J.V.; Goorhuis, A.; Visser, L.G.; Grobusch, M.P.; et al. A Single 17D yellow fever vaccination provides lifelong immunity; characterization of yellow-fever-specific neutralizing antibody and T-cell responses after vaccination. *PLoS ONE* **2016**, *11*, e0149871. [CrossRef] [PubMed]

34. Bodilis, H.C.; Benabdelmoumen, G.; Gergely, A.; Goujon, C.; Pelicot, M.; Poujol, P.; Consigny, P.H. Persistance à long terme des anticorps neutralisants de la fièvre jaune chez les personnes âgées de 60 ans et plus. *Bull. Soc. Pathol. Exot.* **2011**, *104*, 260–265. [CrossRef] [PubMed]

35. Hepburn, M.J.; Kortepeter, M.G.; Pittman, P.R.; Boudreau, E.F.; Mangiafico, J.A.; Buck, P.A.; Norris, S.L.; Anderson, E.L. Neutralizing antibody response to booster vaccination with the 17D yellow fever vaccine. *Vaccine* **2006**, *24*, 2843–2849. [CrossRef] [PubMed]

36. Da Costa Neves, P.C.; de Souza Matos, D.C.; Marcovistz, R.; Galler, R. TLR expression and NK cell activation after human yellow fever vaccination. *Vaccine* **2009**, *27*, 5543–5549. [CrossRef] [PubMed]

37. Gaucher, D.; Therrien, R.; Kettaf, N.; Angermann, B.R.; Boucher, G.; Filali-Mouhim, A.; Moser, J.M.; Mehta, R.S.; Drake, D.R.; Castro, E.; et al. Yellow fever vaccine induces integrated multilineage and polyfunctional immune responses. *J. Exp. Med.* **2008**, *205*, 3119–3131. [CrossRef] [PubMed]

38. Luiza-Silva, M.; Campi-Azevedo, A.C.; Batista, M.A.; Martins, M.A.; Avelar, R.S.; da Silveira Lemos, D.; Bastos Camacho, L.A.; de Menezes Martins, R.; de Lourdes de Sousa Maia, M.; Guedes Farias, R.H.; et al. Cytokine signatures of innate and adaptive immunity in 17DD yellow fever vaccinated children and its association with the level of neutralizing antibody. *J. Infect. Dis.* **2011**, *204*, 873–883. [CrossRef] [PubMed]

39. Silva, M.L.; Martins, M.A.; Espírito-Santo, L.R.; Campi-Azevedo, A.C.; Silveira-Lemos, D.; Ribeiro, J.G.L.; Homma, A.; Kroon, E.G.; Teixeira-Carvalho, A.; Elói-Santos, S.M.; et al. Characterization of main cytokine sources from the innate and adaptive immune responses following primary 17DD yellow fever vaccination in adults. *Vaccine* **2011**, *29*, 583–592. [CrossRef] [PubMed]

40. Querec, T.D.; Akondy, R.S.; Lee, E.K.; Cao, W.; Nakaya, H.I.; Teuwen, D.; Pirani, A.; Gernert, K.; Deng, J.; Marzolf, B.; et al. Systems biology approach predicts immunogenicity of the yellow fever vaccine in humans. *Nat. Immunol.* **2009**, *10*, 116–125. [CrossRef] [PubMed]

41. Querec, T.; Bennouna, S.; Alkan, S.; Laouar, Y.; Gorden, K.; Flavell, R.; Akira, S.; Ahmed, R.; Pulendran, B. Yellow fever vaccine YF-17D activates multiple dendritic cell subsets via TLR2, 7, 8, and 9 to stimulate polyvalent immunity. *J. Exp. Med.* **2006**, *203*, 413–424. [CrossRef] [PubMed]

42. Palmer, D.R.; Fernandez, S.; Bisbing, J.; Peachman, K.K.; Rao, M.; Barvir, D.; Gunther, V.; Burgess, T.; Kohno, Y.; Padmanabhan, R.; et al. Restricted replication and lysosomal trafficking of yellow fever 17D vaccine virus in human dendritic cells. *J. Gen. Virol.* **2007**, *88*, 148–156. [CrossRef] [PubMed]

43. Barba-Spaeth, G.; Longman, R.S.; Albert, M.L.; Rice, C.M. Live attenuated yellow fever 17D infects human DCs and allows for presentation of endogenous and recombinant T cell epitopes. *J. Exp. Med.* **2005**, *202*, 1179–1184. [CrossRef] [PubMed]

44. Akondy, R.S.; Johnson, P.L.F.; Nakaya, H.I.; Edupuganti, S.; Mulligan, M.J.; Lawson, B.; Miller, J.D.; Pulendran, B.; Antia, R.; Ahmed, R. Initial viral load determines the magnitude of the human CD8 T cell response to yellow fever vaccination. *Proc. Natl. Acad. Sci. USA* **2015**, *112*, 3050–3055. [CrossRef] [PubMed]

45. Kohler, S.; Bethke, N.; Böthe, M.; Sommerick, S.; Frentsch, M.; Romagnani, C.; Niedrig, M.; Thiel, A. The early cellular signatures of protective immunity induced by live viral vaccination. *Eur. J. Immunol.* **2012**, *42*, 2363–2373. [CrossRef] [PubMed]

46. Blom, K.; Braun, M.; Ivarsson, M.A.; Gonzalez, V.D.; Falconer, K.; Moll, M.; Ljunggren, H.-G.; Michaëlsson, J.; Sandberg, J.K. Temporal dynamics of the primary human T cell response to yellow fever virus 17D as it matures from an effector- to a memory-type response. *J. Immunol.* **2013**, *190*, 2150–2158. [CrossRef] [PubMed]

47. James, E.A.; LaFond, R.E.; Gates, T.J.; Mai, D.T.; Malhotra, U.; Kwok, W.W. Yellow fever vaccination elicits broad functional CD4+ T cell responses that recognize structural and nonstructural proteins. *J. Virol.* **2013**, *87*, 12794–12804. [CrossRef] [PubMed]

48. Martins, M.A.; Silva, M.L.; Marciano, A.P.V.; Peruhype-Magalhães, V.; Eloi-Santos, S.M.; Ribeiro, J.G.L.; Correa-Oliveira, R.; Homma, A.; Kroon, E.G.; Teixeira-Carvalho, A.; et al. Activation/modulation of adaptive immunity emerges simultaneously after 17DD yellow fever first-time vaccination: Is this the key to prevent severe adverse reactions following immunization? *Clin. Exp. Immunol.* **2007**, *148*, 90–100. [CrossRef] [PubMed]

49. Miller, J.D.; van der Most, R.G.; Akondy, R.S.; Glidewell, J.T.; Albott, S.; Masopust, D.; Murali-Krishna, K.; Mahar, P.L.; Edupuganti, S.; Lalor, S.; et al. Human effector and memory CD8+ T cell responses to smallpox and yellow fever vaccines. *Immunity* **2008**, *28*, 710–722. [CrossRef] [PubMed]

50. Akondy, R.S.; Monson, N.D.; Miller, J.D.; Edupuganti, S.; Teuwen, D.; Wu, H.; Quyyumi, F.; Garg, S.; Altman, J.D.; Del Rio, C.; et al. The yellow fever virus vaccine induces a broad and polyfunctional human memory CD8+ T cell response. *J. Immunol.* **2009**, *183*, 7919–7930. [CrossRef] [PubMed]

51. DeWitt, W.S.; Emerson, R.O.; Lindau, P.; Vignali, M.; Snyder, T.M.; Desmarais, C.; Sanders, C.; Utsugi, H.; Warren, E.H.; McElrath, J.; Makar, K.W.; Wald, A.; Robins, H.S. Dynamics of the cytotoxic T cell response to a model of acute viral infection. *J. Virol.* **2015**, *89*, 4517–4526. [CrossRef] [PubMed]

52. Marraco, S.A.F.; Soneson, C.; Cagnon, L.; Gannon, P.O.; Allard, M.; Maillard, S.A.; Montandon, N.; Rufer, N.; Waldvogel, S.; Delorenzi, M.; et al. Long-lasting stem cell–like memory CD8+ T cells with a naïve-like profile upon yellow fever vaccination. *Sci. Transl. Med.* **2015**, *7*, 282ra48. [CrossRef] [PubMed]

53. Campi-Azevedo, A.C.; Araújo-Porto, L.P.D.; Luiza-Silva, M.; Batista, M.A.; Martins, M.A.; Sathler-Avelar, R.; da Silveira-Lemos, D.; Camacho, L.A.B.; de Menezes Martins, R.; de Lourdes de Sousa Maia, M.; et al. 17DD and 17D-213/77 yellow fever substrains trigger a balanced cytokine profile in primary vaccinated children. *PLoS ONE* **2012**, *7*, e49828. [CrossRef] [PubMed]

54. Crotty, S. Follicular helper CD4 T cells (T_{FH}). *Ann. Rev. Immunol.* **2011**, *29*, 621–663. [CrossRef] [PubMed]

55. Swain, S.L.; McKinstry, K.K.; Strutt, T.M. Expanding roles for CD4+ T cells in immunity to viruses. *Nat. Rev. Immunol.* **2012**, *12*, 136–148. [CrossRef] [PubMed]

56. Bassi, M.R.; Kongsgaard, M.; Steffensen, M.A.; Fenger, C.; Rasmussen, M.; Skjodt, K.; Finsen, B.; Stryhn, A.; Buus, S.; Christensen, J.P.; et al. CD8+ T cells complement antibodies in protecting against yellow fever virus. *J. Immunol.* **2015**, *194*, 1141–1153. [CrossRef] [PubMed]

57. Betts, M.R.; Brenchley, J.M.; Price, D.A.; De Rosa, S.C.; Douek, D.C.; Roederer, M.; Koup, R.A. Sensitive and viable identification of antigen-specific CD8+ T cells by a flow cytometric assay for degranulation. *J. Immunol. Methods* **2003**, *281*, 65–78. [CrossRef]

58. Ter Meulen, J.; Sakho, M.; Koulemou, K.; Magassouba, N.; Bah, A.; Preiser, W.; Daffis, S.; Klewitz, C.; Bae, H.-G.; Niedrig, M.; et al. Activation of the cytokine network and unfavorable outcome in patients with yellow fever. *J. Infect. Dis.* **2004**, *190*, 1821–1827. [CrossRef] [PubMed]

59. Quaresma, J.A.S.; Barros, V.L.R.S.; Pagliari, C.; Fernandes, E.R.; Guedes, F.; Takakura, C.F.H.; Andrade, H.F., Jr.; Vasconcelos, P.F.C.; Duarte, M.I.S. Revisiting the liver in human yellow fever: Virus-induced apoptosis in hepatocytes associated with TGF-β, TNF-α and NK cells activity. *Virology* **2006**, *345*, 22–30. [CrossRef] [PubMed]

60. Quaresma, J.A.S.; Barros, V.L.R.S.; Pagliari, C.; Fernandes, E.R.; Andrade, H.F., Jr.; Vasconcelos, P.F.C.; Duarte, M.I.S. Hepatocyte lesions and cellular immune response in yellow fever infection. *Trans. R. Soc. Trop. Med. Hyg.* **2007**, *101*, 161–168. [CrossRef] [PubMed]

61. Quaresma, J.A.S.; Duarte, M.I.S.; Vasconcelos, P.F.C. Midzonal lesions in yellow fever: A specific pattern of liver injury caused by direct virus action and in situ inflammatory response. *Med. Hypotheses* **2006**, *67*, 618–621. [CrossRef] [PubMed]

62. Mason, R.A.; Tauraso, N.M.; Spertzel, R.O.; Ginn, R.K. Yellow fever vaccine: Direct challenge of monkeys given graded doses of 17D vaccine. *Appl. Microbiol.* **1973**, *25*, 539–544. [PubMed]

63. Bonaldo, M.C.; Martins, M.A.; Rudersdorf, R.; Mudd, P.A.; Sacha, J.B.; Piaskowski, S.M.; Neves, P.C.C.; Santana, M.G.V.D.; Vojnov, L.; Capuano, S.; et al. Recombinant yellow fever vaccine virus 17D expressing simian immunodeficiency virus SIVmac239 Gag induces SIV-specific CD8+ T-cell responses in rhesus macaques. *J. Virol.* **2010**, *84*, 3699–3706. [CrossRef] [PubMed]

64. Mudd, P.A.; Piaskowski, S.M.; Neves, P.C.C.; Rudersdorf, R.; Kolar, H.L.; Eernisse, C.M.; Weisgrau, K.L.; de Santana, M.G.V.; Wilson, N.A.; Bonaldo, M.C.; et al. The live-attenuated yellow fever vaccine 17D induces broad and potent T cell responses against several viral proteins in Indian rhesus macaques—Implications for recombinant vaccine design. *Immunogenetics* **2010**, *62*, 593–600. [CrossRef] [PubMed]

65. Neves, P.C.C.; Rudersdorf, R.A.; Galler, R.; Bonaldo, M.C.; de Santana, M.G.V.; Mudd, P.A.; Martins, M.A.; Rakasz, E.G.; Wilson, N.A.; Watkins, D.I. CD8+ γδ TCR+ and CD4+ T cells produce IFN-γ at 5–7 days after yellow fever vaccination in Indian rhesus macaques, before the induction of classical antigen-specific T cell responses. *Vaccine* **2010**, *28*, 8183–8188. [CrossRef] [PubMed]

66. Meier, K.C.; Gardner, C.L.; Khoretonenko, M.V.; Klimstra, W.B.; Ryman, K.D. A mouse model for studying viscerotropic disease caused by yellow fever virus infection. *PLoS Pathog.* **2009**, *5*, e1000614. [CrossRef] [PubMed]

67. Watson, A.M.; Lam, L.K.M.; Klimstra, W.B.; Ryman, K.D. The 17D-204 Vaccine Strain-Induced Protection against Virulent Yellow Fever Virus Is Mediated by Humoral Immunity and CD4+ but not CD8+ T Cells. *PLoS Pathog.* **2016**, *12*, e1005786. [CrossRef] [PubMed]

68. Liu, T.; Chambers, T.J. Yellow fever virus encephalitis: Properties of the brain-associated T-cell response during virus clearance in normal and γ interferon-deficient mice and requirement for CD4+ lymphocytes. *J. Virol.* **2001**, *75*, 2107–2118. [CrossRef] [PubMed]

69. Vandermost, R.; Harrington, L.; Giuggio, V.; Mahar, P.; Ahmed, R. Yellow fever virus 17D envelope and NS3 proteins are major targets of the antiviral T cell response in mice. *Virology* **2002**, *296*, 117–124. [CrossRef] [PubMed]

70. Bassi, M.R.; Larsen, M.A.B.; Kongsgaard, M.; Rasmussen, M.; Buus, S.; Stryhn, A.; Thomsen, A.R.; Christensen, J.P. Vaccination with replication deficient adenovectors encoding YF-17D antigens induces long-lasting protection from severe yellow fever virus infection in mice. *PLoS Negl. Trop. Dis.* **2016**, *10*, e0004464. [CrossRef] [PubMed]

71. Barrett, A.D.T.; Gould, E.A. Comparison of neurovirulence of different strains of yellow fever virus in mice. *J. Gen. Virol.* **1986**, *67*, 631–637. [CrossRef] [PubMed]

72. Watson, A.M.; Lam, L.K.M.; Klimstra, W.B. CD8+ T cells are Required for the Control of the Live Attenuated 17D-204 Yellow Fever Vaccine. Manuscript under preparation.

73. Haring, J.S.; Badovinac, V.P.; Harty, J.T. Inflaming the CD8+ T cell response. *Immunity* **2006**, *25*, 19–29. [CrossRef] [PubMed]

74. Thompson, L.J.; Kolumam, G.A.; Thomas, S.; Murali-Krishna, K. Innate inflammatory signals induced by various pathogens ddifferentially dictate the IFN-I dependence of CD8 T cells for clonal expansion and memory formation. *J. Immunol.* **2006**, *177*, 1746–1754. [CrossRef] [PubMed]

75. Deauvieau, F.; Sanchez, V.; Balas, C.; Kennel, A.; De Montfort, A.; Lang, J.; Guy, B. Innate immune responses in human dendritic cells upon infection by chimeric yellow-fever dengue vaccine serotypes 1–4. *Am. J. Trop. Med. Hyg.* **2007**, *76*, 144–154. [PubMed]
76. Gandini, M.; Reis, S.R.N.I.; Torrentes-Carvalho, A.; Azeredo, E.L.; Freire, M.D.S.; Galler, R.; Kubelka, C.F. dengue-2 and yellow fever 17DD viruses infect human dendritic cells, resulting in an induction of activation markers, cytokines and chemokines and secretion of different TNF-α and IFN-α profiles. *Memórias do Instituto Oswaldo Cruz* **2011**, *106*, 594–605. [CrossRef] [PubMed]
77. Mandl, J.N.; Akondy, R.; Lawson, B.; Kozyr, N.; Staprans, S.I.; Ahmed, R.; Feinberg, M.B. Distinctive TLR7 signaling, type I IFN production, and attenuated innate and adaptive immune responses to yellow fever virus in a primate reservoir host. *J. Immunol.* **2011**, *186*, 6406–6416. [CrossRef] [PubMed]
78. Bruni, D.; Chazal, M.; Sinigaglia, L.; Chauveau, L.; Schwartz, O.; Desprès, P.; Jouvenet, N. Viral entry route determines how human plasmacytoid dendritic cells produce type I interferons. *Sci. Signal.* **2015**, *8*, ra25. [CrossRef] [PubMed]
79. Thibodeaux, B.A.; Garbino, N.C.; Liss, N.M.; Piper, J.; Blair, C.D.; Roehrig, J.T. A small animal peripheral challenge model of yellow fever using interferon-receptor deficient mice and the 17D-204 vaccine strain. *Vaccine* **2012**, *30*, 3180–3187. [CrossRef] [PubMed]
80. Chen, K.; Liu, J.; Cao, X. Regulation of type I interferon signaling in immunity and inflammation: A comprehensive review. *J. Autoimmun.* **2017**. [CrossRef]
81. Ramana, C.V.; Gil, M.P.; Schreiber, R.D.; Stark, G.R. STAT1-dependent and -independent pathways in IFN-γ-dependent signaling. *Trends Immunol.* **2002**, *23*, 96–101. [CrossRef]
82. Douam, F.; Hrebikova, G.; Albrecht, Y.E.S.; Sellau, J.; Sharon, Y.; Ding, Q.; Ploss, A. Single-cell tracking of flavivirus RNA uncovers species-specific interactions with the immune system dictating disease outcome. *Nat. Commun.* **2017**, *8*, 14781. [CrossRef] [PubMed]
83. Greene, J.M.; Burwitz, B.J.; Blasky, A.J.; Mattila, T.L.; Hong, J.J.; Rakasz, E.G.; Wiseman, R.W.; Hasenkrug, K.J.; Skinner, P.J.; O'Connor, S.L.; et al. Allogeneic lymphocytes persist and traffic in feral MHC-matched Mauritian cynomolgus macaques. *PLoS ONE* **2008**, *3*, e2384. [CrossRef] [PubMed]
84. Ross, C.N.; French, J.A.; Ortí, G. Germ-line chimerism and paternal care in marmosets (*Callithrix kuhlii*). *Proc. Natl. Acad. Sci. USA* **2007**, *104*, 6278–6282. [CrossRef] [PubMed]

Review

Non-Canonical Roles of Dengue Virus Non-Structural Proteins

Julianna D. Zeidler, Lorena O. Fernandes-Siqueira, Glauce M. Barbosa and Andrea T. Da Poian *

Instituto de Bioquímica Médica Leopoldo de Meis, Universidade Federal do Rio de Janeiro,
Av. Carlos Chagas Filho, 373, CCS, Bl. E, sala 18, Rio de Janeiro 21941-90, Brazil;
julianna.zeidler@bioqmed.ufrj.br (J.D.Z.); losiqueira@bioqmed.ufrj.br (L.O.F.-S.);
glaucemb@bioqmed.ufrj.br (G.M.B.)
* Correspondence: dapoian@bioqmed.ufrj.br; Tel.: +55-21-3938-6758

Academic Editor: Michael R. Holbrook
Received: 16 December 2016; Accepted: 8 March 2017; Published: 13 March 2017

Abstract: The Flaviviridae family comprises a number of human pathogens, which, although sharing structural and functional features, cause diseases with very different outcomes. This can be explained by the plurality of functions exerted by the few proteins coded by viral genomes, with some of these functions shared among members of a same family, but others being unique for each virus species. These non-canonical functions probably have evolved independently and may serve as the base to the development of specific therapies for each of those diseases. Here it is discussed what is currently known about the non-canonical roles of dengue virus (DENV) non-structural proteins (NSPs), which may account for some of the effects specifically observed in DENV infection, but not in other members of the Flaviviridae family. This review explores how DENV NSPs contributes to the physiopathology of dengue, evasion from host immunity, metabolic changes, and redistribution of cellular components during infection.

Keywords: dengue virus; non-structural proteins; physiopathology; immunity; metabolism

---

1. Introduction

A large set of arthropod-borne diseases is caused by enveloped viruses of the Flaviviridae family, which includes dengue virus (DENV), yellow fever virus (YFV), Japanese encephalitis virus (JEV), West Nile virus (WNV), Zika virus (ZIKV), and about 70 other members. The Flaviviruses genome is a positive-sense single-strand RNA molecule of about 11 kb translated into a polyprotein that is cleaved to generate three structural (capsid, membrane, and envelope—C, prM, E) and seven non-structural proteins (NSPs—NS1, NS2A, NS2B, NS3, NS4A, NS4B, and NS5), which share functional and structural features among the different members of the family [1]. In addition to the similarities, the different outcomes of the distinct diseases caused by the flaviviruses may be partially explained by non-canonical functions of the viral proteins, which have evolved independently in each virus type. Here we considered the canonical roles those that are shared among all (or most) members of the Flaviviridae family, while the non-canonical are those exclusive to a specific virus or not generalized throughout the family. For instance, flaviviruses' NS1 is known to interfere with the proteins of the complement system, but DENV NS1 has unique interaction partners and mediates this function in a particular manner [2–5]. Viral structural proteins, which dictate targeted-cell type specificity and the architecture of viral particles, may also have non-canonical functions, but this is beyond the scope of this review. As NSPs are dispensed from structural roles in mature virus particles, they are more prone to display multifunctional and non-canonical roles, and more recently have been largely studied probably due to their potential as targets for clinical intervention.

DENV is endemic in more than 100 countries and recent estimates predict about 390 million infections per year [6]. Taking into account the high number of dengue cases and the increased geographical extension of the disease, dengue is considered a global public health problem with deep social and economic implications. The clinical manifestations of dengue vary from a mild fever to life-threatening severe diseases (occurring in a small proportion of cases), known as dengue hemorrhagic fever (DHF) and dengue shock syndrome (DSS), which are characterized by an increase in vascular endothelium permeability leading to plasma leakage, which may evolve into a fatal hypovolemic shock. Although a number of dengue vaccines have been developed, some being in clinical trials, they have shown to be limited with regard to low immunogenicity and partial protection against different DENV serotypes [7]. Among these vaccines, Dengvaxia (developed by Sanofi Pasteur, Lyon, France) has been licensed in several countries, but it is still a matter of concern due to the reported low protection against DENV serotype 2, associated with an increased incidence of hospitalization due to severe dengue of seronegative individuals in the third year after the first dose [7,8]. While vaccination strategies are in development and improvement stages, dengue treatment is still mainly based on supportive clinical interventions that do not always prevent the evolution to the severe forms of the disease [9,10].

2. The Canonical Roles of Flaviviruses' NSPs

Before discussing the particular, non-canonical, roles of DENV proteins in infection and disease establishment, we will summarize, in this topic, the functions carried out by each NSP that are shared among the members of the Flaviviridae family. Most of these NSP canonical functions are related to viral replication, which depends on the assembly of a membrane-bound multi-protein replication complex (RC) [11–14], formed by the association of different NSPs with host co-factors on interconnected lipid vesicles derived from the endoplasmic reticulum (ER) [12,13,15,16]. At these sites, viral RNA is transcribed and translated into a polyprotein, which is cleaved by host and viral proteases to originate the individual viral proteins (Figure 1).

NS1 is a conserved glycosylated protein that may occur in different oligomeric forms during the virus replication cycle, namely immature monomers in the ER lumen, stable hydrophobic homodimers able to interact with membranes, or secreted soluble hexamers harboring a large central 10 nm-diameter open barrel filled with lipids [17–20]. In early stages of infection, the dimeric NS1 associates to the viral RC on the ER membrane, probably interacting with the transmembrane proteins NS4A and NS4B [21–24]. In late infection, the secreted hexamers interact with proteins of the complement system, counteracting the cellular responses to infection [4,22,23,25,26].

NS3 behaves as a protease when it uses NS2B as a co-factor, cleaving the polyprotein at specific sites and host proteins that would impair the establishment of the infection [27–29]. NS3 also plays a role in viral RNA replication, acting as an RNA helicase, nucleoside 5′-triphosphatase (NTPase), and RNA 5′-triphosphatase (RTPase) [24].

NS5 is the most conserved protein among flaviviruses' NSPs, being responsible for the capping, methylation, and replication of the viral genome [30–33]. A NS5 dimer associates with NS3 and NS2B, to form the RC on ER-derived membranes. This trimeric complex is essential for the establishment of the protein-protein and protein-RNA interactions necessary for the polymerization reaction, as also observed for the polymerases of other viruses [34].

The functions carried out by the transmembrane proteins NS2A, NS2B, NS4A, and NS4B are less understood and, for this reason, it is difficult to define the canonical roles of these proteins. They do not have a known enzymatic activity, but act as scaffolds for RC formation [1,11]. For Kunjin virus (KUNV), it was shown that NS2A co-localizes with double-stranded RNA formed during genome replication, binding to the viral RNA 3′UTR, and to NS1, NS3, and NS5 to form the RC [31]. NS2A also participates in the rearrangement of cellular membranes observed during the replication of flaviviruses [32]. NS2B acts as a cofactor for NS3 proteolytic activity. NS4A and NS4B are connected by a transmembrane peptide named 2K, which is cleaved during NS4A maturation. 2K peptide maintains

NS4B associated to ER, even after its cleavage from the polyprotein [33]. NS4A induces the rearranging of ER membranes seen during the infection [35]. In addition, besides participating in viral replication, NS4A promotes autophagy, preventing infection-induced cell death [36]. For WNV, it was observed that NS4A controls the ATPase activity of NS3, while for DENV, NS4B was found to interact with the NS3 helicase domain, assisting its dissociation from the RNA strand. Thus, it seems that NS4A and NS4B work co-operatively during viral replication [37].

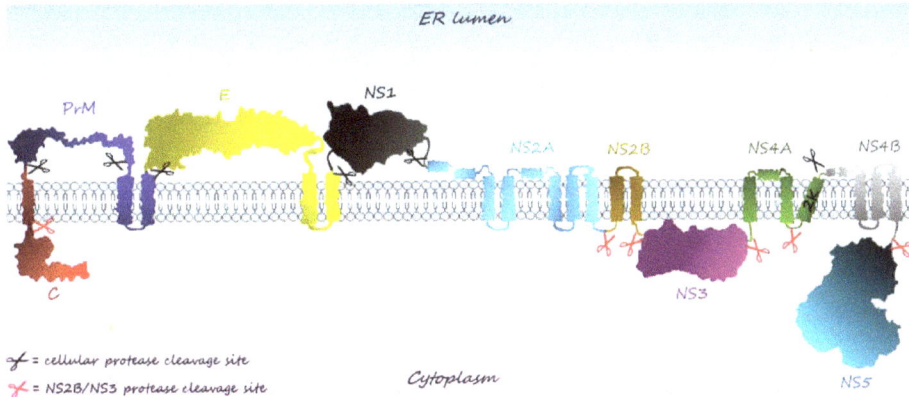

Figure 1. Schematic representation of the flaviviruses' polyprotein. Viral RNA encodes a polyprotein that is co- and post-translationally processed by host proteases (black scissors) or by the viral protease NS2B/NS3 (red scissors) to generate the structural (C, PrM, and E) and non-structural proteins (NS1, NS2A, NS2B, NS3, NS4A, NS4B, and NS5), represented in different colors.

3. Participation of DENV NSPs in Dengue Physiopathology

DHF/DSS pathogenesis encompasses a series of events driven by a cytokine storm triggered during infection that leads to an abrupt increase in endothelial permeability followed by plasma leakage, disseminated intravascular coagulation, and hemorrhage, which may progress to a fatal hypovolemic shock [9]. Among the possible molecular players that mediate these events, DENV NS1 is the more extensively studied. High levels of NS1 are found circulating in the blood of patients [38], and this correlates with the development of DHF [39,40]. Additionally, the detection of NS1 in patients' sera has driven the development of diagnostic tests for dengue [39,40]. Although there are a number of questions regarding this issue that are still not completely answered, some findings start to shed light in the roles of DENV NS1 during infection and disease.

The hexameric form of NS1 circulates associated triglycerides, cholesterols, and phospholipids, a content similar to that present in high-density lipoproteins (HDL) [41]. This, together with the fact that NS1 hexamers were shown to have tropism to the liver when injected in mice [42], and that NS1 internalization was observed in human-cultured hepatocytes, suggested that NS1 would carry lipids in the plasma of dengue patients from tissues to the liver. The implications of these findings are still not understood, but considering that lipoproteins participate in vascular homeostasis, it was hypothesized that the transport of lipids by NS1 may play a role in dengue physiopathology [41].

Many other studies point to a direct role of NS1 in a crucial event of severe dengue physiopathology: the plasma leakage. For instance, antibodies produced against NS1 during infection cross-react with surface antigens of endothelial cells and platelets, possibly contributing to the plasma leakage, dehydration and hypovolemic shock that are observed in the severe forms of dengue [43–45]. NS1 was shown to bind to both prothrombin and thrombin in dengue patients' sera. Thrombin formation is the center of the coagulation cascade, which starts with the release of tissue factor from damaged cells and subsequent activation of an amplifying cascade generating large amounts of factor X from a few initial signaling molecules [46]. Factor X activates prothrombin, generating thrombin, which is able to cleave the fibrinogen to insoluble fibrin fibers, besides being able to cleave prothrombin itself and activating factor XIII, accelerating the coagulation reaction. In vitro assays revealed that recombinant NS1 did not interfere with thrombin activity, but it inhibited prothrombin activation and prolonged the activated partial thromboplastin time (APTT). Thus, NS1 may be responsible for abnormal APTT usually found in the first weeks of dengue onset, and possibly also contribute to plasma leakage by antibody-independent mechanisms that occur in the severe forms of the disease [47]. Additionally, NS1 was found to activate mouse macrophages and human mononuclear cells, inducing the production of proinflammatory cytokines/chemokines via activation of Toll-like receptor 4 (TLR4), which promote the disruption of endothelial cell monolayer integrity and vascular leak, leading the authors to consider NS1 a viral toxin that acts similarly to bacterial endotoxin lipopolysaccharide (LPS) [48]. Accordingly, inoculation of NS1 in mice caused endothelial dysfunction leading to plasma leakage [49]. Another mechanism proposed for NS1-mediated endothelial hyperpermeability includes the induction of sialidases and heparanases expression and activation of the lysosomal protease cathepsin L in endothelial cells, resulting in the degradation of the glycocalyx barrier, a key regulator of vascular permeability [50]. The evidences presented above suggest that NS1-induced plasma leakage is a complex phenomenon resulting from a series of events encompassing endothelial cells, cells from the immune system, and platelets.

Cellular activation and production of large amounts of immune mediators (the "cytokine storm") are known to contribute to dengue pathogenesis and some evidence points to the contribution of DENV NSPs to this scenario. Monomeric NS1 is hydrophilic, but its dimeric form associates to membranes [17,21] and this seems to happen due to the presence of the glycosylphosphatidylinositol (GPI) anchor in NS1 [51], a post-translational modification that targets proteins to membranes [52]. Antibody-induced signal transduction is a common feature of GPI-linked proteins [53–55] and this was found to happen with NS1, being another possible determinant that contributes to cellular activation, cytokine storm, and dengue pathogenesis [51]. Regarding other DENV NSPs, the expression of NS5 was shown to promote the induction of IL-8 expression and secretion in HEK293-transfected cells [56], and both NS5 and NS4B induce the secretion of immunomediators in THP-1 monocytes, being this action enhanced when NS4B is linked to the 2K peptide [57]. In addition, it was observed that the conditioned media of 2K-NS4B-transfected THP1 cells contained IL-8 and TNFα in levels that induce endothelial cell permeability and increased expression of adhesion molecules [58]. NS5 also interacts with death-domain-associate protein (Daxx), a transcription repressor shown to be important to induce expression of the cytokine RANTES [59]. In addition, the ability of NF-κB to bind RANTES promoter is increased in cells expressing NS5 [60], suggesting that NS5 may contribute to DENV pathogenesis by inducing increases of cytokine expression.

4. Evasion from Host Innate Immune Response

To establish a successful infection, viruses need to escape from the complex and robust host immune defenses. The known mechanisms by which DENV evades the host immune system were previously reviewed elsewhere [61,62]. Here, we intend to provide an update of the more recent findings on this issue, focusing on the roles of NSPs. Many DENV NSPs interfere with the host immune response by subverting complement activation, pathways triggered by pattern recognition receptors (PRRs), as well as interferon (IFN)-mediated signaling pathways (Figure 2). It is interesting to note that each DENV NSP acts on multiple points of cellular immune response, revealing the multifunctionality of viral proteins and supporting the idea that viral apparatus evolved to constitute a robust and efficient network against an equally complex system, such as the cellular innate immune system.

The complement system is an important component of the immune response, linking adaptive and innate immunities. Activation of the complement system is triggered by different pathways, as depicted in detail in Figure 2A. Flaviviruses' NS1s are known to interfere with the complement system at different points [2]. In the case of DENV, NS1 inhibits both the classical and the lectin pathways (Figure 2B). NS1 N-linked glycans interact with C4, C4b, C1s proenzyme, and C1s, especially when NS1 is arranged in its hexameric form [3]. The formation of the C4-NS1-C1s/C1 complex results in C4 degradation, impairing complement activation [2]. In addition, DENV NS1 binds and recruits C4BP to the cellular membrane. C4BP is, a negative regulator of complement pathways that promotes dissociation of C4bC2a C3 convertase and acts as a cofactor for cleavage of C4b, so that NS1 binding inactivates C4b on the cell membrane [4,63]. NS1 also interacts with C1q, but the effects of this interaction on inactivation of complement cascade have not been explored so far [5]. Regarding the lectin pathway, MBL has been identified as an NS1 target both in mammalian and mosquito cells [64], protecting infected cells from immune recognition and impairing virus neutralization by complement activation.

Innate immune response against viruses depends on the recognition of pathogen-associated molecular patterns (PAMPs) by PRRs, which trigger signaling pathways that ultimately activate the production of type I IFN and pro-inflammatory cytokines (Figure 2A). Viral RNA recognition is mediated by proteins of the DExD/H-box RNA helicases family, such as the retinoic acid-inducible gene I (RIG-I) and melanoma differentiation-associated gene 5 (MDA5) [65]. Briefly, the interaction between these receptors and the viral RNA allows it to associate to the mitochondrial antiviral signaling protein (MAVS), which activates TANK-protein kinase 1 (TBK1) and IκB kinases (IKK) [66]. These kinases mediate the activation of IFN regulatory factors (IRF), which migrate to the nucleus inducing the expression of type I IFN and pro-inflammatory cytokines. Recently, an adaptor protein named as "stimulator of the interferon gene" (STING) has also been identified as a downstream effector of RIG-I [67]. Mitochondrial-associated membranes (MAM), regions of the ER that are closely juxtaposed to mitochondria and constitute sites of communication and lipid exchange between these organelles, are important for MAVS signaling during an infection, and processes that lead to MAM disruption block the interaction between MAVS and STING, inhibiting the downstream antiviral signaling [67]. With regard to activation of PRR-mediated pathways, our group has shown that DENV infection induces the production of IFN-I and pro-inflammatory cytokines in a manner dependent on the activation of RIG-I [68]. To alleviate host antiviral response, DENV interferes with this pathway at multiple points. It was found that NS2B/NS3 interacts with IKKε, masking the kinase domain and consequently preventing the phosphorylation of IRF3 [69]. Similarly, NS2A and NS4B were shown to inhibit phosphorylation of TANK-binding kinase (TBK1) and its substrate IRF3 [70]. Another member of the DExD/H box helicases family that has a less clear role in innate immune response, DDX21, is a target of DENV NS2B/NS3 protease and its degradation was shown to facilitate DENV replication [71]. NS2B/NS3 also cleaves/inactivates STING, inhibiting the TBK1-mediated IFN expression [28,29]. DENV NS4B was shown to interact with C- and N-terminal domains of MAVS, including its CARD domains, to which RIG-I binds, probably suppressing the oligomerization of MAVS, abrogating its interaction with downstream adaptor proteins and inhibiting IFN production [72]. In addition to participating in the induction of convoluted membranes (CM) in ER (see also the next topic), recently it was shown that NS4B also induces mitochondria elongation at ER-mitochondria contact sites, which was shown to favor DENV replication by impairing translocation of RIG-I to MAMs, impairing the innate immune response [73].

Finally, DENV NSPs interfere with the IFN signaling pathway, an essential host defense against many viruses, including the flaviviruses. Type I IFN (IFNα/β) mediates antiviral responses in an autocrine and paracrine fashion by increasing the expression of hundreds of IFN-stimulated genes (ISGs) [74]. Briefly, binding of type I IFNs to the heterodimeric IFNα receptor (IFNAR, which is composed of the IFNAR1 and IFNAR2 subunits) activates Janus kinase 1 (JAK1) and tyrosine kinase 2 (TYK2), which phosphorylate the signal transducer and activator of transcription (STAT) proteins that, when phosphorylated, dimerize and translocate to nucleus (Figure 2A). STAT1–STAT2 heterodimers associate to IFN regulatory factor 9 (IRF9), and this trimeric complex binds to IFN-stimulated response elements (ISRE), activating the transcription of ISGs. Type I and type II IFNs can also induce homodimerization of STAT1 or STAT3, which translocate to the nucleus and bind to gamma-activated sequences (GAS), stimulating the production of either pro- or anti-inflammatory cytokines. DENV NS1 was identified as a ligand of human STAT3 protein [75], but the implications of this finding are still poorly understood. DENV NS4B, as well as NS4B of other flaviviruses, such as West Nile and yellow fever viruses, has anti-IFN activity [76] by blocking STAT-1 phosphorylation in cells stimulated with IFN [77]. At least for DENV, the 2K peptide is required for the anti-IFN function. In addition, the co-expression of cleaved NS4A and NS4B (mediated by NS2B/NS3 protease) contribute to further ISRE promoter inhibition [76]. So, NS4A, NS4B, and NS2B/NS3 together seem to promote an efficient inhibition of IFN-stimulated signaling pathway during DENV infection. Furthermore, a high-throughput yeast two-hybrid screening showed that flaviviruses' NS5 (including of DENV NS5) also inhibits IFN-mediated signaling [78]. In the case of DENV NS5, it was shown that it mediates STAT2 degradation [79]. Thus, it seems that inhibition of the IFN pathway is important for flaviviruses' infection and different DENV NSPs cooperate to perform anti-INF functions (Figure 2B).

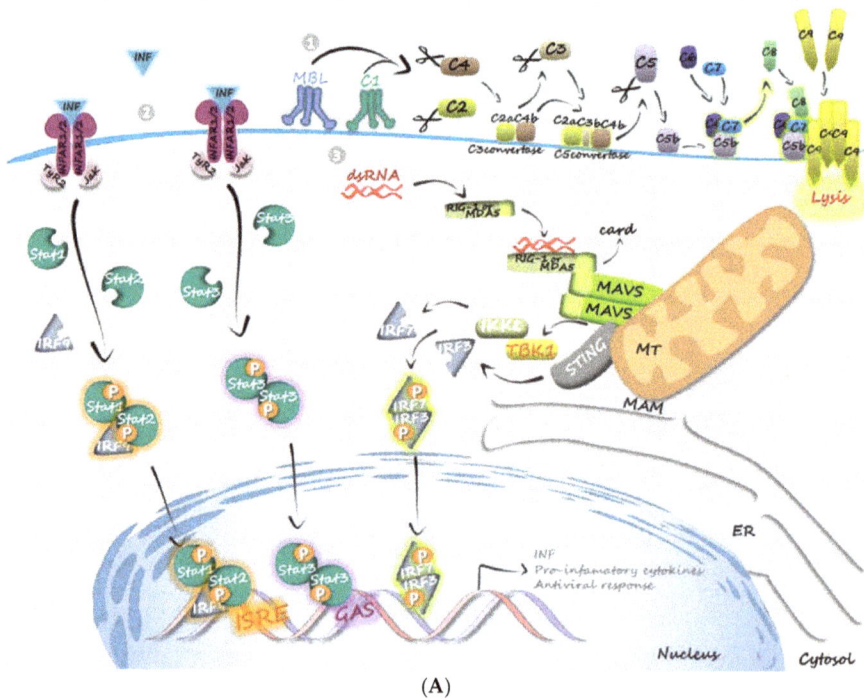

(A)

Figure 2. *Cont.*

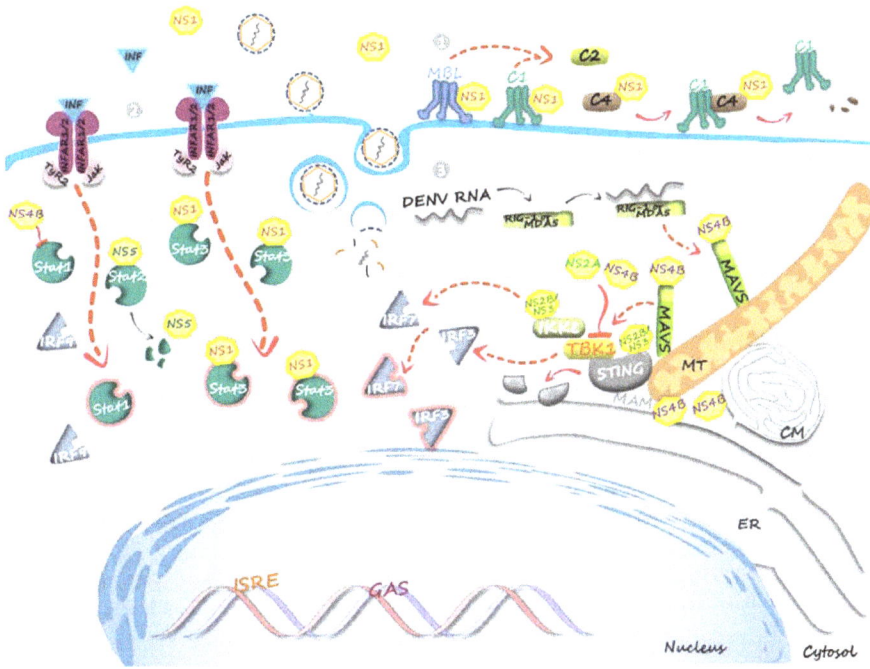

(B)

Figure 2. Involvement of DENV NSPs in the evasion of host innate immune response. (**A**) Pathways of immune response affected by DENV NSPs. (1) Activation of the complement system is triggered by different pathways that converge to the cleavage of factor C3 by the protease C3 convertase. This enzyme formed by the association of two other cleavage products: C4b, a fragment of C4, and C2a, a fragment of C2. The cleavage of C4 and C2 may be catalyzed by two different pathways: the classical pathway, triggered by C1 binding to antigen-antibody complexes, or by the lectin pathway, in which a carbohydrate recognition receptor, such as mannose binding lectin (MBL), associates to a serine protease after binding to carbohydrates. One of the products of C3 cleavage, C3b, binds to C3 convertase changing its substrate specificity, so that the enzyme becomes a C5 convertase. The fragment C5b, generated from the cleavage of C5, binds to the infected cell membrane, initiating the assembly of a complex formed by C6, C7, C8, and C9, which promotes cell lysis. (2) Viral dsRNAs produced during the replication of RNA viruses are recognized by PRRs. Binding of dsRNA leads the cytosolic PRRs to associate with the mitochondrial antiviral signaling protein (MAVS) through its caspase-recruitment domain (CARD), recruiting the TANK-protein kinase 1 (TBK1) and IκB kinase-ε (IKKε). These kinases phosphorylate IRF-3, which forms homodimers or heterodimers with IRF-7, which, in turn, translocate to the nucleus, inducing the expression of type I IFN and pro-inflammatory cytokines. This pathway also includes the participation of the adaptor protein STING, which acts in mitochondrial-associated membrane (MAM) to mediate RIG-I downstream signaling. (3) Type I IFN-mediated antiviral responses occurs via the expression of several IFN-stimulated genes (ISGs). IFN binding to its heterodimeric IFN-α receptor (IFNAR1/2) activates Janus kinase 1 (JAK1) or tyrosine kinase 2 (TYK2), leading to the phosphorylation of the signal transducer and activator of transcription (STAT) proteins, which dimerize and translocate to the nucleus. STAT1–STAT2 heterodimer binds to IFN regulatory factor 9 (IRF9) and migrates to the nucleus, inducing the expression of ISGs through its binding to the IFN-stimulated response elements (ISRE). Type I and type II IFNs can also induce dimerization of STAT3, which translocate to the nucleus, where it binds to gamma-activated sequences (GAS), stimulating the production of both pro- and anti-inflammatory cytokines; (**B**) Participation of DENV NSPs in the evasion of the host immune response. (1) DENV NS1 inhibits complement activation by interacting with different components of the complement system, including C1 proenzyme, C1s, C4, C4b, and MBL. The formation of the complex C4-NS1-C1s/C1 results in degradation of C4, impairing the formation of C3 convertase. NS1 binding to MBL protects DENV against MBL-mediated virus neutralization by the lectin pathway of complement activation. (2) DENV NSPs impair the innate immune response mediated by viral dsRNA recognition. NS4B interacts with the CARD domain of MAVS, impairing its binding to the cytoplasmic PRRs. Moreover, this protein, by inducing the formation of convoluted membranes (CM) and promoting mitochondrial elongation, inhibits the translocation of PRRs to MAMs. NS2B/NS3 interacts with IKKε and cleaves STING and NS2A together with NS4B, inhibiting the phosphorylation of TBK1 and its substrate IRF3. These steps impair the activation of transcription factors IRF-3 and IRF-7. (3) DENV NSPs inhibit INF-stimulated signaling in different points. NS4B interacts with STAT1, blocking its phosphorylation, and NS5 mediates STAT2 degradation, so both proteins inhibit the expression of ISGs by interfering in ISRE activation. Additionally, NS1 interacts with STAT3, inhibiting the formation of its homodimers, thus preventing GAS-induced gene expression. Red arrows represent the events induced by NSPs, while dashed red arrows represent those ones that are blocked by NSPs. ER, endoplasmic retculum; MT, mitochondria.

5. Metabolic Alterations

Alterations in host cell metabolism caused by virus infection have been studied for many years. One of the first investigations in this sense was published in 1928 by Crabtree, where he described changes in oxidative/glycolytic patterns of tissues infected by viruses when compared to the non-infected tissues [80]. Now it is clearer that virus-induced host metabolic re-programming can affect diverse pathways, including not only glycolysis and oxidative phosphorylation, but also the pentose phosphate pathway, pyrimidine, fatty acids, and glutamine metabolisms [81]. These metabolic alterations usually favor viral replication by increasing cellular energy charge and/or supplying substrates required for production of viral progeny.

In the case of DENV, a crescent number of reports on virus-induced metabolic alterations is now appearing in the literature. For instance, our group showed that human hepatocytes infected with DENV displayed morphologically altered and uncoupled mitochondria, decreasing cellular energy charge and causing metabolic stress [82]. Other studies have shown that DENV is able to induce changes in specific metabolic routes in host cells, such as an increase in glycolysis in primary human foreskin fibroblasts [83] and an increased autophagy-mediated mobilization of fatty acids in human hepatocytes [84]. However, the molecular players that mediate these metabolic changes have only started to be elucidated.

A typical cellular alteration observed during flaviviruses infection is the formation of virus-induced ER membrane rearrangements, which, just like a virus factory, creates a subcompartimentalization for the viral replication cycle steps—RNA translation and replication, as well as the virion assembly—allowing proper environments for the coordinated performance of each process without mutual interference. Each flavivirus induces a particular membrane rearrangement, and in the case of DENV, at least three distinct ER membrane-derived structures are formed: single membrane invaginations into the ER lumen (resembling vesicle packets), where the RC is assembled and genome replication takes place; unstructured convoluted membranes (CM), where viral polyprotein is processed and the resulting proteins accumulate to be later used for RC and virion assembly; and membranes associated with the assembly of new viral particles [85,86]. For mounting such a complex structure, which practically consists in an new organelle induced by infection, cellular metabolism should be directed to the mobilization of different substrates, especially lipids, which are essential for membrane formation, besides consisting of an important energy source for virus replication. In this context, an interesting finding was that NS3 interacts with fatty acid synthase (FASN), recruiting this enzyme to sites of viral replication [87] (Figure 3). In addition, interaction between recombinant NS3 and FASN was shown to stimulate FASN activity in vitro, suggesting that this would be a means of increasing de novo FA biosynthesis during DENV infection. The role of NS3 in redistributing FASN to sites of viral replication was later found to be dependent on NS3 interaction with the GTPase Rab18 [88], a protein localized on lipid droplets (LD) which mediate the apposition of these organelles and the ER-derived membranes, enhancing the formation of LD-associated membranes (LAM) [89]. On the other hand, in agreement with the increased requirement of energy supplies during virus replication, DENV infection was shown to increase fatty acid (FA) β-oxidation [84]. At first glace, this would seem contradictory, but it can be explained by differences in subcellular localization of each process: while FA biosynthesis occurs at viral replication sites, providing lipids for membrane synthesis, FA seems to be mobilized from LD to be oxidized in mitochondria, providing energy for virus replication. In addition to activating FA synthesis at the sites of viral replication, NS3/Rab18-induced LAM may facilitate the removal of FA from LD for β-oxidation, which, as mentioned, is increased during infection [84]. So, by managing subcellular localization of specific cellular components, DENV NSP can coordinately stimulate opposite processes that are equally important for formation of new viral particles.

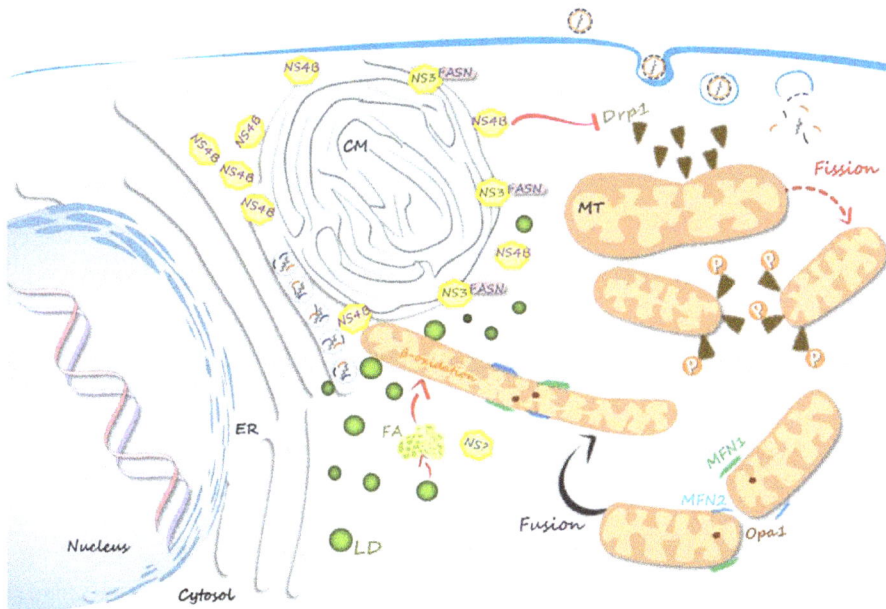

Figure 3. DENV NSPs-induced host metabolic alterations. During infection, DENV NS4B induces the formation of the convoluted membrane (CM), the membranous structure where viral polyprotein is processed and the resulting proteins accumulate. NS3 recruits the enzyme fatty acid synthase (FASN) to the virus replication sites, besides stimulating its enzymatic activity, increasing de novo FA biosynthesis, which provides lipids for inducing the formation of CM, as well as for the assembly of the viral envelope. Fatty acids (FA) are mobilized from lipid droplets (LD) to undergo β-oxidation mitochondria, providing energy to the high-energy demanding virus replication process. NS4B is able to inhibit phosphorylation of cytoplasmic protein dynamin-related protein 1 (Drp1), preventing the mitochondrial fission process. Fused mitochondria accumulate in infected cells, increasing the efficiency of the oxidative metabolism. Red arrows represent the events induced by NSPs, while dashed red arrows represent those ones that are blocked by NSPs. MFN1, mitofusin 1; MFN2, mitofusin 2; Opa1, optic atrophy protein 1; ER, endoplasmic reticulum; MT, mitochondria.

Mitochondria are dynamic organelles that undergo fusion and fission events that ultimately control the rate of oxidative metabolism. Mitochondrial membrane proteins, such as mitofusin 1 (Mfn1), mitofusin 2 (Mfn2), and optic atrophy protein 1 (Opa1), mediate mitochondrial fusion, while phosphorylation the cytoplasmic protein dynamin-related protein 1 (Drp1) promotes its oligomerization on the mitochondrial membrane, inducing the fission of the organelle (Figure 3). While elongated mitochondria are associated to a high respiratory activity, fragmented mitochondria occur predominantly in resting cells or other situations of low energy demand [90].

Among the DENV NSPs, NS4B seems to be the major player in mediating the alterations in mitochondrial morphology. It inhibits the phosphorylation of Drp1, preventing its translocation to the mitochondrial membrane, ultimately inducing the accumulation of elongated mitochondria in the infected cells [73,91]. Mitochondria elongation was shown to be essential for viral replication since the induction of mitochondria fission by overexpression of Drp1 inhibited viral replication [91], while silencing Mfn2 or Drp1 expression inhibited or stimulated DENV replication, respectively [73]. In addition, DENV-infected Huh7 cells, which harbor elongated mitochondria, display increased respiratory rates associated with ATP production [91], indicating that the alterations in mitochondrial morphology induced by infection ensures the high-energy supplies required for virus replication. Then, DENV NSPs seems to promote the assembly of a complex cellular machinery next to sites of virus replication, involving close apposition of the ER with elongated mitochondria and LDs, which may facilitate virus production by providing large amounts of lipids required for viral RNA replication and virion assembly, as well as the energy supply required for virus replication.

6. NSP-Induced Re-Localization of Host Proteins

In addition to recruiting FASN to the RC, DENV NSPs are also involved in the re-localization of other cellular proteins. Positive- and negative-strand RNA viruses usually use nuclear components in their replication [92]. So, localization of RC in the perinuclear region seems to be important in this sense, facilitating the recruitment of nuclear proteins, which allow an efficient replication of the viral genome. A question that is raised in this context is what stabilizes RC in the perinuclear site? At least for DENV infection, this seems to be mediated by the interaction of NS4A with the vimentin scaffold [93], a component of intermediary filaments important for vesicular/organelle transport and positioning [94]. In addition to other evidence, it was shown that vimentin knockdown dissociates DENV RC, which becomes dispersed throughout the cytoplasm, suggesting its role in anchoring RC in the perinuclear region during infection [93].

Furthermore, it was found that the interaction of DENV NS1 with the ribosomal protein RPL-18 and its recruitment to the perinuclear region is required for viral translation and replication [95]. In another study, NS1 was also shown to promote changes in subcellular localization of GAPDH from cytoplasmic to perinuclear regions, resulting in an increase in the enzyme activity [96]. The redistribution of GADPH to sites of viral replication was also described for many other viruses, such as Japanese encephalitis virus [97], parainfluenza virus [98], hepatitis A virus [99], hepatitis C virus [100], and others [101,102]. In all cases, this re-localized pool of GAPDH was shown to bind viral RNA, which seems to be involved in the regulation of viral transcription and/or replication, and not an effect on glycolysis, which would be expected since GAPDH is a glycolysis enzyme. Further studies need to be conducted to determine if GAPDH plays a role in DENV replication, as happens with other viruses.

With a less understood function, DENV2 NS3 subverts nuclear receptor binding protein (NRBP), a protein associated with trafficking between the endoplasmic reticulum (ER) and Golgi, recruiting it to the perinuclear region [103]. It is possible that this is important to regulate the transport of lipids and membranes to the RC, or even play a role in delivering viral particles to the Golgi/secretory pathway, but this should be investigated further.

Acknowledgments: This work was supported by Fundação Carlos Chagas Filho de Amparo à Pesquisa do Estado do Rio de Janeiro, (FAPERJ E-26/201.167/2014) and Conselho Nacional de Desenvolvimento Científico e Tecnológico (MCT/CNPq 306669/2013-7). Julianna D. Zeidler is recipient of a post-doctoral fellowship from Fundação Carlos Chagas Filho de Amparo à Pesquisa do Estado do Rio de Janeiro (FAPERJ E-26/200.460/2015).

Conflicts of Interest: The authors declare no conflict of interest.

References

1. Apte-Sengupta, S.; Sirohi, D.; Kuhn, R.J. Coupling of replication and assembly in flaviviruses. *Curr. Opin. Virol.* **2014**, *9*, 134–142. [CrossRef] [PubMed]
2. Avirutnan, P.; Fuchs, A.; Hauhart, R.E.; Somnuke, P.; Youn, S.; Diamond, M.S.; Atkinson, J.P. Antagonism of the complement component C4 by flavivirus nonstructural protein NS1. *J. Exp. Med.* **2010**, *207*, 793–806. [CrossRef] [PubMed]
3. Somnuke, P.; Hauhart, R.E.; Atkinson, J.P.; Diamond, M.S.; Avirutnan, P. N-linked glycosylation of dengue virus NS1 protein modulates secretion, cell-surface expression, hexamer stability, and interactions with human complement. *Virology* **2011**, *413*, 253–264. [CrossRef] [PubMed]
4. Avirutnan, P.; Hauhart, R.E.; Somnuke, P.; Blom, A.M.; Diamond, M.S.; Atkinson, J.P. Binding of flavivirus nonstructural protein NS1 to C4b binding protein modulates complement activation. *J. Immunol.* **2011**, *187*, 424–433. [CrossRef] [PubMed]
5. Silva, E.M.; Conde, J.N.; Allonso, D.; Nogueira, M.L.; Mohana-Borges, R. Mapping the interactions of dengue virus NS1 protein with human liver proteins using a yeast two-hybrid system: Identification of C1q as an interacting partner. *PLoS ONE* **2013**, *8*, e57514.
6. Bhatt, S.; Gething, P.W.; Brady, O.J.; Messina, J.P.; Farlow, A.W.; Moyes, C.L.; Drake, J.M.; Brownstein, J.S.; Hoen, A.G.; Sankoh, O.; et al. The global distribution and burden of dengue. *Nature* **2013**, *496*, 504–507. [CrossRef] [PubMed]
7. Liu, Y.; Liu, J.; Cheng, G. Vaccines and immunization strategies for dengue prevention. *Emerg. Microbes Infect.* **2016**, *5*. [CrossRef] [PubMed]
8. Halstead, S.B. Critique of World Health Organization Recommendation of a Dengue Vaccine. *J. Infect. Dis.* **2016**, *214*, 1793–1795. [CrossRef] [PubMed]
9. Guzman, M.G.; Harris, E. Dengue. *Lancet* **2015**, *385*, 453–465. [CrossRef]
10. *Dengue: Guidelines for Diagnosis, Treatment, Prevention and Control*, new ed.; World Health Organization: Geneva, Switzerland, 2009.
11. Zou, J.; Xie, X.; Wang, Q.Y.; Dong, H.; Lee, M.Y.; Kang, C.; Yuan, Z.; Shi, P.Y. Characterization of dengue virus NS4A and NS4B protein interaction. *J. Virol.* **2015**, *89*, 3455–3470. [CrossRef] [PubMed]
12. Zou, J.; Lee le, T.; Wang, Q.Y.; Xie, X.; Lu, S.; Yau, Y.H.; Yuan, Z.; Geifman Shochat, S.; Kang, C.; Lescar, J.; et al. Mapping the Interactions between the NS4B and NS3 proteins of dengue virus. *J. Virol.* **2015**, *89*, 3471–3483. [CrossRef] [PubMed]
13. Mackenzie, J. Wrapping things up about virus RNA replication. *Traffic* **2005**, *6*, 967–977. [CrossRef] [PubMed]
14. Salonen, A.; Ahola, T.; Kaariainen, L. Viral RNA replication in association with cellular membranes. *Curr. Top. Microbiol. Immunol.* **2005**, *285*, 139–173. [PubMed]
15. Welsch, S.; Miller, S.; Romero-Brey, I.; Merz, A.; Bleck, C.K.; Walther, P.; Fuller, S.D.; Antony, C.; Krijnse-Locker, J.; Bartenschlager, R. Composition and three-dimensional architecture of the dengue virus replication and assembly sites. *Cell. Host Microbe* **2009**, *5*, 365–375. [CrossRef] [PubMed]
16. Miller, S.; Krijnse-Locker, J. Modification of intracellular membrane structures for virus replication. *Nat. Rev. Microbiol.* **2008**, *6*, 363–374. [CrossRef] [PubMed]
17. Winkler, G.; Maxwell, S.E.; Ruemmler, C.; Stollar, V. Newly synthesized dengue-2 virus nonstructural protein NS1 is a soluble protein but becomes partially hydrophobic and membrane-associated after dimerization. *Virology* **1989**, *171*, 302–305. [CrossRef]
18. Winkler, G.; Randolph, V.B.; Cleaves, G.R.; Ryan, T.E.; Stollar, V. Evidence that the mature form of the flavivirus nonstructural protein NS1 is a dimer. *Virology* **1988**, *162*, 187–196. [CrossRef]
19. Parrish, C.R.; Woo, W.S.; Wright, P.J. Expression of the NS1 gene of dengue virus type 2 using vaccinia virus. Dimerisation of the NS1 glycoprotein. *Arch. Virol.* **1991**, *117*, 279–286. [CrossRef] [PubMed]

20. Flamand, M.; Megret, F.; Mathieu, M.; Lepault, J.; Rey, F.A.; Deubel, V. Dengue virus type 1 nonstructural glycoprotein NS1 is secreted from mammalian cells as a soluble hexamer in a glycosylation-dependent fashion. *J. Virol.* **1999**, *73*, 6104–6110. [PubMed]

21. Mackenzie, J.M.; Jones, M.K.; Young, P.R. Immunolocalization of the dengue virus nonstructural glycoprotein NS1 suggests a role in viral RNA replication. *Virology* **1996**, *220*, 232–240. [CrossRef] [PubMed]

22. Suthar, M.S.; Diamond, M.S.; Gale, M., Jr. West Nile virus infection and immunity. *Nat. Rev. Microbiol.* **2013**, *11*, 115–128. [CrossRef] [PubMed]

23. Muller, D.A.; Young, P.R. The flavivirus NS1 protein: Molecular and structural biology, immunology, role in pathogenesis and application as a diagnostic biomarker. *Antivir. Res.* **2013**, *98*, 192–208. [CrossRef] [PubMed]

24. Li, H.; Clum, S.; You, S.; Ebner, K.E.; Padmanabhan, R. The serine protease and RNA-stimulated nucleoside triphosphatase and RNA helicase functional domains of dengue virus type 2 NS3 converge within a region of 20 amino acids. *J. Virol.* **1999**, *73*, 3108–3116. [PubMed]

25. Akey, D.L.; Brown, W.C.; Dutta, S.; Konwerski, J.; Jose, J.; Jurkiw, T.J.; DelProposto, J.; Ogata, C.M.; Skiniotis, G.; Kuhn, R.J.; et al. Flavivirus NS1 structures reveal surfaces for associations with membranes and the immune system. *Science* **2014**, *343*, 881–885. [CrossRef] [PubMed]

26. Krishna, V.D.; Rangappa, M.; Satchidanandam, V. Virus-specific cytolytic antibodies to nonstructural protein 1 of Japanese encephalitis virus effect reduction of virus output from infected cells. *J. Virol.* **2009**, *83*, 4766–4777. [CrossRef] [PubMed]

27. Ramanathan, M.P.; Chambers, J.A.; Pankhong, P.; Chattergoon, M.; Attatippaholkun, W.; Dang, K.; Shah, N.; Weiner, D.B. Host cell killing by the West Nile Virus NS2B-NS3 proteolytic complex: NS3 alone is sufficient to recruit caspase-8-based apoptotic pathway. *Virology* **2006**, *345*, 56–72. [CrossRef] [PubMed]

28. Yu, C.Y.; Chang, T.H.; Liang, J.J.; Chiang, R.L.; Lee, Y.L.; Liao, C.L.; Lin, Y.L. Dengue virus targets the adaptor protein MITA to subvert host innate immunity. *PLoS Pathog.* **2012**, *8*, e1002780. [CrossRef] [PubMed]

29. Zhong, B.; Yang, Y.; Li, S.; Wang, Y.Y.; Li, Y.; Diao, F.; Lei, C.; He, X.; Zhang, L.; Tien, P.; et al. The adaptor protein MITA links virus-sensing receptors to IRF3 transcription factor activation. *Immunity* **2008**, *29*, 538–550. [CrossRef] [PubMed]

30. Egloff, M.P.; Decroly, E.; Malet, H.; Selisko, B.; Benarroch, D.; Ferron, F.; Canard, B. Structural and functional analysis of methylation and 5′-RNA sequence requirements of short capped RNAs by the methyltransferase domain of dengue virus NS5. *J. Mol. Biol.* **2007**, *372*, 723–736. [CrossRef] [PubMed]

31. Dong, H.; Chang, D.C.; Hua, M.H.; Lim, S.P.; Chionh, Y.H.; Hia, F.; Lee, Y.H.; Kukkaro, P.; Lok, S.M.; Dedon, P.C.; et al. 2′-*O* methylation of internal adenosine by flavivirus NS5 methyltransferase. *PLoS Pathog.* **2012**, *8*, e1002642. [CrossRef] [PubMed]

32. Egloff, M.P.; Benarroch, D.; Selisko, B.; Romette, J.L.; Canard, B. An RNA cap (nucleoside-2′-O-)-methyltransferase in the flavivirus RNA polymerase NS5: Crystal structure and functional characterization. *EMBO J.* **2002**, *21*, 2757–2768. [CrossRef] [PubMed]

33. Ackermann, M.; Padmanabhan, R. De novo synthesis of RNA by the dengue virus RNA-dependent RNA polymerase exhibits temperature dependence at the initiation but not elongation phase. *J. Biol. Chem.* **2001**, *276*, 39926–39937. [CrossRef] [PubMed]

34. Spagnolo, J.F.; Rossignol, E.; Bullitt, E.; Kirkegaard, K. Enzymatic and nonenzymatic functions of viral RNA-dependent RNA polymerases within oligomeric arrays. *RNA* **2010**, *16*, 382–393. [CrossRef] [PubMed]

35. Miller, S.; Kastner, S.; Krijnse-Locker, J.; Buhler, S.; Bartenschlager, R. The non-structural protein 4A of dengue virus is an integral membrane protein inducing membrane alterations in a 2K-regulated manner. *J. Biol. Chem.* **2007**, *282*, 8873–8882. [CrossRef] [PubMed]

36. McLean, J.E.; Wudzinska, A.; Datan, E.; Quaglino, D.; Zakeri, Z. Flavivirus NS4A-induced autophagy protects cells against death and enhances virus replication. *J. Biol. Chem.* **2011**, *286*, 22147–22159. [CrossRef] [PubMed]

37. Umareddy, I.; Chao, A.; Sampath, A.; Gu, F.; Vasudevan, S.G. Dengue virus NS4B interacts with NS3 and dissociates it from single-stranded RNA. *J. Gen. Virol.* **2006**, *87*, 2605–2614. [CrossRef] [PubMed]

38. Young, P.R.; Hilditch, P.A.; Bletchly, C.; Halloran, W. An antigen capture enzyme-linked immunosorbent assay reveals high levels of the dengue virus protein NS1 in the sera of infected patients. *J. Clin. Microbiol.* **2000**, *38*, 1053–1057. [PubMed]

39. Libraty, D.H.; Young, P.R.; Pickering, D.; Endy, T.P.; Kalayanarooj, S.; Green, S.; Vaughn, D.W.; Nisalak, A.; Ennis, F.A.; Rothman, A.L. High circulating levels of the dengue virus nonstructural protein NS1 early in dengue illness correlate with the development of dengue hemorrhagic fever. *J. Infect. Dis.* **2002**, *186*, 1165–1168. [CrossRef] [PubMed]

40. Alcon, S.; Talarmin, A.; Debruyne, M.; Falconar, A.; Deubel, V.; Flamand, M. Enzyme-linked immunosorbent assay specific to Dengue virus type 1 nonstructural protein NS1 reveals circulation of the antigen in the blood during the acute phase of disease in patients experiencing primary or secondary infections. *J. Clin. Microbiol.* **2002**, *40*, 376–381. [CrossRef] [PubMed]

41. Gutsche, I.; Coulibaly, F.; Voss, J.E.; Salmon, J.; d'Alayer, J.; Ermonval, M.; Larquet, E.; Charneau, P.; Krey, T.; Megret, F.; et al. Secreted dengue virus nonstructural protein NS1 is an atypical barrel-shaped high-density lipoprotein. *Proc. Natl. Acad. Sci. USA* **2011**, *108*, 8003–8008. [CrossRef] [PubMed]

42. Alcon-LePoder, S.; Drouet, M.T.; Roux, P.; Frenkiel, M.P.; Arborio, M.; Durand-Schneider, A.M.; Maurice, M.; Le Blanc, I.; Gruenberg, J.; Flamand, M. The secreted form of dengue virus nonstructural protein NS1 is endocytosed by hepatocytes and accumulates in late endosomes: Implications for viral infectivity. *J. Virol.* **2005**, *79*, 11403–11411. [CrossRef] [PubMed]

43. Avirutnan, P.; Punyadee, N.; Noisakran, S.; Komoltri, C.; Thiemmeca, S.; Auethavornanan, K.; Jairungsri, A.; Kanlaya, R.; Tangthawornchaikul, N.; Puttikhunt, C.; et al. Vascular leakage in severe dengue virus infections: A potential role for the nonstructural viral protein NS1 and complement. *J. Infect. Dis.* **2006**, *193*, 1078–1088. [CrossRef] [PubMed]

44. Lin, C.F.; Chiu, S.C.; Hsiao, Y.L.; Wan, S.W.; Lei, H.Y.; Shiau, A.L.; Liu, H.S.; Yeh, T.M.; Chen, S.H.; Liu, C.C.; et al. Expression of cytokine, chemokine, and adhesion molecules during endothelial cell activation induced by antibodies against dengue virus nonstructural protein 1. *J. Immunol.* **2005**, *174*, 395–403. [CrossRef] [PubMed]

45. Lin, C.F.; Lei, H.Y.; Shiau, A.L.; Liu, C.C.; Liu, H.S.; Yeh, T.M.; Chen, S.H.; Lin, Y.S. Antibodies from dengue patient sera cross-react with endothelial cells and induce damage. *J. Med. Virol.* **2003**, *69*, 82–90. [CrossRef] [PubMed]

46. Palta, S.; Saroa, R.; Palta, A. Overview of the coagulation system. *Indian J. Anaesthesia* **2014**, *58*, 515–523. [CrossRef] [PubMed]

47. Lin, S.W.; Chuang, Y.C.; Lin, Y.S.; Lei, H.Y.; Liu, H.S.; Yeh, T.M. Dengue virus nonstructural protein NS1 binds to prothrombin/thrombin and inhibits prothrombin activation. *J. Infect.* **2012**, *64*, 325–334. [CrossRef] [PubMed]

48. Modhiran, N.; Watterson, D.; Muller, D.A.; Panetta, A.K.; Sester, D.P.; Liu, L.; Hume, D.A.; Stacey, K.J.; Young, P.R. Dengue virus NS1 protein activates cells via Toll-like receptor 4 and disrupts endothelial cell monolayer integrity. *Sci. Transl. Med.* **2015**, *7*. [CrossRef] [PubMed]

49. Beatty, P.R.; Puerta-Guardo, H.; Killingbeck, S.S.; Glasner, D.R.; Hopkins, K.; Harris, E. Dengue virus NS1 triggers endothelial permeability and vascular leak that is prevented by NS1 vaccination. *Sci. Transl. Med.* **2015**, *7*. [CrossRef] [PubMed]

50. Puerta-Guardo, H.; Glasner, D.R.; Harris, E. Dengue virus NS1 disrupts the endothelial glycocalyx, leading to hyperpermeability. *PLoS Pathog.* **2016**, *12*, e1005738. [CrossRef] [PubMed]

51. Jacobs, M.G.; Robinson, P.J.; Bletchly, C.; Mackenzie, J.M.; Young, P.R. Dengue virus nonstructural protein 1 is expressed in a glycosyl-phosphatidylinositol-linked form that is capable of signal transduction. *FASEB J.* **2000**, *14*, 1603–1610. [CrossRef] [PubMed]

52. Mayor, S.; Riezman, H. Sorting GPI-anchored proteins. *Nat. Rev. Mol. Cell Biol.* **2004**, *5*, 110–120. [CrossRef] [PubMed]

53. Shenoy-Scaria, A.M.; Kwong, J.; Fujita, T.; Olszowy, M.W.; Shaw, A.S.; Lublin, D.M. Signal transduction through decay-accelerating factor. Interaction of glycosyl-phosphatidylinositol anchor and protein tyrosine kinases p56lck and p59fyn 1. *J. Immunol.* **1992**, *149*, 3535–3541. [PubMed]

54. Lund-Johansen, F.; Olweus, J.; Symington, F.W.; Arli, A.; Thompson, J.S.; Vilella, R.; Skubitz, K.; Horejsi, V. Activation of human monocytes and granulocytes by monoclonal antibodies to glycosylphosphatidylinositol-anchored antigens. *Eur. J. Immunol.* **1993**, *23*, 2782–2791. [CrossRef] [PubMed]

55. Mason, J.C.; Yarwood, H.; Tarnok, A.; Sugars, K.; Harrison, A.A.; Robinson, P.J.; Haskard, D.O. Human Thy-1 is cytokine-inducible on vascular endothelial cells and is a signaling molecule regulated by protein kinase C. *J. Immunol.* **1996**, *157*, 874–883. [PubMed]

56. Medin, C.L.; Fitzgerald, K.A.; Rothman, A.L. Dengue virus nonstructural protein NS5 induces interleukin-8 transcription and secretion. *J. Virol.* **2005**, *79*, 11053–11061. [CrossRef] [PubMed]
57. Kelley, J.F.; Kaufusi, P.H.; Volper, E.M.; Nerurkar, V.R. Maturation of dengue virus nonstructural protein 4B in monocytes enhances production of dengue hemorrhagic fever-associated chemokines and cytokines. *Virology* **2011**, *418*, 27–39. [CrossRef] [PubMed]
58. Kelley, J.F.; Kaufusi, P.H.; Nerurkar, V.R. Dengue hemorrhagic fever-associated immunomediators induced via maturation of dengue virus nonstructural 4B protein in monocytes modulate endothelial cell adhesion molecules and human microvascular endothelial cells permeability. *Virology* **2012**, *422*, 326–337. [CrossRef] [PubMed]
59. Khunchai, S.; Junking, M.; Suttitheptumrong, A.; Kooptiwut, S.; Haegeman, G.; Limjindaporn, T.; Yenchitsomanus, P.T. NF-κB is required for dengue virus NS5-induced RANTES expression. *Virus Res.* **2015**, *197*, 92–100. [CrossRef] [PubMed]
60. Khunchai, S.; Junking, M.; Suttitheptumrong, A.; Yasamut, U.; Sawasdee, N.; Netsawang, J.; Morchang, A.; Chaowalit, P.; Noisakran, S.; Yenchitsomanus, P.T.; et al. Interaction of dengue virus nonstructural protein 5 with Daxx modulates RANTES production. *Biochem. Biophys. Res. Commun.* **2012**, *423*, 398–403. [CrossRef] [PubMed]
61. Makhluf, H.; Shresta, S. Innate antiviral immunity against dengue virus. *Crit. Rev. Immunol.* **2015**, *35*, 253–260. [CrossRef] [PubMed]
62. Morrison, J.; Aguirre, S.; Fernandez-Sesma, A. Innate immunity evasion by Dengue virus. *Viruses* **2012**, *4*, 397–413. [CrossRef] [PubMed]
63. Stoermer, K.A.; Morrison, T.E. Complement and viral pathogenesis. *Virology* **2011**, *411*, 362–373. [CrossRef] [PubMed]
64. Thiemmeca, S.; Tamdet, C.; Punyadee, N.; Prommool, T.; Songjaeng, A.; Noisakran, S.; Puttikhunt, C.; Atkinson, J.P.; Diamond, M.S.; Ponlawat, A.; et al. Secreted NS1 protects dengue virus from mannose-binding lectin-mediated neutralization. *J. Immunol.* **2016**, *197*, 4053–4065. [CrossRef] [PubMed]
65. Fullam, A.; Schroder, M. DExD/H-box RNA helicases as mediators of anti-viral innate immunity and essential host factors for viral replication. *Biochim. Biophys. Acta* **2013**, *1829*, 854–865. [CrossRef] [PubMed]
66. Wu, J.; Chen, Z.J. Innate immune sensing and signaling of cytosolic nucleic acids. *Annu. Rev. Immunol.* **2014**, *32*, 461–488. [CrossRef] [PubMed]
67. West, A.P.; Shadel, G.S.; Ghosh, S. Mitochondria in innate immune responses. *Nat. Rev. Immunol.* **2011**, *11*, 389–402. [CrossRef] [PubMed]
68. Da Conceicao, T.M.; Rust, N.M.; Berbel, A.C.; Martins, N.B.; do Nascimento Santos, C.A.; Da Poian, A.T.; de Arruda, L.B. Essential role of RIG-I in the activation of endothelial cells by dengue virus. *Virology* **2013**, *435*, 281–292. [CrossRef] [PubMed]
69. Anglero-Rodriguez, Y.I.; Pantoja, P.; Sariol, C.A. Dengue virus subverts the interferon induction pathway via NS2B/3 protease-IkappaB kinase epsilon interaction. *Clin. Vaccine Immunol.* **2014**, *21*, 29–38. [CrossRef] [PubMed]
70. Dalrymple, N.A.; Cimica, V.; Mackow, E.R. Dengue virus NS proteins inhibit RIG-I/MAVS signaling by blocking TBK1/IRF3 phosphorylation: Dengue virus serotype 1 NS4A is a unique interferon-regulating virulence determinant. *mBio* **2015**, *6*, e00553-15. [CrossRef] [PubMed]
71. Dong, Y.; Ye, W.; Yang, J.; Han, P.; Wang, Y.; Ye, C.; Weng, D.; Zhang, F.; Xu, Z.; Lei, Y. DDX21 translocates from nucleus to cytoplasm and stimulates the innate immune response due to dengue virus infection. *Biochem. Biophys. Res. Commun.* **2016**, *473*, 648–653. [CrossRef] [PubMed]
72. He, Z.; Zhu, X.; Wen, W.; Yuan, J.; Hu, Y.; Chen, J.; An, S.; Dong, X.; Lin, C.; Yu, J.; et al. Dengue virus subverts host innate immunity by targeting adaptor protein MAVS. *J. Virol.* **2016**, *90*, 7219–7230. [CrossRef] [PubMed]
73. Chatel-Chaix, L.; Cortese, M.; Romero-Brey, I.; Bender, S.; Neufeldt, C.J.; Fischl, W.; Scaturro, P.; Schieber, N.; Schwab, Y.; Fischer, B.; et al. Dengue virus perturbs mitochondrial morphodynamics to dampen innate immune responses. *Cell. Host Microbe* **2016**, *20*, 342–356. [CrossRef] [PubMed]
74. Ivashkiv, L.B.; Donlin, L.T. Regulation of type I interferon responses. *Nat. Rev. Immunol.* **2014**, *14*, 36–49. [CrossRef] [PubMed]
75. Chua, J.J.; Bhuvanakantham, R.; Chow, V.T.; Ng, M.L. Recombinant non-structural 1 (NS1) protein of dengue-2 virus interacts with human STAT3beta protein. *Virus Res.* **2005**, *112*, 85–94. [CrossRef] [PubMed]
76. Munoz-Jordan, J.L.; Laurent-Rolle, M.; Ashour, J.; Martinez-Sobrido, L.; Ashok, M.; Lipkin, W.I.; Garcia-Sastre, A. Inhibition of alpha/beta interferon signaling by the NS4B protein of flaviviruses. *J. Virol.* **2005**, *79*, 8004–8013. [CrossRef] [PubMed]

77. Munoz-Jordan, J.L.; Sanchez-Burgos, G.G.; Laurent-Rolle, M.; Garcia-Sastre, A. Inhibition of interferon signaling by dengue virus. *Proc. Natl. Acad. Sci. USA* **2003**, *100*, 14333–14338. [CrossRef] [PubMed]

78. Le Breton, M.; Meyniel-Schicklin, L.; Deloire, A.; Coutard, B.; Canard, B.; de Lamballerie, X.; Andre, P.; Rabourdin-Combe, C.; Lotteau, V.; Davoust, N. Flavivirus NS3 and NS5 proteins interaction network: A high-throughput yeast two-hybrid screen. *BMC Microbiol.* **2011**, *11*. [CrossRef] [PubMed]

79. Morrison, J.; Laurent-Rolle, M.; Maestre, A.M.; Rajsbaum, R.; Pisanelli, G.; Simon, V.; Mulder, L.C.; Fernandez-Sesma, A.; Garcia-Sastre, A. Dengue virus co-opts UBR4 to degrade STAT2 and antagonize type I interferon signaling. *PLoS Pathog.* **2013**, *9*, e1003265. [CrossRef] [PubMed]

80. Crabtree, H.G. The carbohydrate metabolism of certain pathological overgrowths. *Biochem. J.* **1928**, *22*, 1289–1298. [CrossRef] [PubMed]

81. El-Bacha, T.; Da Poian, A.T. Virus-induced changes in mitochondrial bioenergetics as potential targets for therapy. *Int. J. Biochem. Cell Biol.* **2013**, *45*, 41–46. [CrossRef] [PubMed]

82. El-Bacha, T.; Midlej, V.; Pereira da Silva, A.P.; Silva da Costa, L.; Benchimol, M.; Galina, A.; Da Poian, A.T. Mitochondrial and bioenergetic dysfunction in human hepatic cells infected with dengue 2 virus. *Biochim. Biophys. Acta* **2007**, *1772*, 1158–1166. [CrossRef] [PubMed]

83. Fontaine, K.A.; Sanchez, E.L.; Camarda, R.; Lagunoff, M. Dengue virus induces and requires glycolysis for optimal replication. *J. Virol.* **2015**, *89*, 2358–2366. [CrossRef] [PubMed]

84. Heaton, N.S.; Randall, G. Dengue virus-induced autophagy regulates lipid metabolism. *Cell. Host Microbe* **2010**, *8*, 422–432. [CrossRef] [PubMed]

85. Paul, D.; Bartenschlager, R. Flaviviridae replication organelles: Oh, what a tangled web we weave. *Annu. Rev. Virol.* **2015**, *2*, 289–310. [CrossRef] [PubMed]

86. Den Boon, J.A.; Diaz, A.; Ahlquist, P. Cytoplasmic viral replication complexes. *Cell. Host Microbe* **2010**, *8*, 77–85. [CrossRef] [PubMed]

87. Heaton, N.S.; Perera, R.; Berger, K.L.; Khadka, S.; Lacount, D.J.; Kuhn, R.J.; Randall, G. Dengue virus nonstructural protein 3 redistributes fatty acid synthase to sites of viral replication and increases cellular fatty acid synthesis. *Proc. Natl. Acad. Sci. USA* **2010**, *107*, 17345–17350. [CrossRef] [PubMed]

88. Tang, W.C.; Lin, R.J.; Liao, C.L.; Lin, Y.L. Rab18 facilitates dengue virus infection by targeting fatty acid synthase to sites of viral replication. *J. Virol.* **2014**, *88*, 6793–6804. [CrossRef] [PubMed]

89. Ozeki, S.; Cheng, J.; Tauchi-Sato, K.; Hatano, N.; Taniguchi, H.; Fujimoto, T. Rab18 localizes to lipid droplets and induces their close apposition to the endoplasmic reticulum-derived membrane. *J. Cell Sci.* **2005**, *118*, 2601–2611. [CrossRef] [PubMed]

90. Westermann, B. Bioenergetic role of mitochondrial fusion and fission. *Biochim. Biophys. Acta* **2012**, *1817*, 1833–1838. [CrossRef] [PubMed]

91. Barbier, V.; Lang, D.; Valois, S.; Rothman, A.L.; Medin, C.L. Dengue virus induces mitochondrial elongation through impairment of Drp1-triggered mitochondrial fission. *Virology* **2017**, *500*, 149–160. [CrossRef]

92. Hiscox, J.A. The interaction of animal cytoplasmic RNA viruses with the nucleus to facilitate replication. *Virus Res.* **2003**, *95*, 13–22. [CrossRef]

93. Teo, C.S.; Chu, J.J. Cellular vimentin regulates construction of dengue virus replication complexes through interaction with NS4A protein. *J. Virol.* **2014**, *88*, 1897–1913. [CrossRef] [PubMed]

94. Styers, M.L.; Salazar, G.; Love, R.; Peden, A.A.; Kowalczyk, A.P.; Faundez, V. The endo-lysosomal sorting machinery interacts with the intermediate filament cytoskeleton. *Mol. Biol. Cell* **2004**, *15*, 5369–5382. [CrossRef] [PubMed]

95. Cervantes-Salazar, M.; Angel-Ambrocio, A.H.; Soto-Acosta, R.; Bautista-Carbajal, P.; Hurtado-Monzon, A.M.; Alcaraz-Estrada, S.L.; Ludert, J.E.; Del Angel, R.M. Dengue virus NS1 protein interacts with the ribosomal protein RPL18: This interaction is required for viral translation and replication in Huh-7 cells. *Virology* **2015**, *484*, 113–126. [CrossRef] [PubMed]

96. Allonso, D.; Andrade, I.S.; Conde, J.N.; Coelho, D.R.; Rocha, D.C.; da Silva, M.L.; Ventura, G.T.; Silva, E.M.; Mohana-Borges, R. Dengue virus NS1 protein modulates cellular energy metabolism by increasing glyceraldehyde-3-phosphate dehydrogenase activity. *J. Virol.* **2015**, *89*, 11871–11883. [CrossRef] [PubMed]

97. Yang, S.H.; Liu, M.L.; Tien, C.F.; Chou, S.J.; Chang, R.Y. Glyceraldehyde-3-phosphate dehydrogenase (GAPDH) interaction with 3′ ends of Japanese encephalitis virus RNA and colocalization with the viral NS5 protein. *J. Biomed. Sci.* **2009**, *16*. [CrossRef] [PubMed]

98. De, B.P.; Gupta, S.; Zhao, H.; Drazba, J.A.; Banerjee, A.K. Specific interaction in vitro and in vivo of glyceraldehyde-3-phosphate dehydrogenase and LA protein with cis-acting RNAs of human parainfluenza virus type 3. *J. Biol. Chem.* **1996**, *271*, 24728–24735. [CrossRef] [PubMed]

99. Schultz, D.E.; Hardin, C.C.; Lemon, S.M. Specific interaction of glyceraldehyde 3-phosphate dehydrogenase with the 5′-nontranslated RNA of hepatitis A virus. *J. Biol. Chem.* **1996**, *271*, 14134–14142. [PubMed]

100. Petrik, J.; Parker, H.; Alexander, G.J. Human hepatic glyceraldehyde-3-phosphate dehydrogenase binds to the poly(U) tract of the 3′ non-coding region of hepatitis C virus genomic RNA. *J. Gen. Virol.* **1999**, *80*, 3109–3113. [CrossRef] [PubMed]

101. Sirover, M.A. New insights into an old protein: The functional diversity of mammalian glyceraldehyde-3-phosphate dehydrogenase. *Biochim. Biophys. Acta* **1999**, *1432*, 159–184. [CrossRef]

102. Prasanth, K.R.; Huang, Y.W.; Liou, M.R.; Wang, R.Y.; Hu, C.C.; Tsai, C.H.; Meng, M.; Lin, N.S.; Hsu, Y.H. Glyceraldehyde 3-phosphate dehydrogenase negatively regulates the replication of Bamboo mosaic virus and its associated satellite RNA. *J. Virol.* **2011**, *85*, 8829–8840. [CrossRef] [PubMed]

103. Chua, J.J.; Ng, M.M.; Chow, V.T. The non-structural 3 (NS3) protein of dengue virus type 2 interacts with human nuclear receptor binding protein and is associated with alterations in membrane structure. *Virus Res.* **2004**, *102*, 151–163. [CrossRef] [PubMed]

viruses

MDPI

Review

Zika Virus: Recent Advances towards the Development of Vaccines and Therapeutics

Monica A. McArthur

Center for Vaccine Development, University of Maryland School of Medicine, Baltimore, MD 21201, USA;
mmcarthu@som.umaryland.edu; Tel.: +1-410-706-5328

Academic Editor: Eric Freed
Received: 18 April 2017; Accepted: 8 June 2017; Published: 13 June 2017

Abstract: Zika is a rapidly emerging public health threat. Although clinical infection is frequently mild, significant neurological manifestations have been demonstrated in infants born to Zika virus (ZIKV) infected mothers. Due to the substantial ramifications of intrauterine infection, effective counter-measures are urgently needed. In order to develop effective anti-ZIKV vaccines and therapeutics, improved animal models and a better understanding of immunological correlates of protection against ZIKV are required. This review will summarize what is currently known about ZIKV, the clinical manifestations and epidemiology of Zika as well as, the development of animal models to study ZIKV infection, host immune responses against ZIKV, and the current state of development of vaccines and therapeutics against ZIKV.

Keywords: Zika virus; animal models; immunology; vaccines; congenital Zika syndrome

1. Introduction

As of March 2017, 84 countries and territories have reported evidence of mosquito-borne Zika virus (ZIKV) transmission, including the United States of America [1]. Furthermore, many of these countries have reported microcephaly and other central nervous system (CNS) malformations potentially associated with ZIKV infection, or suggestive of congenital infection as well as increased incidence of Guillain–Barré syndrome (GBS) and/or laboratory confirmation of ZIKV infection among GBS cases [1]. The temporal association of ZIKV transmission with clusters of microcephaly and GBS was declared a Public Health Emergency of International Concern (PHEIC) by the World Health Organization (WHO) on 1 February 2016, a designation which was subsequently ended at the fifth meeting of the Emergency Committee on ZIKV, microcephaly and other neurological disorders on 18 November 2016 [2,3]. According to the WHO, the decision was made to escalate the coordination and response to ZIKV to a more sustained program to address the long-term nature of the disease and its consequences [2]. As ZIKV continues to spread, many "unknowns" remain and considerable research is needed to advance our understanding of this important pathogen. The current literature pertaining to the clinical manifestations and epidemiology of ZIKV as well as, the development of animal models to study ZIKV infection, host immune responses against ZIKV, and the current state of development of vaccines and therapeutics against ZIKV will be discussed.

2. Isolation and Characterization

ZIKV was first isolated from a sentinel rhesus monkey in Uganda, near the Zika forest, in 1947 [4]; however, there was limited documentation of human infection prior to the 2007 outbreak on Yap Island in the Federated States of Micronesia [5]. ZIKV belongs to the family *Flaviviridae*, genus *Flavivirus* and is a mosquito-borne virus in the Spondweni group [6]. Flaviviruses are among the most medically significant arboviruses and include, in addition to ZIKV, pathogens such as yellow fever virus (YFV),

Japanese encephalitis virus (JEV), West Nile Virus (WNV), tick-borne encephalitis virus (TBEV), and the four dengue viruses (DENV1–4).

Flaviviruses are enveloped RNA viruses that contain the viral genome complexed with multiple copies of the capsid protein (C) and surrounded by an icosahedral shell composed of 180 copies of the envelope (E) glycoprotein and the membrane (M) or precursor membrane (prM) proteins. The E and M/prM proteins are anchored in a lipid membrane [7]. The full-length genome of ZIKV is 10,794 nucleotides and encodes 3419 amino acids, which, in addition to the structural proteins (C, prM, and E), constitute seven non-structural proteins (NS1, NS2a, NS2b, NS3, NS4a, NS4b, and NS5) [6,7]. The non-structural proteins are involved in replication and assembly of the virus as well as antagonizing the host innate immune response [7]. Specifically, NS3 and NS5 are large, multifunctional proteins which contain several enzymatic activities involved in polyprotein processing (NS3) and RNA replication (NS3 and NS5) [8]. Additionally, NS5 has been demonstrated to antagonize the interferon (IFN) response [9].The crystal structure of ZIKV has been solved and demonstrates that the mature ZIKV structure is similar to mature WNV and DENV structures [7]. However, there are notable differences in the E protein structure which may be responsible for cellular tropism and contribute to disease outcomes [7].

Phylogenetic analyses of ZIKV strains isolated from 1947 to 2016 identified two distinct clades (lineages), African and Asian [10]. The causative agent of the current ZIKV epidemic belongs to the Asian lineage, which, while not closely related to the African lineage, shares a common ancestor [10]. Comparisons of the E protein sequences from the two lineages found that the Asian lineage contains insertions in the E protein glycosylation motif which are not present in the African lineage [10]. When the amino acid sequences of the E protein of ZIKV isolates from human (2015/2016 epidemic), monkey (pre-epidemic), and mosquito (pre-epidemic) were compared, a total of sixteen amino acid substitutions were identified, resulting in subtle structural changes [10]. These changes, although subtle, may impact ZIKV virulence and host tropism [10].

3. Epidemiology

Despite relatively limited reports of human cases of ZIKV infection from 1947 until the 2007 outbreak on Yap Island, ZIKV has now spread dramatically to include over 80 countries and territories with vector-borne transmission [1]. Following the introduction of ZIKV to Yap Island, sporadic cases of ZIKV infection continued to be reported in Southeast Asia during the mid-2010s [11]. A major epidemic of ZIKV infection ensued in French Polynesia in 2013–2014 and some severe cases were associated with neurological complications such as GBS [12,13]. ZIKV was first reported in Brazil in 2015 with large numbers of suspected cases and the initial observation that the number of newborn infants with microcephaly was increased in ZIKV-affected areas later that year [11]. ZIKV has since spread to include much of Central and South America and the Caribbean. Furthermore, vector-borne transmission of ZIKV has been documented in the US in both Texas and Florida [14].

The primary mode of ZIKV transmission is through the bite of infected mosquitoes, with *Aedes aegypti* and *Aedes albopictus* being the predominant vectors [6,15]. While *Ae. aegypti* mosquitoes are confined to tropical and sub-tropical regions, *Ae. albopictus*, are distributed throughout tropical, sub-tropical, and temperate regions [16]. Since 2015, ZIKV has spread rapidly through the range occupied by *Aedes* mosquitos in the Americas [15]. In addition to vector-borne transmission, sexual transmission is a contributor to ZIKV spread [17–20]. Additional modes of transmission have also been reported, including transmission from mother to child, blood transfusion-related transmission, laboratory transmission, and transmission by physical contact [16].

It is likely that ZIKV will continue to spread; however, it is unclear at this time whether ZIKV transmission will remain epidemic, eventually becoming episodic with intervening periods of relative inactivity, or if it will become endemic with seasonal transmission patterns [21]. Characterization of the epidemiology of ZIKV is complicated by inadequate diagnostic assays (discussed below) and co-circulation with viruses demonstrating similar clinical presentations (i.e., Chikungunya virus (CHIKV) and DENV) [22].

4. Clinical Manifestations

Most individuals, ~80%, infected with ZIKV are asymptomatic [23]. When symptoms are present, they are generally mild and may include pruritic maculopapular rash, fever (typically low-grade), arthritis, arthralgia, non-purulent conjunctivitis, and edema of the extremities [5,23–25]. Additional symptoms of headache, myalgia, retro-orbital pain, low-back pain, lymphadenopathy, and vomiting have also been reported [5,24]. While disease is frequently mild, fatal cases have been reported, particularly in individuals with underlying medical conditions [26–31].

4.1. Neurological Manifestations of ZIKV Infection

A major concern during the current ZIKV epidemic has been the association of ZIKV with neurological complications. GBS is the most frequent neurological sequela reported [12,13,25,32]. Although GBS is commonly thought to have a good prognosis, up to 20% of GBS patients may remain severely disabled and even with treatment, approximately 5% of GBS cases are fatal [33]. GBS is characterized by progressive bilateral and relatively symmetric weakness of the limbs frequently with hyporeflexia or areflexia. There is considerable evidence suggesting GBS has an autoimmune etiology [33]. GBS has previously been associated with preceding infections such as *Campylobacter jejuni*, cytomegalovirus (CMV), Epstein–Barr virus (EBV), varicella zoster virus (VZV) and *Mycoplasma pneumoniae*. Additionally, there has been a link to influenza vaccination [34]. The mechanism by which ZIKV infection is associated with GBS remains unknown. It is therefore of critical importance to monitor for GBS in large scale vaccine trials as well as natural infection.

Other neurological complications associated with ZIKV infection including encephalitis, meningoencephalitis, and acute myelitis have also been reported [30,35,36].

4.2. Congenital Zika Syndrome

The Brazil Ministry of Health reported an unusual increase in cases of microcephaly in October and November 2015, the Pan American Health Organization (PAHO) and WHO requested Member States to monitor for similar events [37]. The identification of an increased number of cases of microcephaly potentially associated with ZIKV prompted the declaration of a PHEIC in February 2016 [3]. Subsequently, multiple reports of microcephaly and other birth defects following ZIKV infection during pregnancy have been made [38–43]. Following recognition of increases in the number of infants with microcephaly in Brazil [44] and French Polynesia [40] which were spatially and temporally associated with outbreaks of ZIKV infection, a causal link was suspected [45]. Subsequently, Rassmussen et al. applied Shepard's criteria for causality and suggested that sufficient evidence had been accumulated to infer a causal relationship between ZIKV infection and microcephaly as well as other brain anomalies [46]. Evidence to support the role of ZIKV infection during pregnancy in causing microcephaly and other neurological malformations includes: (a) temporal association (with a six-month delay) of increases in microcephaly with increased ZIKV transmission [44]; (b) data modeling indicating the first trimester is the primary risk period of pregnancy [40]; (c) identification of ZIKV in the brains of fetuses and infants (who died) with microcephaly [47,48]; (d) attenuation of human neural progenitor cell growth following in vitro ZIKV infection [49]; (e) in vitro infection of placental macrophages and cytotrophoblasts [50]; and (f) evidence of microcephaly associated with ZIKV infection in mouse and non-human primate models [51–54].

5. Animal Models

Animal models play an important role in understanding viral pathogenesis, identifying promising vaccine candidates, and testing therapeutics. The scientific community has exerted considerable effort in developing animal models which recapitulate various aspects of human ZIKV infection and disease.

For a comprehensive review of animal models for ZIKV infection and Zika disease, see Morrison and Diamond [55].

5.1. Rodent Models

Initial attempts to infect wild-type (wt) mice with ZIKV resulted in infection only by the intra-cerebral (ic) route [4]; however, passage through mouse brain resulted in adaptation to the mouse model [56]. Similarly, guinea pigs and rabbits did not show signs of disease following inoculation with ZIKV [56]. Although adult immunocompetent mice do not typically develop overt signs of clinical disease, wt mice have been used to demonstrate intrauterine infection [52,57]. Additionally, treatment of wt mice with antibodies blocking aspects of the interferon (IFN) response have been used as described below [54,58,59]. More recently immunocompromised mouse models, mostly lacking IFN responses, have been utilized to study systemic infection with ZIKV [54,59–62]. A variety of mouse models which lack various components of the IFN response have been employed, including A129 (lack IFN α/β receptor and are, therefore, incapable of responding to Type I IFN) [61,62], AG129 (lack Type I and II IFN responsiveness) [60,62], $Irf3^{-/-}$ $Irf5^{-/-}$ $Irf7^{-/-}$ (triple knockouts which produce very little IFN α/β) [59], $Ifnar1^{-/-}$ (lack IFN-α/β receptor) [54,59]. Interestingly, while Lazear et al., reported severe neurological disease in $Irf3^{-/-}$ $Irf5^{-/-}$ $Irf7^{-/-}$ triple knock-out (TKO) and $Ifnar1^{-/-}$ mice following subcutaneous inoculation with either a contemporary human isolate from French Polynesia (H/PF/2013) or the original Ugandan ZIKV strain (MR766), they found no overt disease in $Irf3^{-/-}$, $Irf^{-/-}$, and $Mavs^{-/-}$ single knock-out mice [59]. Despite the fact that both H/PF/2013 and MR766 elicited neurological disease in these models, H/PF/2013 demonstrated greater pathogenicity than the original Ugandan strain, MR766. Following intravenous inoculation, $Irf3^{-/-}$ $Irf5^{-/-}$ $Irf7^{-/-}$ TKO were more susceptible to ZIKV infection than $Ifnar1^{-/-}$ mice indicating a potential role for IRF-3-dependent, IFN-α/β-independent restriction mechanisms [59]. Furthermore, wt mice treated with MAR1-5A3 (a monoclonal antibody (mAb) that blocks the IFN-α/β receptor (IFNAR)) were susceptible to ZIKV infection but did not demonstrate a same neurological phenotype as $Ifnar1^{-/-}$ mice [59]. Additionally, mice immunosuppressed by treatment with dexamethasone developed ZIKV infection when inoculated with a clinical ZIKV isolate from Puerto Rico (PRVABC59) and treatment with type I IFN resulted in improved clinical outcome [63]. Taken together, these studies indicate an important role for the IFN response in protection against ZIKV infection.

Perhaps even more important than models of systemic ZIKV infection, are models of ZIKV infection during pregnancy. Multiple mouse models for ZIKV infection during pregnancy have been proposed. The $Ifnar1^{-/-}$ and MAR1-5A3 blocking models were used by Miner et al. to investigate ZIKV replication and trans-placental transmission in pregnant dams following infection with a clinical isolate from French Polynesia (H/PF/2013) [54]. Infection of $Ifnar1^{-/-}$ dams mated with wt males resulted in fetal demise, while when pregnant wt dams were given anti-IFNAR mAb prior to and during infection, there was mild intrauterine growth restriction (IUGR) and viral infection within the fetal head during a key period of neurodevelopment [54]. Another study using wt mice (SJL and C57BL/6 strains) found that infection of pregnant SJL mice with a Brazilian strain of ZIKV intravenously (iv) resulted in IUGR and ZIKV RNA was identified in multiple fetal tissues, particularly the brain [52]. Furthermore, surviving fetuses demonstrated cortical malformations with reduced cell number and cortical layer thickness (both signs associated with microcephaly in humans) as well as ocular abnormalities [52]. Adult C57 mice inoculated intra-peritoneally (ip) with a contemporary clinical ZIKV isolate were also able to transmit virus to their unborn fetuses and resulted in infection of primary neural progenitor cells responsible for cortex development as well as reduction of these cortex founder cells in the fetuses [57]. Intravaginal exposure to a Cambodian ZIKV strain, FSS13025, in both wt and $Ifnar1^{-/-}$ mice demonstrated replication of the virus in the vaginal mucosa (higher levels were identified in $Ifnar1^{-/-}$ mice) which resulted in IUGR and fetal brain infection (wt mice) and severe IUGR and fetal death ($Ifnar1^{-/-}$ mice) [64]. Direct infection of embryonic brains has also been demonstrated, and an Asian ZIKV strain, SZ01, was found to infect neural progenitor cells and cause cell death resulting in microcephaly [53]. Additionally, transcriptomic analyses following ZIKV infection demonstrated enrichment of up-regulated immune-response-related and apoptosis pathways providing further evidence that cytokines may play a role in ZIKV pathogenesis [53]. In sum, studies in mice have demonstrated that the interferon response is likely to play a critical role in protection from or susceptibility to ZIKV infection. While these studies support a causal role for ZIKV infection during pregnancy and microcephaly/neurological malformations, it is important to note that there are considerable differences in the morphological, spatial, and temporal placentation between mice and humans as well as differences in in utero brain development [65,66]. Further efforts to develop novel and enhance existing animal models to better recapitulate human disease are needed.

5.2. Non-Human Primate (NHP)Models

While mice represent a relatively easy to study small animal model for ZIKV, the previously mentioned differences between humans and mice necessitate additional models of disease. Immunocompetent macaque monkeys have similar gestation and fetal development to humans providing an animal model that more faithfully recapitulates human disease [51,67–69]. In non-pregnant rhesus macaques, the duration of detectable viral RNA in plasma is similar to that in humans, 6–7 days post infection, and the decline in plasma levels of viral RNA is coincident with increases in anti-ZIKV neutralizing antibody titers [69]. Dudley et al., have recently demonstrated prolonged viremia following infection with an Asian lineage ZIKV in pregnant rhesus macaques compared to non-pregnant animals [67]. Furthermore, they identified ZIKV-specific proliferation of natural killer (NK) cells and adaptive immune responses, including proliferation of CD4+ and CD8+ T cells, circulating plasmablasts, and ZIKV-specific IFN-γ production (by Elispot) [67]. ZIKV RNA has also been detected in placenta, fetal brain and liver as well as maternal tissues following infection of a pigtail macaque with an Asian lineage ZIKV [51]. Additionally, Magnetic Resonance Imaging (MRI) abnormalities in the fetal brain were identified 10 days after inoculation suggesting that fetal brain injury begins shortly after infection [51]. In a recent study by Osuna et al., both rhesus and cynomolgus macaques were shown to be highly susceptible to ZIKV infection following subcutaneous (sc) inoculation with isolates of Thai and Puerto Rican origin [68]. Activation of T and B cells was reported and multiple animals developed T cells specific for ZIKV peptides as evidenced by cytokine production following in vitro stimulation [68]. Studies utilizing non-human primate models to test anti-ZIKV vaccine candidates will be discussed below.

6. Immunology

6.1. Innate Immunity

Early responses by the innate immune system are the first line of host defense for the suppression of viral infections. IFN production is a major component of the innate response and the transcriptional regulation of numerous IFN-regulated genes leads to an antiviral environment [70]. As previously discussed, mice deficient in various components of the IFN response have increased susceptibility to ZIKV infection compared to wt mice [54,59–62,64]. It has been shown that human primary trophoblast cells, isolated from full-term placentas, constitutively release antiviral type III IFNλ1 which acts in both autocrine and paracrine fashion to protect cells from ZIKV infection [71]. These in vitro data suggest that ZIKV may not access the fetal compartment by direct replication in placental syncytiotrophoblasts during later stages of pregnancy [71]. However, it remains to be determined whether first-trimester trophoblasts are more permissive to ZIKV infection than late pregnancy villous trophoblasts and/or the antiviral effects of IFNλs [71]. Importantly, there are significant differences in the type III IFN pathway between mice and humans potentially complicating the interpretation of mouse models of ZIKV during pregnancy [71]. Flaviviruses have previously been shown to antagonize IFN signaling through multiple mechanisms [72]. In vitro studies have demonstrated that ZIKV NS5 binds to and targets the human IFN-regulated transcriptional activator STAT2 for proteasomal degradation; in contrast, mouse

STAT2 is refractory to ZIKV NS5 [73]. NS5-mediated degradation of STAT2 has also been demonstrated for DENV, but the mechanisms of degradation differ between DENV and ZIKV [73]. Interestingly, although Type I and Type III IFN use different cell-surface receptors they both signal through the Janus kinases-signal transducers and activators of transcription (Jak-STAT) pathway, including STAT2 [73,74]. For this reason, it is possible that while, Type III IFN are produced by villous trophoblasts [71], ZIKV may be able to evade Type III IFN signaling by degrading STAT2 mediated by NS5 [73]. In a mouse model of intravaginal ZIKV infection, it was demonstrated that antiviral Type I and III IFN and other inflammatory mediators were poorly induced and that there was robust viral replication in the vaginal mucosa [75]. Viral replication has also been shown in the vaginal mucosa of wt pregnant mice with even higher levels of viral replication demonstrated in *Ifnar1$^{-/-}$* mice [64]. Furthermore, Khan et al. found that if the dampened innate immune response was augmented (either by systemic infection with an unrelated pathogen or vaginal administration of acitretin) ZIKV replication was inhibited [75]. Taken together with studies using various strains of IFN-deficient mice, these data indicate a critical role for IFN in resistance to ZIKV infection.

6.2. Humoral Immunity

Neutralizing antibodies are thought to be a major factor in the protection against ZIKV infection [76–78]. In fact, broadly neutralizing human monoclonal antibodies directed against the E protein were protective against ZIKV-infection in type I/II interferon receptor-knockout mice [78]. However, these mAb were initially derived from DENV-infected individuals, not ZIKV-infected individuals. In another study, Sapparapu et al. isolated a panel of human mAb from individuals infected with ZIKV and identified a potent neutralizing antibody recognizing a quaternary epitope of the E protein dimer-dimer interface [76]. This mAb was also found to protect anti-IFNAR1 treated C57BL/6 mice from challenge with a mouse-adapted strain of ZIKV [76]. Furthermore, passive transfer of anti-ZIKV antibodies elicited by immunization has been shown to be protective in mice [79,80].

Antibodies against flaviviruses tend to be highly cross-reactive [81] and ZIKV is no exception. In fact, serological diagnosis has been complicated by a lack of ZIKV-specific antibody based assays. Both ELISA and Plaque Reduction Neutralization Tests (PRNT) demonstrate significant cross-reactivity with related viruses, including DENV which co-circulates in many of the areas affected by the current ZIKV epidemic [82]. Multiple groups have identified antibodies from DENV patients, which cross-react, and in some cases, neutralize ZIKV [78,83–86]. Barba-Spaeth et al. identified a subset of ZIKV-neutralizing antibodies that target a conformational epitope and reported the crystal structure of two of these antibodies in complex with the ZIKV E protein to reveal an epitope conserved between DENV and ZIKV [83]. These structural details of a conserved quaternary epitope may provide an important antigenic target for the development of cross-protective vaccines [83].

Although neutralizing antibodies may provide protection against ZIKV infection, there is also concern for antibody dependent enhancement (ADE) which may occur due to cross-reactive, poorly neutralizing antibodies. ADE is an immunological phenomenon in which non-neutralizing or weakly neutralizing antibodies facilitate viral entrance into Fc-receptor bearing cells such as monocytes and macrophages and has been described extensively in the context of secondary DENV infection [87–90]. While there has, to date, been no indication in epidemiological studies that prior DENV infection results in ADE of ZIKV infection or disease, in vitro analyses of ZIKV cross-reactive antibodies from DENV-infected individuals have demonstrated enhanced ZIKV infection through an Fc receptor-mediated process [84,86]. In AG129 (immunocompromised) mice, a DENV cross-reactive anti-ZIKV mAb enhanced DENV2 infection and disease; however, the LALA version of the mAb, which lacks the ability to bind to the Fc receptor, did not enhance disease [77]. While evidence of ADE has been demonstrated in vitro and in mouse models, extensive epidemiological studies are required to determine the clinical significance of ADE in areas where ZIKV and DENV co-circulate.

6.3. Cell Mediated Immunity

While significant information is emerging regarding humoral responses against ZIKV, there have been relatively few studies that explore cell mediated immune (CMI) responses against ZIKV. Data from DENV infection have demonstrated that T cell responses contribute to protection and/or disease enhancement [91–95]. Similarly, strong CD4+ and CD8+ T cell responses against the live-attenuated YFV vaccine, 17D have been established [96–98]. CMI responses against flaviviruses including JEV and WNV have also been reported [99,100]. Rivino et al. presented the argument that due to the high degree of sequence homology among DENV and ZIKV, some of the HLA-restricted CD8+ T cell epitopes may be conserved [92].

ZIKV-specific T cell responses have been elaborated in animal models [67,68,80,101–103] and humans [77]. In the NHP model, both CD4+ and CD8+ T cells were expanded as plasma viral RNA loads decreased following infection of rhesus macaques with an Asian lineage ZIKV strain. Additionally, IFN-γ secretion was detected in response to in vitro stimulation of peripheral blood mononuclear cells (PBMC) isolated from infected animals with overlapping peptides from the ZIKV NS5 [67]. Furthermore, increased expression of CD69 (a marker of early T cell activation) by T cells occurred following ZIKV infection and peaked between Days 2 and 5 post-infection [68]. ZIKV-specific cytokine production by CD4+ and CD8+ T cells was identified in both PBMC and lymph nodes of ZIKV-infected NHP at multiple time-points post-infection with peak responses noted on day 28 post-infection [68]. Elong Ngono et al. recently used mouse models to begin to dissect the CD8+ T cell responses against ZIKV [58]. They identified increases in antigen-experienced CD8+ T cells in mice infected with both African (MR766) and Asian (FSS13025) strains of ZIKV [58]. Moreover, they identified ZIKV-derived epitopes recognized by CD8+ T cells which encompassed peptides from the majority of the ZIKV proteins with a predominance of E protein-derived epitopes [58]. These ZIKV-responsive CD8+ T cells were multifunctional and also found to mediate cytotoxicity [58]. Furthermore, depletion of CD8+ T cells resulted in increased levels of infectious ZIKV in the serum and multiple tissues, including the brain [58]. In humans, CD4+ T cell responses against the ZIKV NS1 protein have been demonstrated [77]. These T cell responses were primarily in the CXCR3+ compartment and were poorly cross-reactive with DENV [77]. Although much remains to be learned about the T cell responses against ZIKV, current data support an important role for T cells in protection against ZIKV.

7. Diagnostics

The development of diagnostic tests for ZIKV has proven challenging due, in part, to serological cross-reactivity among flaviviruses [104,105]. There are currently no US Food and Drug Administration (FDA) approved ZIKV diagnostic tests; however, both serologic and nucleic acid amplification tests (NAAT) including the immunoglobulin (Ig)M class capture enzyme-linked immunosorbent assay (MAC-ELISA) and the Trioplex reverse transcription polymerase chain reaction (RT-PCR) have been given Emergency Use Authorization (EUA) by the FDA [106]. In addition to these diagnostic tests given EUA, there are numerous research-based assays; however, these assays do not provide validation using the most recent viral strains and fully documented clinical specimens [104].

7.1. Serological Tests

Due to the complexities of serological diagnostic testing for ZIKV infection, interim guidance for the interpretation of ZIKV antibody test results has been published [107]. Huzly et al. demonstrated high specificity of the Eurimmun ZIKV ELISA in European individuals with previous flavivirus exposure (either from infection or vaccination) including TBEV, DENV, and YFV [108]. However, a very recent study demonstrated high levels of cross-reactivity using commercially available DENV ELISA kits [109]. In this study, 100% of convalescent samples from NAAT confirmed ZIKV-infected individuals demonstrated cross-reactivity with DENV [109]. Although there is still some risk of cross-reactivity, the plaque reduction neutralization test remains the most specific serological method for diagnosing flaviviruses, including ZIKV [104]. An ELISA using the ZIKV NS1 antigen has been proposed as a more specific serological test [110]. In an initial study of sera from individuals with either RT-PCR confirmed or suspected ZIKV infection, sensitivity of combined IgG and IgM ELISA was 100% if samples were obtained ≥6 days post-symptom onset [110]. Furthermore, there was extremely limited cross-reactivity with sera from known DENV, WNV, JEV infected or YFV vaccinated individuals (specificity of 99.8%) [110]. The significant cross-reactivity demonstrated by most serological tests has hindered diagnosis in areas were multiple flaviviruses co-circulate. Development of serological tests with improved specificity is essential for assessing prevalence of ZIKV infection and aiding in diagnosis.

7.2. NAAT

NAAT to detect ZIKV RNA in the serum, whole blood, urine, and cerebrospinal fluid (CSF) are available [105]. While NAAT are more specific than serological tests, they are limited by the short duration of RNAemia and RNAuria. In general, ZIKV RNA is detectable from serum for approximately seven days post-infection and in urine for approximately 15–20 days post-infection [111]. The sensitivity of different NAAT varies. In a comparison of seven published and two new RT-PCR assays, Corman et al. found that some published RT-PCR assays may be of limited value for diagnostics in the current outbreak due to lack of sensitivity or difficulty in obtaining necessary reagents [112]. Currently, 12 molecular assays are available through EUA [106].

8. Vaccines

It is widely accepted that vaccines provide a cost-effective method of preventing infectious diseases [113]. Given the rapid spread of and severe outcomes associated with ZIKV infection, the development of a safe and efficacious vaccine is critical. The existence of successful vaccines against other flaviviral diseases (YFV, JEV, DENV, and TBEV) indicates that it is possible to develop a vaccine against ZIKV. Forty-five ZIKV vaccine candidates consisting of multiple vaccine platforms are currently under consideration and at various stages of development as summarized in the WHO vaccine pipeline tracker [114]. Five candidate vaccines, including inactivated whole organism, DNA, synthetic peptide, and mRNA platforms are already in Phase I clinical trials (Table 1) and larger Phase II and III studies are planned pending the results of the Phase I trials [21]. Due to the large number of vaccine candidates under investigation, this review will focus on those in clinical trials or for which animal model data is available.

Table 1. Summary of anti-Zika virus (ZIKV) vaccine candidates currently in clinical trials.

Type of Vaccine	Developers/Collaborators	Candidate Vaccine Name (If Available)	Stage of Development	Clinical Trial Registration Number
Inactivated whole organism	WRAIR/BIDMC/Harvard/ NIAID/Sanofi Pasteur		Clinical (Phase 1)	NCT02963909 NCT02952833 NCT02937233
DNA	GeneOne Life Science, Inc/Inovio Pharmaceuticals	GLS-5700	Clinical (Phase I)	NCT02809443 NCT02887482
DNA	VRC/NIAID	VRC ZIKV DNA	Clinical (Phase I)	NCT02840487 NCT02996461
Synthetic peptide	NIAID	AGS-v	Clinical (Phase I)	NCT03055000
Measles-vectored	Themis Bioscience	MV-ZIKA	Clinical (Phase I)	NCT02996890
mRNA	Valera (Moderna)	mRNA-1325	Clinical (Phase I)	NCT03014089

WRAIR: Walter Reed Army Institute of Research, BIDMC: Beth Israel Deaconess Medical Center, NIAID: National Institutes of Allergy and Infectious Diseases, VRC: Vaccine Research Center.

8.1. Inactivated Whole Organism (with or without Adjuvant)

Several developers are focusing on inactivated whole organism vaccine candidates [114]. Inactivated vaccines against flaviviruses including JEV and TBEV have been used successfully providing support for this method [115,116]. Benefits of an inactivated whole organism vaccine include multiple antigenic targets and non-replicating virus, which may improve safety. Purified inactivated virus (PIV) derived from a Puerto Rican strain of ZIKV was tested in Balb/c mice which received 1 µg of PIV vaccine with alum by either the intramuscular or subcutaneous route or sham inoculated with alum alone [117]. The PIV vaccine was shown to induce ZIKV-specific neutralizing antibodies after a single immunization and complete protection against ZIKV viremia was observed in those mice that received intramuscular injection [117]. It is important, however, as mentioned above, to note the limitations of this mouse model in recapitulating human disease. In recent studies by Abbink et al., 16 rhesus monkeys were immunized subcutaneously with 5 µg of ZIKV PIV with alum as an adjuvant or sham inoculated. It was demonstrated that all PIV inoculated animals developed both E protein-specific antibodies as well as neutralizing antibodies [79]. Furthermore, all PIV-inoculated monkeys were protected against challenge with wild-type ZIKV and had no detectable virus by RT-PCR in blood, urine, CSF, colorectal, or cervicovaginal secretions [79]. All sham inoculated monkeys had detectable viremia and in the majority, virus was detected in other body fluids as well [79]. In order to improve vaccine accessibility and reduce manufacturing costs, Yang et al. very recently proposed a cDNA clone-launched platform for high yield production of inactivated ZIKV vaccines [118]. Currently, Phase I trials of anti-ZIKV PIV vaccine candidates are underway (Table 1) [114,119].

8.2. DNA

DNA vaccines have been in development since the early 1990s and consist of a selected gene sequence cloned into a plasmid backbone. The plasmid is injected allowing DNA to be taken up by antigen presenting cells, which then express the plasmid-encoded genes to generate the target antigen(s) [120]. A candidate DNA vaccine against WNV was previously tested in humans and demonstrated excellent safety and immunogenicity [121,122]. A monovalent (DENV1) DNA vaccine candidate was also found to have an excellent safety profile in a Phase I clinical trial; however, it did not induce high levels of neutralizing antibodies [123]. Multiple ZIKV DNA vaccine candidates are in Phase I clinical trials (Table 1) [114,124]. These DNA vaccine candidates encode the *prM/E* genes of ZIKV. The E protein, in particular, is thought to be a major antigen against which neutralizing antibodies are produced and is considered important for protective efficacy in flaviviral vaccines [21]. In mice, a single immunization with DNA vaccine candidate encoding prM/E elicited higher E-specific antibody titers than did a DNA vaccine candidate encoding only E indicating the importance of prM in immunogenicity [117]. Furthermore, the prM/E DNA vaccine induced CD4+ and CD8+ T lymphocyte responses against E as demonstrated by increased IFN-γ production by intracellular cytokine staining [117]. Multiple anti-ZIKV DNA vaccine candidates induce both binding and neutralizing anti-ZIKV antibodies in rhesus monkeys [79,80,125].

8.3. RNA

RNA vaccines contain an open reading frame encoding the antigen of interest which is then translated by the host cellular machinery [126]. Because there is not the potential for genome integration, RNA-based vaccines may have a safety advantage over DNA vaccines [102]. Immunogenicity of several RNA vaccine candidates in animal models have been reported in the literature [102,103,127]. A lipid nanoparticle encapsulated modified mRNA encoding prM/E from an Asian lineage ZIKV strain was shown to protect both AG129 mice as well as C57BL/6 mice treated with blocking anti-IFNAR1 antibody (to create a lethal challenge model) [127]. Richner et al., further modified the mRNA to delete an immunodominant epitope within the E domain II fusion loop and demonstrated that the fusion loop mutant elicited serum antibody responses and protected against ZIKV challenge in mice.

Furthermore, the antibodies elicited by the fusion loop mutant caused less ADE of DENV1 infection in cell culture and less immune enhancement of DENV2 infection in AG129 mice [127]. Another recently reported anti-ZIKV RNA nanoparticle vaccine candidate encoding the prM/E proteins of an Asian lineage ZIKV induced both antibody and CD8+ T cell responses in C57BL/6 mice [102]. Furthermore, in the rhesus macaque model a single immunization of a different ZIKV prM/E encoding mRNA lipid nanoparticle vaccine demonstrated neutralizing antibody titers that were fifty times greater than those induced by a single immunization of a DNA vaccine and more than twice as high as those induced by two immunizations of a DNA vaccine measured using the same assay in the same laboratory [103,125]. These results indicate promise for RNA anti-ZIKV vaccine candidates. There is currently a Phase I clinical trial of an mRNA vaccine candidate underway (Table 1) [114].

8.4. Recombinant Viral Vector

In rhesus monkeys, a rhesus adenovirus serotype 52 (RhAd52) vector-based vaccine elicited ZIKV-specific neutralizing antibodies following a single immunization which demonstrated a substantial breadth of antibody responses against linear ZIKV E protein epitopes (peptide microarray assays) [79]. Furthermore, this vaccine candidate protected against challenge with wt ZIKV as demonstrated by lack of detectable viral RNA in plasma [79]. Very recently, a recombinant vesicular stomatitis virus (rVSV) anti-ZIKV vaccine was tested in mice and demonstrated maternal protective immunity in challenged newborn mice born to vaccinated mothers [101]. Betancourt et al. investigated multiple rVSV expressing ZIKV E or prM/E constructs in mice and identified an attenuated VSV with mutated matrix protein expressing prM/E (VSVm-ZprME) that induced high neutralizing anti-ZIKV antibody titers as well as IFN-γ production by CD8+ T cells [101]. Furthermore, in a neonatal mouse challenge model, seven-day-old mice born to VSVm-ZprME vaccinated mothers were partially protected against neurological manifestations of ZIKV infection following challenge [101]. Additional recombinant viral-vectored anti-ZIKV vaccine candidates are in the pre-clinical stages of development [114].

Additional vaccine candidate platforms in pre-clinical studies include live-attenuated vaccines, recombinant subunit vaccines, peptide vaccines, and ZIKV exosome vaccines [114].

9. Therapeutics

There are currently no drugs approved for the treatment of ZIKV-infection. The aim of drug development is primarily to reduce viral load, reduce symptoms, and protect the unborn fetus from neurological sequelae [128]. Multiple studies have focused on "re-purposing" existing compounds for the treatment of ZIKV [129–134]. Zmurko et al. tested multiple compounds, including ribavirin and polymerase inhibitor 7-deaza-2'-C-methyladenosine (7DMA), to identify potential anti-ZIKV therapeutics [134]. Of the compounds tested, 7DMA inhibited ZIKV replication in vitro, and, when administered for 10 consecutive days (beginning 1 h prior to infection), reduced viremia as well as delayed time to disease progression in ZIKV-infected AG129 mice [134]. While 7DMA was well tolerated by the mice, there was only modest reduction in viremia and the initiation of treatment prior to infection is impractical in non-research settings. Rapid, high-throughput screening of drug/compound libraries has also been utilized in an attempt to identify compounds with in vitro anti-ZIKV activity [129,131]. In a large screen of 727 compounds using a high-throughput cell-based assay to screen for anti-ZIKV activity, ZIKV was found to be sensitive to pyrimidine synthesis inhibitors (e.g., brequinar) [129]. Furthermore, Barrows et al. screened 774 FDA-approved drugs for anti-ZIKV activity and identified over 20 compounds, including mycophenolate mofetil, daptomycin, and sertraline, that reduced viral infection in vitro [131]. In an even larger screen, ~6000 compounds, including approved drugs, clinical trial drug candidates, and pharmacologically active compounds, were tested to determine their ability to either inhibit ZIKV infection or suppress infection-induced caspase-3 activity in neuronal cells [133]. Of the compounds screened in this study, emricasan, a pan-caspase inhibitor, was the most potent anti-cell-death compound and it demonstrated neuroprotective activity for human neuronal progenitor cells, but did not suppress ZIKV replication [133]. Additionally, several anti-malarial compounds were identified as having anti-DENV2 and anti-ZIKV properties using a cell-based cytotoxicity assay [130]. Bullard-Feibelman et al. demonstrated that an FDA-approved hepatitis C virus (HCV) anti-viral, sofosbuvir, inhibited ZIKV replication and infection in tissue culture as well as protected mice from ZIKV-induced death [132]. However, it is important to note that in this study mice were treated one day after inoculation with ZIKV, a time-line that is not practical in human infections.

Although anti-ZIKV therapeutic agents would likely be a welcome addition to the armamentarium, there are multiple challenges that face their development. Because clinical infection with ZIKV is typically mild, the primary populations for whom treatment would be indicated are pregnant women and those at increased risk for neurological complications, such as GBS. There are multiple ethical considerations in the development of therapeutics to be used during pregnancy. In general, pregnant women are excluded from clinical trials of new investigational compounds. In this setting, the inclusion of pregnant women would be warranted; however, the agent would need to be low risk to the mother, low risk to the fetus (not teratogenic), effective in preventing adverse fetal outcomes, and practical for use in resource-limited settings [128].

10. Conclusions

ZIKV is a rapidly emerging virus with a complex clinical picture. To date, much progress has been made in understanding the epidemiology and pathogenesis as well as developing vaccines which are essential to addressing the spread of ZIKV. However, many questions remain and a sustained global effort is required to battle this public health threat. Among the major issues requiring attention are the need for: (a) an improved understanding of the epidemiology including the potential interaction of ZIKV infection with other flaviviral infections; (b) an improved understanding of the clinical risk factors and mechanisms of severe disease; (c) identification of immunological correlates of protection which may aid in the rapid development of vaccines; and (d) the development of additional animal models which faithfully recapitulate human disease to better understand ZIKV pathogenesis and facilitate the development of countermeasures. In sum, an improved understanding of the epidemiology, pathogenesis, clinical risk factors for severe manifestations, and immunological correlates of protection are critical in developing effective countermeasures against this pathogen, including vaccines and therapeutics.

Acknowledgments: This work was supported, in part, by the Robert Wood Johnson Foundation, Harold Amos Medical Faculty Development Program (73307) and the National Institute of Allergy and Infectious Diseases Cooperative Centers on Human Immunity (NIAID, U19 AI-110820).

Conflicts of Interest: The authors declare no conflict of interest.

References

1. WHO Situation Report 10 March 2017. Available online: http://apps.who.int/iris/bitstream/10665/254714/1/zikasitrep10Mar17-eng.pdf?ua=1 (accessed on 2 April 2017).
2. WHO Situation Report 24 November 2016. Available online: http://www.who.int/emergencies/zika-virus/situation-report/24-november-2016/en/ (accessed on 27 December 2016).
3. WHO Zika Situation Report 5 February 2016. Available online: http://apps.who.int/iris/bitstream/10665/204348/1/zikasitrep_5Feb2016_eng.pdf?ua=1 (accessed on 7 April 2017).
4. Dick, G.W.; Kitchen, S.F.; Haddow, A.J. Zika virus. I. Isolations and serological specificity. *Trans. R. Soc. Trop. Med. Hyg.* **1952**, *46*, 509–520. [CrossRef]
5. Duffy, M.R.; Chen, T.H.; Hancock, W.T.; Powers, A.M.; Kool, J.L.; Lanciotti, R.S.; Pretrick, M.; Marfel, M.; Holzbauer, S.; Dubray, C.; et al. Zika virus outbreak on Yap Island, Federated States of Micronesia. *N. Engl. J. Med.* **2009**, *360*, 2536–2543. [CrossRef] [PubMed]
6. Kuno, G.; Chang, G.J. Full-length sequencing and genomic characterization of Bagaza, Kedougou, and Zika viruses. *Arch. Virol.* **2007**, *152*, 687–696. [CrossRef] [PubMed]

7. Sirohi, D.; Chen, Z.; Sun, L.; Klose, T.; Pierson, T.C.; Rossmann, M.G.; Kuhn, R.J. The 3.8 A resolution cryo-EM structure of Zika virus. *Science* **2016**, *352*, 467–470. [CrossRef] [PubMed]
8. Lindenbach, B.D.; Rice, C.M. Molecular biology of flaviviruses. *Adv. Virus Res.* **2003**, *59*, 23–61. [PubMed]
9. Best, S.M. The Many Faces of the Flavivirus NS5 Protein in Antagonism of Type I Interferon Signaling. *J. Virol.* **2017**, *91*. [CrossRef] [PubMed]
10. Ramaiah, A.; Dai, L.; Contreras, D.; Sinha, S.; Sun, R.; Arumugaswami, V. Comparative analysis of protein evolution in the genome of pre-epidemic and epidemic Zika virus. *Infect. Genet. Evol.* **2017**, *51*, 74–85. [CrossRef] [PubMed]
11. Song, B.H.; Yun, S.I.; Woolley, M.; Lee, Y.M. Zika virus: History, epidemiology, transmission, and clinical presentation. *J. Neuroimmunol.* **2017**. [CrossRef] [PubMed]
12. Cao-Lormeau, V.M.; Blake, A.; Mons, S.; Lastere, S.; Roche, C.; Vanhomwegen, J.; Dub, T.; Baudouin, L.; Teissier, A.; Larre, P.; et al. Guillain-Barre Syndrome outbreak associated with Zika virus infection in French Polynesia: A case-control study. *Lancet* **2016**, *387*, 1531–1539. [CrossRef]
13. Oehler, E.; Watrin, L.; Larre, P.; Leparc-Goffart, I.; Lastere, S.; Valour, F.; Baudouin, L.; Mallet, H.; Musso, D.; Ghawche, F. Zika virus infection complicated by Guillain-Barre syndrome—Case report, French Polynesia, December 2013. *Euro Surveill.* **2014**, *19*, 20720. [CrossRef] [PubMed]
14. Zika Virus Cases in the US. Available online: https://www.cdc.gov/zika/geo/united-states.html (accessed on 7 April 2017).
15. Kindhauser, M.K.; Allen, T.; Frank, V.; Santhana, R.S.; Dye, C. Zika: The origin and spread of a mosquito-borne virus. *Bull. World Health Organ.* **2016**, *94*, 675–686. [CrossRef] [PubMed]
16. Sharma, A.; Lal, S.K. Zika Virus: Transmission, Detection, Control, and Prevention. *Front. Microbiol.* **2017**, *8*, 110. [CrossRef] [PubMed]
17. D'Ortenzio, E.; Matheron, S.; Yazdanpanah, Y.; de Lamballerie, X.; Hubert, B.; Piorkowski, G.; Maquart, M.; Descamps, D.; Damond, F.; Leparc-Goffart, I. Evidence of Sexual Transmission of Zika Virus. *N. Engl. J. Med.* **2016**, *374*, 2195–2198. [CrossRef] [PubMed]
18. Foy, B.D.; Kobylinski, K.C.; Chilson Foy, J.L.; Blitvich, B.J.; Travassos da Rosa, A.; Haddow, A.D.; Lanciotti, R.S.; Tesh, R.B. Probable non-vector-borne transmission of Zika virus, Colorado, USA. *Emerg. Infect. Dis.* **2011**, *17*, 880–882. [CrossRef] [PubMed]
19. Frank, C.; Cadar, D.; Schlaphof, A.; Neddersen, N.; Gunther, S.; Schmidt-Chanasit, J.; Tappe, D. Sexual transmission of Zika virus in Germany, April 2016. *Euro Surveill.* **2016**, *21*. [CrossRef] [PubMed]
20. Freour, T.; Mirallie, S.; Hubert, B.; Splingart, C.; Barriere, P.; Maquart, M.; Leparc-Goffart, I. Sexual transmission of Zika virus in an entirely asymptomatic couple returning from a Zika epidemic area, France, April 2016. *Euro Surveill.* **2016**, *21*. [CrossRef] [PubMed]
21. Barouch, D.H.; Thomas, S.J.; Michael, N.L. Prospects for a Zika Virus Vaccine. *Immunity* **2017**, *46*, 176–182. [CrossRef] [PubMed]
22. Waggoner, J.J.; Gresh, L.; Vargas, M.J.; Ballesteros, G.; Tellez, Y.; Soda, K.J.; Sahoo, M.K.; Nunez, A.; Balmaseda, A.; Harris, E.; et al. Viremia and Clinical Presentation in Nicaraguan Patients Infected With Zika Virus, Chikungunya Virus, and Dengue Virus. *Clin. Infect. Dis.* **2016**, *63*, 1584–1590. [CrossRef] [PubMed]
23. Grossi-Soyster, E.N.; LaBeaud, A.D. Clinical aspects of Zika virus. *Curr. Opin. Pediatr.* **2016**. [CrossRef] [PubMed]
24. Brasil, P.; Calvet, G.A.; Siqueira, A.M.; Wakimoto, M.; de Sequeira, P.C.; Nobre, A.; Quintana Mde, S.; Mendonca, M.C.; Lupi, O.; de Souza, R.V.; et al. Zika Virus Outbreak in Rio de Janeiro, Brazil: Clinical Characterization, Epidemiological and Virological Aspects. *PLoS Negl. Trop. Dis.* **2016**, *10*, e0004636. [CrossRef] [PubMed]
25. Calvet, G.A.; Santos, F.B.; Sequeira, P.C. Zika virus infection: Epidemiology, clinical manifestations and diagnosis. *Curr. Opin. Infect. Dis.* **2016**, *29*, 459–466. [CrossRef] [PubMed]
26. Arzuza-Ortega, L.; Polo, A.; Perez-Tatis, G.; Lopez-Garcia, H.; Parra, E.; Pardo-Herrera, L.C.; Rico-Turca, A.M.; Villamil-Gomez, W.; Rodriguez-Morales, A.J. Fatal Sickle Cell Disease and Zika Virus Infection in Girl from Colombia. *Emerg. Infect. Dis.* **2016**, *22*, 925–927. [CrossRef] [PubMed]
27. Azevedo, R.S.; Araujo, M.T.; Martins Filho, A.J.; Oliveira, C.S.; Nunes, B.T.; Cruz, A.C.; Nascimento, A.G.; Medeiros, R.C.; Caldas, C.A.; Araujo, F.C.; et al. Zika virus epidemic in Brazil. I. Fatal disease in adults: Clinical and laboratorial aspects. *J. Clin. Virol.* **2016**, *85*, 56–64. [CrossRef] [PubMed]

28. Dirlikov, E.; Ryff, K.R.; Torres-Aponte, J.; Thomas, D.L.; Perez-Padilla, J.; Munoz-Jordan, J.; Caraballo, E.V.; Garcia, M.; Segarra, M.O.; Malave, G.; et al. Update: Ongoing Zika Virus Transmission—Puerto Rico, 1 November 2015–14 April 2016. *MMWR Morb. Mortal. Wkly. Rep.* **2016**, *65*, 451–455. [CrossRef] [PubMed]

29. Sarmiento-Ospina, A.; Vasquez-Serna, H.; Jimenez-Canizales, C.E.; Villamil-Gomez, W.E.; Rodriguez-Morales, A.J. Zika virus associated deaths in Colombia. *Lancet Infect. Dis.* **2016**, *16*, 523–524. [CrossRef]

30. Soares, C.N.; Brasil, P.; Carrera, R.M.; Sequeira, P.; de Filippis, A.B.; Borges, V.A.; Theophilo, F.; Ellul, M.A.; Solomon, T. Fatal encephalitis associated with Zika virus infection in an adult. *J. Clin. Virol.* **2016**, *83*, 63–65. [CrossRef] [PubMed]

31. Swaminathan, S.; Schlaberg, R.; Lewis, J.; Hanson, K.E.; Couturier, M.R. Fatal Zika Virus Infection with Secondary Nonsexual Transmission. *N. Engl. J. Med.* **2016**, *375*, 1907–1909. [CrossRef] [PubMed]

32. Brasil, P.; Sequeira, P.C.; Freitas, A.D.; Zogbi, H.E.; Calvet, G.A.; de Souza, R.V.; Siqueira, A.M.; de Mendonca, M.C.; Nogueira, R.M.; de Filippis, A.M.; et al. Guillain-Barre syndrome associated with Zika virus infection. *Lancet* **2016**, *387*, 1482. [CrossRef]

33. Yuki, N.; Hartung, H.P. Guillain-Barre syndrome. *N. Engl. J. Med.* **2012**, *366*, 2294–2304. [CrossRef] [PubMed]

34. Sandhu, S.K.; Hua, W.; MaCurdy, T.E.; Franks, R.L.; Avagyan, A.; Kelman, J.; Worrall, C.M.; Ball, R.; Nguyen, M. Near real-time surveillance for Guillain-Barre syndrome after influenza vaccination among the Medicare population, 2010/11 to 2013/14. *Vaccine* **2017**, *35*, 2986–2992. [CrossRef] [PubMed]

35. Carteaux, G.; Maquart, M.; Bedet, A.; Contou, D.; Brugieres, P.; Fourati, S.; Cleret de Langavant, L.; de Broucker, T.; Brun-Buisson, C.; Leparc-Goffart, I.; et al. Zika Virus Associated with Meningoencephalitis. *N. Engl. J. Med.* **2016**, *374*, 1595–1596. [CrossRef] [PubMed]

36. Mecharles, S.; Herrmann, C.; Poullain, P.; Tran, T.H.; Deschamps, N.; Mathon, G.; Landais, A.; Breurec, S.; Lannuzel, A. Acute myelitis due to Zika virus infection. *Lancet* **2016**, *387*, 1481. [CrossRef]

37. Epidemiological Alert Increase of Microcephaly in the Northeast of Brazil. Available online: http://www.paho.org/hq/index.php?option=com_docman&task=doc_view&Itemid=270&gid=32285&lang=en (accessed on 7 April 2017).

38. Boeuf, P.; Drummer, H.E.; Richards, J.S.; Scoullar, M.J.; Beeson, J.G. The global threat of Zika virus to pregnancy: Epidemiology, clinical perspectives, mechanisms, and impact. *BMC Med.* **2016**, *14*, 112. [CrossRef] [PubMed]

39. Carvalho, M.D.; Miranda-Filho, D.B.; van der Linden, V.; Sobral, P.F.; Ramos, R.C.; Rocha, M.A.; Cordeiro, M.T.; de Alencar, S.P.; Nunes, M.L. Sleep EEG patterns in infants with congenital Zika virus syndrome. *Clin. Neurophysiol.* **2016**, *128*, 204–214. [CrossRef] [PubMed]

40. Cauchemez, S.; Besnard, M.; Bompard, P.; Dub, T.; Guillemette-Artur, P.; Eyrolle-Guignot, D.; Salje, H.; van Kerkhove, M.D.; Abadie, V.; Garel, C.; et al. Association between Zika virus and microcephaly in French Polynesia, 2013–2015: A retrospective study. *Lancet* **2016**, *387*, 2125–2132. [CrossRef]

41. Costello, A.; Dua, T.; Duran, P.; Gulmezoglu, M.; Oladapo, O.T.; Perea, W.; Pires, J.; Ramon-Pardo, P.; Rollins, N.; Saxena, S. Defining the syndrome associated with congenital Zika virus infection. *Bull. World Health Organ.* **2016**, *94*, 406A. [CrossRef] [PubMed]

42. De Paula Freitas, B.; de Oliveira Dias, J.R.; Prazeres, J.; Sacramento, G.A.; Ko, A.I.; Maia, M.; Belfort, R. Ocular Findings in Infants With Microcephaly Associated With Presumed Zika Virus Congenital Infection in Salvador, Brazil. *JAMA Ophthalmol.* **2016**. [CrossRef] [PubMed]

43. De Paula Freitas, B.; Ko, A.I.; Khouri, R.; Mayoral, M.; Henriques, D.F.; Maia, M.; Belfort, R. Glaucoma and Congenital Zika Syndrome. *Ophthalmology* **2016**. [CrossRef] [PubMed]

44. Kleber de Oliveira, W.; Cortez-Escalante, J.; de Oliveira, W.T.; do Carmo, G.M.; Henriques, C.M.; Coelho, G.E.; Araujo de Franca, G.V. Increase in Reported Prevalence of Microcephaly in Infants Born to Women Living in Areas with Confirmed Zika Virus Transmission During the First Trimester of Pregnancy—Brazil, 2015. *MMWR Morb. Mortal. Wkly. Rep.* **2016**, *65*, 242–247. [CrossRef] [PubMed]

45. Frank, C.; Faber, M.; Stark, K. Causal or not: Applying the Bradford Hill aspects of evidence to the association between Zika virus and microcephaly. *EMBO Mol. Med.* **2016**, *8*, 305–307. [CrossRef] [PubMed]

46. Rasmussen, S.A.; Jamieson, D.J.; Honein, M.A.; Petersen, L.R. Zika Virus and Birth Defects—Reviewing the Evidence for Causality. *N. Engl. J. Med.* **2016**. [CrossRef] [PubMed]

47. Martines, R.B.; Bhatnagar, J.; de Oliveira Ramos, A.M.; Davi, H.P.; Iglezias, S.D.; Kanamura, C.T.; Keating, M.K.; Hale, G.; Silva-Flannery, L.; Muehlenbachs, A.; et al. Pathology of congenital Zika syndrome in Brazil: A case series. *Lancet* **2016**, *388*, 898–904. [CrossRef]

48. Mlakar, J.; Korva, M.; Tul, N.; Popovic, M.; Poljsak-Prijatelj, M.; Mraz, J.; Kolenc, M.; Resman Rus, K.; Vesnaver Vipotnik, T.; Fabjan Vodusek, V.; et al. Zika Virus Associated with Microcephaly. *N. Engl. J. Med.* **2016**, *374*, 951–958. [CrossRef] [PubMed]

49. Tang, H.; Hammack, C.; Ogden, S.C.; Wen, Z.; Qian, X.; Li, Y.; Yao, B.; Shin, J.; Zhang, F.; Lee, E.M.; et al. Zika Virus Infects Human Cortical Neural Progenitors and Attenuates Their Growth. *Cell Stem Cell* **2016**, *18*, 587–590. [CrossRef] [PubMed]

50. Quicke, K.M.; Bowen, J.R.; Johnson, E.L.; McDonald, C.E.; Ma, H.; O'Neal, J.T.; Rajakumar, A.; Wrammert, J.; Rimawi, B.H.; Pulendran, B.; et al. Zika Virus Infects Human Placental Macrophages. *Cell Host Microbe* **2016**, *20*, 83–90. [CrossRef] [PubMed]

51. Adams Waldorf, K.M.; Stencel-Baerenwald, J.E.; Kapur, R.P.; Studholme, C.; Boldenow, E.; Vornhagen, J.; Baldessari, A.; Dighe, M.K.; Thiel, J.; Merillat, S.; et al. Fetal brain lesions after subcutaneous inoculation of Zika virus in a pregnant nonhuman primate. *Nat. Med.* **2016**, *22*, 1256–1259. [CrossRef] [PubMed]

52. Cugola, F.R.; Fernandes, I.R.; Russo, F.B.; Freitas, B.C.; Dias, J.L.; Guimaraes, K.P.; Benazzato, C.; Almeida, N.; Pignatari, G.C.; Romero, S.; et al. The Brazilian Zika virus strain causes birth defects in experimental models. *Nature* **2016**, *534*, 267–271. [CrossRef] [PubMed]

53. Li, C.; Xu, D.; Ye, Q.; Hong, S.; Jiang, Y.; Liu, X.; Zhang, N.; Shi, L.; Qin, C.F.; Xu, Z. Zika Virus Disrupts Neural Progenitor Development and Leads to Microcephaly in Mice. *Cell Stem Cell* **2016**, *19*, 120–126. [CrossRef] [PubMed]

54. Miner, J.J.; Cao, B.; Govero, J.; Smith, A.M.; Fernandez, E.; Cabrera, O.H.; Garber, C.; Noll, M.; Klein, R.S.; Noguchi, K.K.; et al. Zika Virus Infection during Pregnancy in Mice Causes Placental Damage and Fetal Demise. *Cell* **2016**, *165*, 1081–1091. [CrossRef] [PubMed]

55. Morrison, T.E.; Diamond, M.S. Animal Models of Zika Virus Infection, Pathogenesis, and Immunity. *J. Virol.* **2017**, *91*. [CrossRef] [PubMed]

56. Dick, G.W. Zika virus. II. Pathogenicity and physical properties. *Trans. R. Soc. Trop. Med. Hyg.* **1952**, *46*, 521–534. [CrossRef]

57. Wu, K.Y.; Zuo, G.L.; Li, X.F.; Ye, Q.; Deng, Y.Q.; Huang, X.Y.; Cao, W.C.; Qin, C.F.; Luo, Z.G. Vertical transmission of Zika virus targeting the radial glial cells affects cortex development of offspring mice. *Cell Res.* **2016**, *26*, 645–654. [CrossRef] [PubMed]

58. Elong Ngono, A.; Vizcarra, E.A.; Tang, W.W.; Sheets, N.; Joo, Y.; Kim, K.; Gorman, M.J.; Diamond, M.S.; Shresta, S. Mapping and Role of the CD8+ T Cell Response During Primary Zika Virus Infection in Mice. *Cell Host Microbe* **2017**, *21*, 35–46. [CrossRef] [PubMed]

59. Lazear, H.M.; Govero, J.; Smith, A.M.; Platt, D.J.; Fernandez, E.; Miner, J.J.; Diamond, M.S. A Mouse Model of Zika Virus Pathogenesis. *Cell Host Microbe* **2016**, *19*, 720–730. [CrossRef] [PubMed]

60. Aliota, M.T.; Caine, E.A.; Walker, E.C.; Larkin, K.E.; Camacho, E.; Osorio, J.E. Characterization of Lethal Zika Virus Infection in AG129 Mice. *PLoS Negl. Trop. Dis.* **2016**, *10*, e0004682. [CrossRef] [PubMed]

61. Dowall, S.D.; Graham, V.A.; Rayner, E.; Atkinson, B.; Hall, G.; Watson, R.J.; Bosworth, A.; Bonney, L.C.; Kitchen, S.; Hewson, R. A Susceptible Mouse Model for Zika Virus Infection. *PLoS Negl. Trop. Dis.* **2016**, *10*, e0004658. [CrossRef] [PubMed]

62. Rossi, S.L.; Tesh, R.B.; Azar, S.R.; Muruato, A.E.; Hanley, K.A.; Auguste, A.J.; Langsjoen, R.M.; Paessler, S.; Vasilakis, N.; Weaver, S.C. Characterization of a Novel Murine Model to Study Zika Virus. *Am. J. Trop. Med. Hyg.* **2016**, *94*, 1362–1369. [CrossRef] [PubMed]

63. Chan, J.F.; Zhang, A.J.; Chan, C.C.; Yip, C.C.; Mak, W.W.; Zhu, H.; Poon, V.K.; Tee, K.M.; Zhu, Z.; Cai, J.P.; et al. Zika Virus Infection in Dexamethasone-immunosuppressed Mice Demonstrating Disseminated Infection with Multi-organ Involvement Including Orchitis Effectively Treated by Recombinant Type I Interferons. *EBioMedicine* **2016**, *14*, 112–122. [CrossRef] [PubMed]

64. Yockey, L.J.; Varela, L.; Rakib, T.; Khoury-Hanold, W.; Fink, S.L.; Stutz, B.; Szigeti-Buck, K.; Van den Pol, A.; Lindenbach, B.D.; Horvath, T.L.; et al. Vaginal Exposure to Zika Virus during Pregnancy Leads to Fetal Brain Infection. *Cell* **2016**, *166*, 1247–1256. [CrossRef] [PubMed]

65. Arck, P.C.; Hecher, K. Fetomaternal immune cross-talk and its consequences for maternal and offspring's health. *Nat. Med.* **2013**, *19*, 548–556. [CrossRef] [PubMed]

66. Mysorekar, I.U.; Diamond, M.S. Modeling Zika Virus Infection in Pregnancy. *N. Engl. J. Med.* **2016**, *375*, 481–484. [CrossRef] [PubMed]

67. Dudley, D.M.; Aliota, M.T.; Mohr, E.L.; Weiler, A.M.; Lehrer-Brey, G.; Weisgrau, K.L.; Mohns, M.S.; Breitbach, M.E.; Rasheed, M.N.; Newman, C.M.; et al. A *Rhesus macaque* model of Asian-lineage Zika virus infection. *Nat. Commun.* **2016**, *7*, 12204. [CrossRef] [PubMed]

68. Osuna, C.E.; Lim, S.Y.; Deleage, C.; Griffin, B.D.; Stein, D.; Schroeder, L.T.; Omange, R.; Best, K.; Luo, M.; Hraber, P.T.; et al. Zika viral dynamics and shedding in rhesus and cynomolgus macaques. *Nat. Med.* **2016**, *22*, 1448–1455. [CrossRef] [PubMed]

69. Coffey, L.L.; Pesavento, P.A.; Keesler, R.I.; Singapuri, A.; Watanabe, J.; Watanabe, R.; Yee, J.; Bliss-Moreau, E.; Cruzen, C.; Christe, K.L.; et al. Zika Virus Tissue and Blood Compartmentalization in Acute Infection of *Rhesus macaques*. *PLoS ONE* **2017**, *12*, e0171148. [CrossRef] [PubMed]

70. Schneider, W.M.; Chevillotte, M.D.; Rice, C.M. Interferon-stimulated genes: A complex web of host defenses. *Annu. Rev. Immunol.* **2014**, *32*, 513–545. [CrossRef] [PubMed]

71. Bayer, A.; Lennemann, N.J.; Ouyang, Y.; Bramley, J.C.; Morosky, S.; Marques, E.T., Jr.; Cherry, S.; Sadovsky, Y.; Coyne, C.B. Type III Interferons Produced by Human Placental Trophoblasts Confer Protection against Zika Virus Infection. *Cell Host Microbe* **2016**, *19*, 705–712. [CrossRef] [PubMed]

72. Versteeg, G.A.; Garcia-Sastre, A. Viral tricks to grid-lock the type I interferon system. *Curr. Opin. Microbiol.* **2010**, *13*, 508–516. [CrossRef] [PubMed]

73. Grant, A.; Ponia, S.S.; Tripathi, S.; Balasubramaniam, V.; Miorin, L.; Sourisseau, M.; Schwarz, M.C.; Sanchez-Seco, M.P.; Evans, M.J.; Best, S.M.; et al. Zika Virus Targets Human STAT2 to Inhibit Type I Interferon Signaling. *Cell Host Microbe* **2016**, *19*, 882–890. [CrossRef] [PubMed]

74. Kotenko, S.V.; Gallagher, G.; Baurin, V.V.; Lewis-Antes, A.; Shen, M.; Shah, N.K.; Langer, J.A.; Sheikh, F.; Dickensheets, H.; Donnelly, R.P. IFN-lambdas mediate antiviral protection through a distinct class II cytokine receptor complex. *Nat. Immunol.* **2003**, *4*, 69–77. [CrossRef] [PubMed]

75. Khan, S.; Woodruff, E.M.; Trapecar, M.; Fontaine, K.A.; Ezaki, A.; Borbet, T.C.; Ott, M.; Sanjabi, S. Dampened antiviral immunity to intravaginal exposure to RNA viral pathogens allows enhanced viral replication. *J. Exp. Med.* **2016**, *213*, 2913–2929. [CrossRef] [PubMed]

76. Sapparapu, G.; Fernandez, E.; Kose, N.; Bin, C.; Fox, J.M.; Bombardi, R.G.; Zhao, H.; Nelson, C.A.; Bryan, A.L.; Barnes, T.; et al. Neutralizing human antibodies prevent Zika virus replication and fetal disease in mice. *Nature* **2016**, *540*, 443–447. [CrossRef] [PubMed]

77. Stettler, K.; Beltramello, M.; Espinosa, D.A.; Graham, V.; Cassotta, A.; Bianchi, S.; Vanzetta, F.; Minola, A.; Jaconi, S.; Mele, F.; et al. Specificity, cross-reactivity, and function of antibodies elicited by Zika virus infection. *Science* **2016**, *353*, 823–826. [CrossRef] [PubMed]

78. Swanstrom, J.A.; Plante, J.A.; Plante, K.S.; Young, E.F.; McGowan, E.; Gallichotte, E.N.; Widman, D.G.; Heise, M.T.; de Silva, A.M.; Baric, R.S. Dengue Virus Envelope Dimer Epitope Monoclonal Antibodies Isolated from Dengue Patients Are Protective against Zika Virus. *MBio* **2016**, *7*. [CrossRef] [PubMed]

79. Abbink, P.; Larocca, R.A.; De La Barrera, R.A.; Bricault, C.A.; Moseley, E.T.; Boyd, M.; Kirilova, M.; Li, Z.; Ng'ang'a, D.; Nanayakkara, O.; et al. Protective efficacy of multiple vaccine platforms against Zika virus challenge in rhesus monkeys. *Science* **2016**, *353*, 1129–1132. [CrossRef] [PubMed]

80. Muthumani, K.; Griffin, B.D.; Agarwal, S.; Kudchodkar, S.B.; Reuschel, E.L.; Choi, H.; Kraynyak, K.A.; Duperret, E.K.; Keaton, A.A.; Chung, C.; et al. In vivo protection against ZIKV infection and pathogenesis through passive antibody transfer and active immunisation with a prMEnv DNA vaccine. *Npj Vaccines* **2016**, *1*. [CrossRef]

81. De Madrid, A.T.; Porterfield, J.S. The flaviviruses (group B arboviruses): A cross-neutralization study. *J. Gen. Virol.* **1974**, *23*, 91–96. [CrossRef] [PubMed]

82. Fauci, A.S.; Morens, D.M. Zika Virus in the Americas—Yet Another Arbovirus Threat. *N. Engl. J. Med.* **2016**, *374*, 601–604. [CrossRef] [PubMed]

83. Barba-Spaeth, G.; Dejnirattisai, W.; Rouvinski, A.; Vaney, M.C.; Medits, I.; Sharma, A.; Simon-Loriere, E.; Sakuntabhai, A.; Cao-Lormeau, V.M.; Haouz, A.; et al. Structural basis of potent Zika-dengue virus antibody cross-neutralization. *Nature* **2016**, *536*, 48–53. [CrossRef] [PubMed]

84. Dejnirattisai, W.; Supasa, P.; Wongwiwat, W.; Rouvinski, A.; Barba-Spaeth, G.; Duangchinda, T.; Sakuntabhai, A.; Cao-Lormeau, V.M.; Malasit, P.; Rey, F.A.; et al. Dengue virus sero-cross-reactivity drives

antibody-dependent enhancement of infection with Zika virus. *Nat. Immunol.* **2016**, *17*, 1102–1108. [CrossRef] [PubMed]

85. Harrison, S.C. Immunogenic cross-talk between dengue and Zika viruses. *Nat. Immunol.* **2016**, *17*, 1010–1012. [CrossRef] [PubMed]

86. Priyamvada, L.; Quicke, K.M.; Hudson, W.H.; Onlamoon, N.; Sewatanon, J.; Edupuganti, S.; Pattanapanyasat, K.; Chokephaibulkit, K.; Mulligan, M.J.; Wilson, P.C.; et al. Human antibody responses after dengue virus infection are highly cross-reactive to Zika virus. *Proc. Natl. Acad. Sci. USA* **2016**, *113*, 7852–7857. [CrossRef] [PubMed]

87. Halstead, S.B. Neutralization and antibody-dependent enhancement of dengue viruses. *Adv. Virus Res.* **2003**, *60*, 421–467. [PubMed]

88. Halstead, S.B. Antibodies determine virulence in dengue. *Ann. N. Y. Acad. Sci.* **2009**, *1171*, E48–E56. [CrossRef] [PubMed]

89. Kliks, S.C.; Nisalak, A.; Brandt, W.E.; Wahl, L.; Burke, D.S. Antibody-dependent enhancement of dengue virus growth in human monocytes as a risk factor for dengue hemorrhagic fever. *Am. J. Trop. Med. Hyg.* **1989**, *40*, 444–451. [CrossRef] [PubMed]

90. Wahala, W.M.; Silva, A.M. The human antibody response to dengue virus infection. *Viruses* **2011**, *3*, 2374–2395. [CrossRef] [PubMed]

91. Duangchinda, T.; Dejnirattisai, W.; Vasanawathana, S.; Limpitikul, W.; Tangthawornchaikul, N.; Malasit, P.; Mongkolsapaya, J.; Screaton, G. Immunodominant T-cell responses to dengue virus NS3 are associated with DHF. *Proc. Natl. Acad. Sci. USA* **2010**, *107*, 16922–16927. [CrossRef] [PubMed]

92. Rivino, L.; Lim, M.Q. CD4+ and CD8+ T-cell immunity to Dengue—Lessons for the study of Zika virus. *Immunology* **2016**. [CrossRef] [PubMed]

93. Simmons, C.P.; Dong, T.; Chau, N.V.; Dung, N.T.; Chau, T.N.; Thao le, T.T.; Dung, N.T.; Hien, T.T.; Rowland-Jones, S.; Farrar, J. Early T-cell responses to dengue virus epitopes in Vietnamese adults with secondary dengue virus infections. *J. Virol.* **2005**, *79*, 5665–5675. [CrossRef] [PubMed]

94. Weiskopf, D.; Angelo, M.A.; Sidney, J.; Peters, B.; Shresta, S.; Sette, A. Immunodominance changes as a function of the infecting dengue virus serotype and primary versus secondary infection. *J. Virol.* **2014**, *88*, 11383–11394. [CrossRef] [PubMed]

95. Weiskopf, D.; Sette, A. T-cell immunity to infection with dengue virus in humans. *Front. Immunol.* **2014**, *5*, 93. [CrossRef] [PubMed]

96. Akondy, R.S.; Monson, N.D.; Miller, J.D.; Edupuganti, S.; Teuwen, D.; Wu, H.; Quyyumi, F.; Garg, S.; Altman, J.D.; Del Rio, C.; et al. The yellow fever virus vaccine induces a broad and polyfunctional human memory CD8+ T cell response. *J. Immunol.* **2009**, *183*, 7919–7930. [CrossRef] [PubMed]

97. Cong, Y.; McArthur, M.A.; Cohen, M.; Jahrling, P.B.; Janosko, K.B.; Josleyn, N.; Kang, K.; Zhang, T.; Holbrook, M.R. Characterization of Yellow Fever Virus Infection of Human and Non-human Primate Antigen Presenting Cells and Their Interaction with CD4+ T Cells. *PLoS Negl. Trop. Dis.* **2016**, *10*, e0004709. [CrossRef] [PubMed]

98. Miller, J.D.; van der Most, R.G.; Akondy, R.S.; Glidewell, J.T.; Albott, S.; Masopust, D.; Murali-Krishna, K.; Mahar, P.L.; Edupuganti, S.; Lalor, S.; et al. Human effector and memory CD8+ T cell responses to smallpox and yellow fever vaccines. *Immunity* **2008**, *28*, 710–722. [CrossRef] [PubMed]

99. Kim, J.H.; Patil, A.M.; Choi, J.Y.; Kim, S.B.; Uyangaa, E.; Hossain, F.M.; Park, S.Y.; Lee, J.H.; Eo, S.K. CCR5 ameliorates Japanese encephalitis via dictating the equilibrium of regulatory CD4(+)Foxp3(+) T and IL-17(+)CD4(+) Th17 cells. *J. Neuroinflamm.* **2016**, *13*, 223. [CrossRef] [PubMed]

100. Netland, J.; Bevan, M.J. CD8 and CD4 T cells in West Nile virus immunity and pathogenesis. *Viruses* **2013**, *5*, 2573–2584. [CrossRef] [PubMed]

101. Betancourt, D.; de Queiroz, N.M.; Xia, T.; Ahn, J.; Barber, G.N. Cutting Edge: Innate Immune Augmenting Vesicular Stomatitis Virus Expressing Zika Virus Proteins Confers Protective Immunity. *J. Immunol.* **2017**, *198*, 3023–3028. [CrossRef] [PubMed]

102. Chahal, J.S.; Fang, T.; Woodham, A.W.; Khan, O.F.; Ling, J.; Anderson, D.G.; Ploegh, H.L. An RNA nanoparticle vaccine against Zika virus elicits antibody and CD8+ T cell responses in a mouse model. *Sci. Rep.* **2017**, *7*, 252. [CrossRef] [PubMed]

103. Pardi, N.; Hogan, M.J.; Pelc, R.S.; Muramatsu, H.; Andersen, H.; DeMaso, C.R.; Dowd, K.A.; Sutherland, L.L.; Scearce, R.M.; Parks, R.; et al. Zika virus protection by a single low-dose nucleoside-modified mRNA vaccination. *Nature* **2017**, *543*, 248–251. [CrossRef] [PubMed]

104. Charrel, R.N.; Leparc-Goffart, I.; Pas, S.; de Lamballerie, X.; Koopmans, M.; Reusken, C. Background review for diagnostic test development for Zika virus infection. *Bull. World Health Organ.* **2016**, *94*, 574D–584D. [CrossRef] [PubMed]

105. Landry, M.L.; St George, K. Laboratory Diagnosis of Zika Virus Infection. *Arch. Pathol. Lab. Med.* **2017**, *141*, 60–67. [CrossRef] [PubMed]

106. US Food & Drug Administration Emergency Use Authorizations. Available online: https://www.fda.gov/MedicalDevices/Safety/EmergencySituations/ucm161496.htm#zika (accessed on 7 March 2017).

107. Rabe, I.B.; Staples, J.E.; Villanueva, J.; Hummel, K.B.; Johnson, J.A.; Rose, L.; Hills, S.; Wasley, A.; Fischer, M.; Powers, A.M.; et al. Interim Guidance for Interpretation of Zika Virus Antibody Test Results. *MMWR Morb. Mortal. Wkly. Rep.* **2016**, *65*, 543–546. [CrossRef] [PubMed]

108. Huzly, D.; Hanselmann, I.; Schmidt-Chanasit, J.; Panning, M. High specificity of a novel Zika virus ELISA in European patients after exposure to different flaviviruses. *Euro Surveill.* **2016**, *21*. [CrossRef] [PubMed]

109. Felix, A.C.; Souza, N.C.; Figueiredo, W.M.; Costa, A.A.; Inenami, M.; da Silva, R.M.; Levi, J.E.; Pannuti, C.S.; Romano, C.M. Cross reactivity of commercial anti-dengue immunoassays in patients with acute Zika virus infection. *J. Med. Virol.* **2017**. [CrossRef] [PubMed]

110. Steinhagen, K.; Probst, C.; Radzimski, C.; Schmidt-Chanasit, J.; Emmerich, P.; van Esbroeck, M.; Schinkel, J.; Grobusch, M.P.; Goorhuis, A.; Warnecke, J.M.; et al. Serodiagnosis of Zika virus (ZIKV) infections by a novel NS1-based ELISA devoid of cross-reactivity with dengue virus antibodies: A multicohort study of assay performance, 2015 to 2016. *Euro Surveill.* **2016**, *21*. [CrossRef] [PubMed]

111. Moulin, E.; Selby, K.; Cherpillod, P.; Kaiser, L.; Boillat-Blanco, N. Simultaneous outbreaks of dengue, chikungunya and Zika virus infections: Diagnosis challenge in a returning traveller with nonspecific febrile illness. *New Microbes New Infect.* **2016**, *11*, 6–7. [CrossRef] [PubMed]

112. Corman, V.M.; Rasche, A.; Baronti, C.; Aldabbagh, S.; Cadar, D.; Reusken, C.B.; Pas, S.D.; Goorhuis, A.; Schinkel, J.; Molenkamp, R.; et al. Assay optimization for molecular detection of Zika virus. *Bull. World Health Organ.* **2016**, *94*, 880–892. [CrossRef] [PubMed]

113. Levine, M.M.; Sztein, M.B. Vaccine development strategies for improving immunization: The role of modern immunology. *Nat. Immunol.* **2004**, *5*, 460–464. [CrossRef] [PubMed]

114. WHO Vaccine Pipeline Tracker Zika. Available online: http://www.who.int/immunization/research/vaccine_pipeline_tracker_spreadsheet/en/ (accessed on 14 March 2017).

115. Lehrer, A.T.; Holbrook, M.R. Tick-borne Encephalitis Vaccines. *J. Bioterror. Biodef.* **2011**, *2011*, 3. [CrossRef] [PubMed]

116. McArthur, M.A.; Holbrook, M.R. Japanese Encephalitis Vaccines. *J. Bioterror. Biodef.* **2011**, 2. [CrossRef] [PubMed]

117. Larocca, R.A.; Abbink, P.; Peron, J.P.; Zanotto, P.M.; Iampietro, M.J.; Badamchi-Zadeh, A.; Boyd, M.; Ng'ang'a, D.; Kirilova, M.; Nityanandam, R.; et al. Vaccine protection against Zika virus from Brazil. *Nature* **2016**, *536*, 474–478. [CrossRef] [PubMed]

118. Yang, Y.; Shan, C.; Zou, J.; Muruato, A.E.; Bruno, D.N.; de Almeida Medeiros Daniele, B.; Vasconcelos, P.F.; Rossi, S.L.; Weaver, S.C.; Xie, X.; et al. A cDNA Clone-Launched Platform for High-Yield Production of Inactivated Zika Vaccine. *EBioMedicine* **2017**, *17*, 145–156. [CrossRef] [PubMed]

119. Abbasi, J. First Inactivated Zika Vaccine Trial. *JAMA* **2016**, *316*, 2588. [CrossRef] [PubMed]

120. Danko, J.R.; Beckett, C.G.; Porter, K.R. Development of dengue DNA vaccines. *Vaccine* **2011**, *29*, 7261–7266. [CrossRef] [PubMed]

121. Ledgerwood, J.E.; Pierson, T.C.; Hubka, S.A.; Desai, N.; Rucker, S.; Gordon, I.J.; Enama, M.E.; Nelson, S.; Nason, M.; Gu, W.; et al. A West Nile virus DNA vaccine utilizing a modified promoter induces neutralizing antibody in younger and older healthy adults in a phase I clinical trial. *J. Infect. Dis.* **2011**, *203*, 1396–1404. [CrossRef] [PubMed]

122. Martin, J.E.; Pierson, T.C.; Hubka, S.; Rucker, S.; Gordon, I.J.; Enama, M.E.; Andrews, C.A.; Xu, Q.; Davis, B.S.; Nason, M.; et al. A West Nile virus DNA vaccine induces neutralizing antibody in healthy adults during a phase 1 clinical trial. *J. Infect. Dis.* **2007**, *196*, 1732–1740. [CrossRef] [PubMed]

123. Beckett, C.G.; Tjaden, J.; Burgess, T.; Danko, J.R.; Tamminga, C.; Simmons, M.; Wu, S.J.; Sun, P.; Kochel, T.; Raviprakash, K.; et al. Evaluation of a prototype dengue-1 DNA vaccine in a Phase 1 clinical trial. *Vaccine* **2011**, *29*, 960–968. [CrossRef] [PubMed]

124. Abbasi, J. Zika Vaccine Enters Clinical Trials. *JAMA* **2016**, *316*, 1249. [CrossRef] [PubMed]

125. Dowd, K.A.; Ko, S.Y.; Morabito, K.M.; Yang, E.S.; Pelc, R.S.; DeMaso, C.R.; Castilho, L.R.; Abbink, P.; Boyd, M.; Nityanandam, R.; et al. Rapid development of a DNA vaccine for Zika virus. *Science* **2016**, *354*, 237–240. [CrossRef] [PubMed]

126. Kramps, T.; Elbers, K. Introduction to RNA Vaccines. *Methods Mol. Biol.* **2017**, *1499*, 1–11. [CrossRef] [PubMed]

127. Richner, J.M.; Himansu, S.; Dowd, K.A.; Butler, S.L.; Salazar, V.; Fox, J.M.; Julander, J.G.; Tang, W.W.; Shresta, S.; Pierson, T.C.; et al. Modified mRNA Vaccines Protect against Zika Virus Infection. *Cell* **2017**, *169*, 176. [CrossRef] [PubMed]

128. Salam, A.P.; Rojek, A.; Dunning, J.; Horby, P.W. Clinical Trials of Therapeutics for the Prevention of Congenital Zika Virus Disease: Challenges and Potential Solutions. *Ann. Intern. Med.* **2017**. [CrossRef] [PubMed]

129. Adcock, R.S.; Chu, Y.K.; Golden, J.E.; Chung, D.H. Evaluation of anti-Zika virus activities of broad-spectrum antivirals and NIH clinical collection compounds using a cell-based, high-throughput screen assay. *Antivir. Res.* **2017**, *138*, 47–56. [CrossRef] [PubMed]

130. Balasubramanian, A.; Teramoto, T.; Kulkarni, A.A.; Bhattacharjee, A.K.; Padmanabhan, R. Antiviral activities of selected antimalarials against dengue virus type 2 and Zika virus. *Antivir. Res.* **2017**, *137*, 141–150. [CrossRef] [PubMed]

131. Barrows, N.J.; Campos, R.K.; Powell, S.T.; Prasanth, K.R.; Schott-Lerner, G.; Soto-Acosta, R.; Galarza-Munoz, G.; McGrath, E.L.; Urrabaz-Garza, R.; Gao, J.; et al. A Screen of FDA-Approved Drugs for Inhibitors of Zika Virus Infection. *Cell Host Microbe* **2016**, *20*, 259–270. [CrossRef] [PubMed]

132. Bullard-Feibelman, K.M.; Govero, J.; Zhu, Z.; Salazar, V.; Veselinovic, M.; Diamond, M.S.; Geiss, B.J. The FDA-approved drug sofosbuvir inhibits Zika virus infection. *Antivir. Res.* **2017**, *137*, 134–140. [CrossRef] [PubMed]

133. Xu, M.; Lee, E.M.; Wen, Z.; Cheng, Y.; Huang, W.K.; Qian, X.; Tcw, J.; Kouznetsova, J.; Ogden, S.C.; Hammack, C.; et al. Identification of small-molecule inhibitors of Zika virus infection and induced neural cell death via a drug repurposing screen. *Nat. Med.* **2016**, *22*, 1101–1107. [CrossRef] [PubMed]

134. Zmurko, J.; Marques, R.E.; Schols, D.; Verbeken, E.; Kaptein, S.J.; Neyts, J. The Viral Polymerase Inhibitor 7-Deaza-2′-C-Methyladenosine Is a Potent Inhibitor of In Vitro Zika Virus Replication and Delays Disease Progression in a Robust Mouse Infection Model. *PLoS Negl. Trop. Dis.* **2016**, *10*, e0004695. [CrossRef] [PubMed]

viruses

MDPI

Commentary

The New High Resolution Crystal Structure of NS2B-NS3 Protease of Zika Virus

Syed Lal Badshah [1,*], Abdul Naeem [2] and Yahia Mabkhot [3,*]

[1] Department of Chemistry, Islamia College University, Peshawar 25120, Khyber Pukhtoonkhwa, Pakistan
[2] National Center of Excellence in Physical Chemistry, University of Peshawar, Peshawar 25120,
   Khyber Pukhtoonkhwa, Pakistan; naeeem64@yahoo.com
[3] Department of Chemistry, College of Science, King Saud University, Riyadh 11451, Saudi Arabia
* Correspondence: shahbiochemist@gmail.com (S.L.B.); yahia@ksu.edu.sa (Y.M.);
   Tel.: +92-331-921-6672 (S.L.B.); +966-11-467-5898 (Y.M.)

Academic Editor: Michael Holbrook
Received: 19 December 2016; Accepted: 1 January 2017; Published: 10 January 2017

Abstract: Zika virus (ZIKV) is the cause of a significant viral disease affecting humans, which has spread throughout many South American countries and has also become a threat to Southeastern Asia. This commentary discusses the article "Crystal structure of unlinked NS2B-NS3 protease from Zika virus" published recently in the journal *Science* by Zhang et al. of Nanyang Technological University, Singapore. They resolved a 1.58 Å resolution structure of the NS2B-NS3 protease of ZIKV and demonstrated how peptide and non-peptide inhibitors interact with this structure, along with the different conformational states that were observed. This protease crystal structure offers new opportunities for the design and development of novel antiviral drugs used for the treatment and control of ZIKV.

Keywords: Zika virus; microcephaly; non-structural protein; protease; inhibitors

1. Introduction

Zika virus (ZIKV) is a mosquito-borne virus [1], initially isolated from a monkey in the Zika forest of Uganda [2]. The *Aedes aegypti* (*Ae. aegypti*) mosquito species serves as the principal vector for the virus. ZIKV is a member of the *Flaviviridae* family, which includes other highly pathogenic viruses such as dengue virus, West Nile virus, yellow fever virus, Japanese encephalitis virus, and tick-borne encephalitis virus [3–6]. ZIKV became endemic in 2015 in South and Central America, with an especially high concentration in Brazil, and from there the virus quickly spread to other parts of the Americas [7,8]. ZIKV crosses the placental barrier in pregnant women, and can cause teratogenic effects, such as microcephaly in newborns, although the exact mechanism is still not fully understood. Due to the dramatic rise in microcephaly cases caused by ZIKV, the World Health Organization (WHO) declared the virus to be a public health emergency [7,9–12]. The main factors leading to the spread of the virus—and thus increased incidence of microcephaly in newborns—are thought to be the increased mobility of humans and the wide distribution of the *Ae. aegypti* mosquito vector [13,14].

2. NS2B-NS3 Protease of Zika Virus (ZIKV)

The genome of ZIKV encodes a single polyprotein that is co- and post-translationally cleaved to generate three structural proteins and seven non-structural proteins [15,16]. Several of the non-structural proteins function as enzymes for the virus [17]. Among these is the protease NS2B-NS3, whose function is to cleave the virus polyprotein at proper sites, and is required for ZIKV replication. Similar to most viruses, the non-structural proteins of ZIKV are suitable drug targets, and it is therefore highly desirable to understand the crystal structure of these non-structural proteins [18].

Viruses **2017**, *9*, 7

In their recent article, Zhang et al. resolved a 1.58 Å resolution structure of the NS2B-NS3 protease without a linker [19]. Prior to this, they had also published work on a slightly lower resolution structure with a linker and with different ligands in different states [20,21]. The new unlinked NS2B-NS3 structure has an established binding pocket that does not show prominent conformational changes when a substrate or an inhibitor binds with it. This preformed binding cavity is shaped like a cross and contains sub-compartments, where the different residues of the substrate peptide can bind during catalysis. The NS3 N-terminal tetrapeptide group—which contains lysine 14,15, glutamate 16, and glycine 17 ($K_{14}K_{15}E_{16}G_{17}$)—folds into a hairpin structure and occupies this active site or binding cavity. This tetrapeptide forms several different kinds of interaction within the binding pocket, which includes hydrogen bonding and a pi-stacking interaction. Several of the protein intramolecular hydrogen bondings are with the backbone, and that is why it is called the reverse peptide. The formation of the reverse peptide bond is believed to be an ideal area of exploitation for drug design.

In order to understand the full catalytic activity of NS2B-NS3 protease, in vitro activities were performed, and the C-terminal part of the ZIKV NS2B was observed to be quite flexible. When the inhibitor is removed from the C-terminus of NS2B, it then becomes structurally disordered, and is thusly labeled as an open conformation. On the other hand, the ligand-bound protease is a compact structure, and through folding shows close contact with the NS3, and is labeled a closed form conformation. The previously resolved crystal structure of NS2B-NS3 has a long glycine linker which prohibits ligand binding due to steric clashes of different residues.

The structural dynamics of the NS2B-NS3 protease in solution form were also observed through nuclear magnetic resonance (NMR) spectroscopy, which showed a properly folded form of the protein. The different conformational states of the protease enzyme were explored by titrating it with a bipeptide of acetyl lysine-arginine (AcKR) [19]. The AcKR has been previously shown to act as an inhibitor of the West Nile virus (WNV) protease with an IC_{50} of more than 100 μM [19]. The ^{1}H-^{15}N-HSQC spectra of ZIKV protease showed different conformational changes upon the dipeptide binding in the catalytic triad and the protease in general [19]. The ^{15}N-R_1, R_2 and heteronuclear Overhauser effect (NOE) showed stable conformation for the N-terminal region and for the C-terminal β-hairpin region of NS2B. Zhang et al. also showed that small molecule inhibitors can bind to the ZIKV protease [19] when they resolved a crystal structure of the ZIKV NS2B-NS3 protease with the EN300 molecule ((1H-benzo [*d*] imidazole-1-yl) methanol). Although the EN300 did not directly bind with NS2B, it made several interactions of pi-stacking and hydrogen bonding in one of the pockets of the NS3 part of the protein, and there were no major conformational changes in the protein.

3. Concluding Remarks

These results show that the ZIKV protease can be further explored for other active binding pockets, which will be targets for inhibitor molecules. Although Zhang et al. focused on the need to design peptide-based drugs (especially cyclic peptide molecules that act as inhibitors), every option of drug design from small organic molecules to large peptide-based inhibitors should be evaluated for inhibitory interactions with such an influential protease enzyme [19]. Secondly, computational drug design, virtual screening, and molecular dynamic simulation-based approaches are required to determine suitable drug candidates for NS2B-NS3 protease of ZIKV. Similarly, the vaccine development for ZIKV should also be expedited, and the academic and industrial partnership regarding vaccine and therapeutic development needs to be strengthened.

Acknowledgments: The authors appreciate the Deanship of Scientific Research at King Saud University for the funding of this Prolific Research group (PRG-1437-29).

Author Contributions: All the authors contributed equally.

Conflicts of Interest: The authors declare no conflict of interest.

References

1. Maharajan, M.K.; Ranjan, A.; Chu, J.F.; Foo, W.L.; Chai, Z.X.; Lau, E.Y.Y.; Ye, H.M.; Theam, X.J.; Lok, Y.L. Zika Virus Infection: Current Concerns and Perspectives. *Clin. Rev. Allergy Immunol.* **2016**, *51*, 383–394. [CrossRef] [PubMed]

2. Kirya, B.G.; Mukwaya, L.G.; Sempala, S.D.K. A yellow fever epizootic in Zika forest, Uganda, during 1972: Part 1: Virus isolation and sentinel monkeys. *Trans. R. Soc. Trop. Med. Hyg.* **1977**, *71*, 254–260. [CrossRef]

3. Anderson, K.B.; Thomas, S.J.; Endy, T.P. The emergence of Zika virus: A narrative review. *Ann. Intern. Med.* **2016**, *165*, 175–183. [CrossRef] [PubMed]

4. Musso, D.; Gubler, D.J. Zika virus. *Clin. Microbiol. Rev.* **2016**, *29*, 487–524. [CrossRef] [PubMed]

5. Coyne, C.B.; Lazear, H.M. Zika virus—Reigniting the TORCH. *Nat. Rev. Microbiol.* **2016**, *14*, 707–715. [CrossRef] [PubMed]

6. Leyssen, P.; De Clercq, E.; Neyts, J. Perspectives for the treatment of infections with Flaviviridae. *Clin. Microbiol. Rev.* **2000**, *13*, 67–82. [CrossRef] [PubMed]

7. Mlakar, J.; Korva, M.; Tul, N.; Popovic, M.; Poljsak-Prijatelj, M.; Mraz, J.; Kolenc, M.; Resman Rus, K.; Vesnaver Vipotnik, T.; Fabjan Vodusek, V.; et al. Zika Virus Associated with Microcephaly. *N. Engl. J. Med.* **2016**, *374*, 951–958. [CrossRef] [PubMed]

8. Petersen, E.; Wilson, M.E.; Touch, S.; McCloskey, B.; Mwaba, P.; Bates, M.; Dar, O.; Mattes, F.; Kidd, M.; Ippolito, G.; et al. Rapid Spread of Zika Virus in The Americas—Implications for Public Health Preparedness for Mass Gatherings at the 2016 Brazil Olympic Games. *Int. J. Infect. Dis.* **2016**, *44*, 11–15. [CrossRef] [PubMed]

9. Adibi, J.J.; Marques, E.T.A.; Cartus, A.; Beigi, R.H. Teratogenic effects of the Zika virus and the role of the placenta. *Lancet* **2016**, *387*, 1587–1590. [CrossRef]

10. White, M.K.; Wollebo, H.S.; Beckham, J.D.; Tyler, K.L.; Khalili, K. Zika virus: An emergent neuropathological agent. *Ann. Neurol.* **2016**. [CrossRef] [PubMed]

11. World Health Organization. *Zika Virus Situation Report*; WHO: Geneva, Switzerland, 2016.

12. Barjas-Castro, M.L.; Angerami, R.N.; Cunha, M.S.; Suzuki, A.; Nogueira, J.S.; Rocco, I.M.; Maeda, A.Y.; Vasami, F.G.S.; Katz, G.; Boin, I.F.S.F.; et al. Probable transfusion-transmitted Zika virus in Brazil. *Transfusion* **2016**, *56*, 1684–1688. [CrossRef] [PubMed]

13. Attar, N. Zika virus circulates in new regions. *Nat. Rev. Microbiol.* **2016**, *14*, 62. [CrossRef]

14. Gatherer, D.; Kohl, A. Zika virus: A previously slow pandemic spreads rapidly through the Americas. *J. Gen. Virol.* **2016**, *97*, 269–273. [CrossRef] [PubMed]

15. Cunha, M.S.; Esposito, D.L.A.; Rocco, I.M.; Maeda, A.Y.; Vasami, F.G.S.; Nogueira, J.S.; de Souza, R.P.; Suzuki, A.; Addas-Carvalho, M.; de Lourdes Barjas-Castro, M.; et al. First Complete Genome Sequence of Zika Virus (Flaviviridae, Flavivirus) from an Autochthonous Transmission in Brazil. *Genome Announc.* **2016**, *4*, e00032-16. [CrossRef] [PubMed]

16. Lindenbach, B.D.; Rice, C.M. Flaviviridae: The Viruses and Their Replication. In *Fields Virology*; Lippincott Williams & Wilkins: Philadelphia, PA, USA, 2007; pp. 1101–1152.

17. King, A.M.Q.; Lefkowitz, E.; Adams, M.J.; Carstens, E.B. Family Flaviviridae. In *Virus Taxonomy. Ninth Report of the International Committee on Taxonomy of Viruses*; Elsevier: Amsterdam, The Netherlands, 2012; pp. 1003–1020.

18. Sironi, M.; Forni, D.; Clerici, M.; Cagliani, R. Nonstructural Proteins Are Preferential Positive Selection Targets in Zika Virus and Related Flaviviruses. *PLoS Negl. Trop. Dis.* **2016**, *10*, e0004978. [CrossRef] [PubMed]

19. Zhang, Z.; Li, Y.; Loh, Y.R.; Phoo, W.W.; Hung, A.W.; Kang, C.; Luo, D. Crystal structure of unlinked NS2B-NS3 protease from Zika virus. *Science* **2016**. [CrossRef] [PubMed]

20. Lei, J.; Hansen, G.; Nitsche, C.; Klein, C.D.; Zhang, L.; Hilgenfeld, R. Crystal structure of Zika virus NS2B-NS3 protease in complex with a boronate inhibitor. *Science* **2016**, *353*, 503–505. [CrossRef] [PubMed]

21. Phoo, W.W.; Li, Y.; Zhang, Z.; Lee, M.Y.; Loh, Y.R.; Tan, Y.B.; Ng, E.Y.; Lescar, J.; Kang, C.; Luo, D. Structure of the NS2B-NS3 protease from Zika virus after self-cleavage. *Nat. Commun.* **2016**, *7*, 13410. [CrossRef] [PubMed]

viruses

MDPI

Article

Chloroquine, an Endocytosis Blocking Agent, Inhibits Zika Virus Infection in Different Cell Models

Rodrigo Delvecchio [1,†], Luiza M. Higa [1,†], Paula Pezzuto [1,†], Ana Luiza Valadão [1,†], Patrícia P. Garcez [2,3], Fábio L. Monteiro [1], Erick C. Loiola [3], André A. Dias [4], Fábio J. M. Silva [2], Matthew T. Aliota [5], Elizabeth A. Caine [5], Jorge E. Osorio [5], Maria Bellio [4], David H. O'Connor [6], Stevens Rehen [2,3], Renato Santana de Aguiar [1], Andrea Savarino [7], Loraine Campanati [2,*] and Amilcar Tanuri [1,*]

[1] Department of Genetics, Institute of Biology, Federal University of Rio de Janeiro, Rio de Janeiro 21941-902, Brazil; digodelvecchio@gmail.com (R.D.); luizahiga@gmail.com (L.M.H.); paulapezzuto81@gmail.com (P.P.); analuvaladao@gmail.com (A.L.V.); fabiolimonte@gmail.com (F.L.M.); santanarnt@gmail.com (R.S.d.A.)

[2] Institute of Biomedical Sciences, Federal University of Rio de Janeiro, Rio de Janeiro 21941-902, Brazil; ppgarcez@gmail.com (P.P.G.); fabiojms@icb.ufrj.br (F.J.M.S.); srehen@lance-ufrj.org (S.R.)

[3] D'Or Institute for Research and Education (IDOR), Rio de Janeiro 22281-100, Brazil; erickloiola@lance-ufrj.org

[4] Department of Immunology, Federal University of Rio de Janeiro, Rio de Janeiro 21941-902, Brazil; aadias2005@yahoo.com.br (A.A.D.); mariabellioufrj@gmail.com (M.B.)

[5] Department of Pathobiological Sciences, University of Wisconsin-Madison, Madison, WI 53706, USA; mtaliota@wisc.edu (M.T.A.); eacaine@wisc.edu (E.A.C.); jorge.osorio@wisc.edu (J.E.O.)

[6] Department of Pathology and Laboratory Medicine, University of Wisconsin-Madison, Madison, WI 53706, USA; dhoconno@wisc.edu

[7] Istituto Superiore di Sanità, Deptartment of Infectious Diseases, 299 Viale Regina Elena, 00161 Rome, Italy; ansavari@yahoo.com

* Correspondence: lcampanati@gmail.com (L.C.); atanuri@biologia.ufrj.br (A.T.); Tel.: +55-213-938-6384 (L.C.)

† These authors contributed equally to this work.

Academic Editor: Michael R. Holbrook
Received: 16 October 2016; Accepted: 18 November 2016; Published: 29 November 2016

Abstract: Zika virus (ZIKV) infection in utero might lead to microcephaly and other congenital defects. Since no specific therapy is available thus far, there is an urgent need for the discovery of agents capable of inhibiting its viral replication and deleterious effects. Chloroquine is widely used as an antimalarial drug, anti-inflammatory agent, and it also shows antiviral activity against several viruses. Here we show that chloroquine exhibits antiviral activity against ZIKV in Vero cells, human brain microvascular endothelial cells, human neural stem cells, and mouse neurospheres. We demonstrate that chloroquine reduces the number of ZIKV-infected cells in vitro, and inhibits virus production and cell death promoted by ZIKV infection without cytotoxic effects. In addition, chloroquine treatment partially reveres morphological changes induced by ZIKV infection in mouse neurospheres.

Keywords: Zika virus; chloroquine; antiviral; microcephaly; neural stem cell

1. Introduction

Zika virus (ZIKV) is an arthropod-borne virus, transmitted by *Aedes* mosquitoes, that belongs to the *Flavivirus* genus, which also includes other pathogens such as West Nile virus (WNV), yellow fever virus (YFV), Japanese encephalitis virus (JEV), and dengue virus (DENV). Zika virus was first isolated from a sentinel monkey in the Zika forest in Uganda in 1947 [1]. Since then, ZIKV has been isolated from humans and mosquitoes throughout Africa and Southeast Asian countries. Phylogenetic analysis of the nonstructural protein 5 encoding region has disclosed three ZIKV lineages: East African, West African, and Asian [2]. Sequences from Uganda and Senegal belong to the East African cluster, while sequences from Nigeria and also some sequences from Senegal are clustered on the West African lineage [3]. Most of the viral sequences recovered from recent outbreaks, from 2007 until now, belong to the Asian lineage, such as strains isolated in American, South Asian, and Pacific countries. In utero exposure to Asian lineage ZIKV might lead to microcephaly and other developmental malformations including calcifications, arthrogryposis, ventriculomegaly, lissencephaly, cerebellar atrophy, and ocular abnormalities [4–8]. Although all ZIKV lineages can infect humans, these severe manifestations reported after in utero infection have only been associated with Asian lineages, including Brazilian isolates [6,9].

Asian ZIKV strains were detected in the brain and amniotic fluid of newborns and stillborns with microcephaly [4–6,8,9] and the African MR766 strain was shown to kill human neuroprogenitor cells in vitro as well as decrease the growth rate of brain organoids [10,11].

Symptoms of infections with the Asian ZIKV include low-grade fever, headache, rash, conjunctivitis, arthritis, and myalgia [4,12]. Mild symptoms such as headache and low-grade fever were reported by a human volunteer during infection with the African ZIKV strain [13]. However, in rare instances, infection with the Asian ZIKV is associated with cases of Guillain–Barré syndrome [14] and meningoencephalitis [15]. Currently, there is no vaccine or specific therapeutic approaches to prevent or treat ZIKV infections. With the alarming increase in the number of countries affected and the potential for viral spread through global travel and sexual transmission [16,17], there is an urgent need to find a treatment capable of lessening the effects of the disease and inhibiting further transmission.

Chloroquine, a 4-aminoquinoline, is a weak base that is rapidly imported into acidic vesicles, increasing their pH [18]. It is approved by the Food and Drug Administration (FDA) to treat malaria and has long been prophylactically prescribed to pregnant women at risk of exposure to *Plasmodium* parasites [19]. Chloroquine, through the inhibition of pH-dependent steps of viral replication, restricts human immunodeficiency virus (HIV) [20], influenza virus [21], DENV [22], JEV [23], and WNV infection [24]. Here we investigated the antiviral effects of chloroquine on both Asian (using a Brazilian isolate) and African ZIKV infections in different cell types.

2. Materials and Methods

2.1. Cell Culture

Vero cells (ATCC, Manassas, VA, USA) are derived from the kidney of African green monkey and were grown in DMEM High Glucose (GIBCO, Thermo Fisher Scientific, Waltham, MA, USA) supplemented with 5% fetal bovine serum (FBS) (GIBCO). Human brain microvascular endothelial cells (hBMEC) were a kind gift from Dr. Julio Scharfstein (Federal University of Rio de Janeiro, Rio de Janeiro, Brazil), and hBMEC isolation was performed as previously described [25]. These cells were cultured in DMEM High Glucose supplemented with 20% FBS. The C6/36 cell line is derived from *Aedes albopictus*. C6/36 cells (ATCC, Manassas, VA, USA) were grown in Leibovitz L-15 medium (GIBCO) supplemented with 2.95 g/L tryptose phosphate broth (Sigma Aldrich, Boston, MA, USA), 2 mM glutamine (GIBCO), 0.075% sodium bicarbonate (GIBCO), 1X non-essential amino acids (GIBCO) and 5% FBS. Neural stem cells (NSCs) were derived from human induced pluripotent stem cells (iPSCs). iPSCs were provided by the Biobank of iPSCs of the Brazilian Ministry of Health (CONEP B-027 # 25000.111598/2014-04). According to the supplier, fibroblasts were reprogrammed using the protocol developed by Paulsen et al. [26], and transduced with the CytoTune-iPS Sendai kit (Thermo Fisher Scientific, Waltham, MA, USA). iPSCs presented a normal karyotype and the expression of pluripotency markers. These cells were cultured with E8 culture media (GIBCO) on a Matrigel (BD Biosciences, San Jose, CA, USA) coated surface. iPSC colonies were manually passaged every 5–7 days until they reached 70%–80% confluence and were maintained at 37 °C in humidified air with 5% CO_2. To produce NSCs, human iPSCs were exposed to the serum-free neural induction medium (GIBCO), containing Neurobasal medium (GIBCO) and the pluripotent stem cell neural induction supplement (GIBCO), according to the manufacturer's protocol [27]. Briefly, the medium was changed every other day until day 7, when initial NSCs are split and grown on neural expansion medium (Advanced DMEM/F12 and neurobasal medium (1:1) with neural induction supplement; GIBCO). NSCs were used after four passages in neural expansion media. Mouse central nervous system (CNS) cells were harvested from Swiss mouse embryos at embryo day 14 (E14) and grown for 72 h as free floating neurospheres in neurobasal culture media (GIBCO) supplemented with 1X B27 (GIBCO).

2.2. Compounds

Chloroquine diphosphate was kindly supplied by FarManguinhos (Fiocruz, Rio de Janeiro, Brazil). The lyophilized powder was diluted in double distilled water to 20 mM. The chloroquine solution was filtered through a 0.22 μm membrane and stored at −80 °C.

2.3. Viruses

ZIKV strain MR 766 (Uganda/Africa, accession no.: NC012532.1, kindly provided by Dr. Davis Ferreira, Federal University of Rio de Janeiro, Rio de Janeiro, Brazil) was isolated from a rhesus monkey and injected intracerebrally on Swiss mice for several passages [1] and ZIKV BR (Recife/Brazil, ZIKV PE/243, accession no: KX197192.1, kindly provided by Dr. Marli Tenório Cordeiro, Centro de Pesquisas Aggeu Magalhães, Recife, Brazil) was isolated from a patient presenting classical symptoms of ZIKV infection [28]. These viruses were propagated in Vero and C6/36 cells, respectively. Briefly, the cells were infected with ZIKV at a multiplicity of infection (MOI) of 0.01 and incubated at 37 °C. After 1 h, the inoculum was removed and replaced with DMEM high-glucose (Vero) and Leibovitz L-15 (C6/36) growth media supplemented with 2% FBS. After 4 to 6 days, the conditioned medium was harvested, centrifuged at $300 \times g$, and sterile-filtered to remove cells and cellular debris. Virus stocks were stored at −80 °C.

2.4. Plaque Assay

Virus titers were determined by plaque assay performed on Vero cells. Virus stocks or samples were serially diluted and adsorbed to confluent monolayers. After 1 h, the inoculum was removed and cells were overlaid with semisolid medium constituted of alpha-MEM (GIBCO) containing 1% carboxymethyl cellulose (Sigma Aldrich) and 1% FBS (GIBCO). Cells were further incubated for 5 days when cells were fixed in 4% formaldehyde. Cells were stained with 1% crystal violet in 20% ethanol for plaque visualization. Titers were expressed as plaque forming units (PFU) per milliliter.

2.5. Quantification of ZIKV-Infected Cells by Flow Cytometry

Vero cells, hBMEC, and NSC were infected with ZIKV MR 766 or ZIKV BR strain at an MOI of 2. After 1 h, the inoculum was removed and medium containing chloroquine (6.25 to 50 μM) was added to the cells. After 4 to 5 days, cells were fixed with 4% paraformaldehyde (Sigma Aldrich) in phosphate buffered saline (PBS) for 15 min at room temperature and washed with PBS. Cells were permeabilized with 0.1% Triton X-100 (Sigma Aldrich) in PBS, washed with PBS, and blocked with PBS with 5% FBS. Cells were incubated with 4G2, a pan-flavivirus antibody raised against the ZIKV envelope E protein produced in 4G2-4-15 hybridoma (ATCC), diluted 1:5 in PBS with 5% FBS. Cells were labeled with donkey anti-mouse Alexa Fluor 488 antibody (Thermo Scientific, Waltham, MA, USA) diluted 1:1000 in PBS with 5% FBS, and were analyzed by flow cytometry in a BD Accuri C6 (Becton, Dickinson and Company, Franklin Lakes, NJ, USA) for ZIKV infection.

2.6. Immunofluorescence Assay

Vero, hBMEC, and NSC were seeded on black 96-well plates with clear bottoms and infected with ZIKV MR 766 or ZIKV BR strain at an MOI of 2 for 1 h. Neurospheres were seeded on coverslips and infected with ZIKV MR766 with 2.5×10^5 PFU After infection, the viral inoculum was removed and cells were incubated with medium containing chloroquine (1.56 to 50 μM) for 4 to 5 days, depending on cell type. Cells and neurospheres were fixed with 4% paraformaldehyde in PBS for 20 min at room temperature. The fixative was removed and the samples were washed three times with PBS. Blocking of unspecific binding of the antibody and permeabilization were performed with PBS supplemented with 3% bovine serum albumin (BSA, Sigma Aldrich) and 0.1% Triton X-100 for 40 min at room temperature. Incubation with anti-Map2 antibody was performed following the manufacturer's instructions (ABCAM, Cambridge, Cambridgeshire, UK, #32454; 1:1000) and the 4G2 antibody was diluted 1:4 in PBS with 3% BSA and incubated with cells for 1 h. After washing three times with PBS, cells were incubated with secondary antibodies coupled to Alexa fluorochromes (Thermo Fisher Scientific) for 40 min and washed five times. Coverslips with neurospheres were mounted with Prolong Gold mounting medium (Thermo Fisher Scientific). Samples were imaged using either Leica SP5 or Leica SPE (Leica Biosystems, Wetzlar, Hesse, DE) confocal microscopes and a Nikon TE300 (Tokyo, Japan) inverted microscope coupled to a Leica DFC310FX camera (Leica Biosystems, Wetzlar, Germany).

2.7. Virus Infection Inhibition Through Cell Viability Assay

Vero, hBMEC, and NSC were exposed to ZIKV MR 766 at an MOI of 2. After 1 h, inoculum was removed and chloroquine-containing medium (6.25 to 50 μM) was added to the cells. Five days post-infection, 20 μL of CellTiter Blue reagent (Promega, Madison, WI, USA) was added in each well, incubated for 12–16 h, and fluorescence was measured (560/590 nm), except for NSCp14 cells when CellTiter Blue was added at three days post-infection. Mean fluorescence intensity (MFI) and standard deviation were displayed. In order to calculate the half maximum effective concentration (EC50) that protects cells from death caused by ZIKV infection, MFI values from ZIKV MR766 control were subtracted from every condition and then these values were normalized over the mock control. These data were plotted and Hill 4 parameter sigmoidal regression was performed on Sigma Plot v.12.0 (Systat Software Inc., San Jose, CA, USA).

2.8. Drug Cytotoxicity Assay

Vero cells, hBMEC, and NSC were incubated with medium containing chloroquine (1.56 to 200 μM) for 3 days (NSC) or 5 days (Vero and hBMEC). CellTiter Blue reagent (Promega) was added in each well, incubated for 12–16 h, and fluorescence was measured (560/590 nm). In order to calculate the 50% cytotoxicity concentration (CC50), MFI values were normalized over the untreated control. These data were plotted and Hill 4 parameter sigmoidal regression was performed on Sigma Plot v.12.0 (Systat Software Inc.).

2.9. Time-of-Addition Assay

Vero cells were inoculated with ZIKV MR 766 at an MOI of 10 for 1 h at 4 °C. Cells were washed three times with cold PBS to remove unbound virus and treated with 50 μM chloroquine that was added at different time points: 0, 0.5, 3, 12, and 24 h post-infection. Conditioned media were collected at 30 h post-infection to analyze the production of virus particles through viral RNA content or the amount of infectious virus particles. Viral RNA was extracted and quantitative reverse transcription polymerase chain reaction (RT-qPCR) was performed. The titer of infectious virus particles was determined by plaque assay.

2.10. ZIKV Inhibition Assessed by Virus Production

Vero cells were infected with ZIKV MR766 or ZIKV BR strain at an MOI of 2 for 1 h at 4 °C. Virus input was washed three times with cold PBS and cells were treated with chloroquine (6.25 to 50 μM) for 48 h, and then the supernatant was collected and the RNA was extracted and analyzed by relative quantification by RT-qPCR. The supernatant was also evaluated by plaque assay to quantitate the infectious virus particles.

2.11. Viral RNA Extraction and RT-qPCR

Viral RNA was extracted from 200 μL supernatant of infected cells using QIAmp MiniElute Virus Spin (QIAgen, Hilden, Düsseldorf ,DE), following the manufacturer's recommendations. ZIKV detection was performed using One Step Taqman RT-qPCR (Thermo Fisher Scientific) on a 7500 Real-Time PCR System (Applied Biosystems) with primers and the probe described elsewhere [2]. Threshold cycle (CT) was determined and ΔCT (CT chloroquine treated − CT untreated) was calculated. The fold reduction of virus particles' release, including defective viral particles, were calculated by $2^{\Delta Ct}$.

2.12. Statistical Analysis

Mean and standard deviation (SD) were calculated for each assay. One way analysis of variance (ANOVA) was conducted using the non-parametric test (Kruskal–Wallis) followed by Dunn's multiple comparisons test. A *p*-value of <0.05 was considered significant. All analyses were performed on GraphPad Prism v.7 (GraphPad Software, San Diego, CA, USA). The sample size is provided in the respective figure legends.

3. Results

3.1. Chloroquine Inhibits ZIKV Infection in Vero Cells

We characterized the antiviral properties of chloroquine in Vero cells, a model widely used to study viral infections. Vero cells were infected with ZIKV MR766 at an MOI of 2 (i.e., 2 PFU/cell) and were then treated for 5 days with chloroquine in concentrations ranging from 6.25 to 50 μM. Viral infectivity was assessed using the 4G2 antibody, which detects flavivirus envelope E protein. We observed that chloroquine treatment decreased the number of ZIKV-infected cells in a dose-dependent manner. Flow cytometry analysis showed a reduction of 35% and 65% in ZIKV-infected cells when cultures were treated with 25 μM and 50 μM chloroquine, respectively, compared to untreated infected cells (Figure 1A). Immunofluorescence staining corroborated these results (Figure 1B) and additionally, chloroquine decreased the production of infectious (Figure 1C) and total (Figure 1D) virus particles, including defective viral particles, by ZIKV-infected cells. To confirm that viral inhibition is independent of chloroquine cytotoxicity, the viability of uninfected cells treated with chloroquine (1.56 to 200 μM) for 5 days was analyzed. Chloroquine did not impact cell viability at concentrations of 50 μM or lower (Figure 1E). We further analyzed whether chloroquine treatment could protect Vero cells from ZIKV infection as assessed by cell viability. Chloroquine, ranging from 12.5 to 50 μM, increased cell viability from 55% up to 100% (Figure 1F).

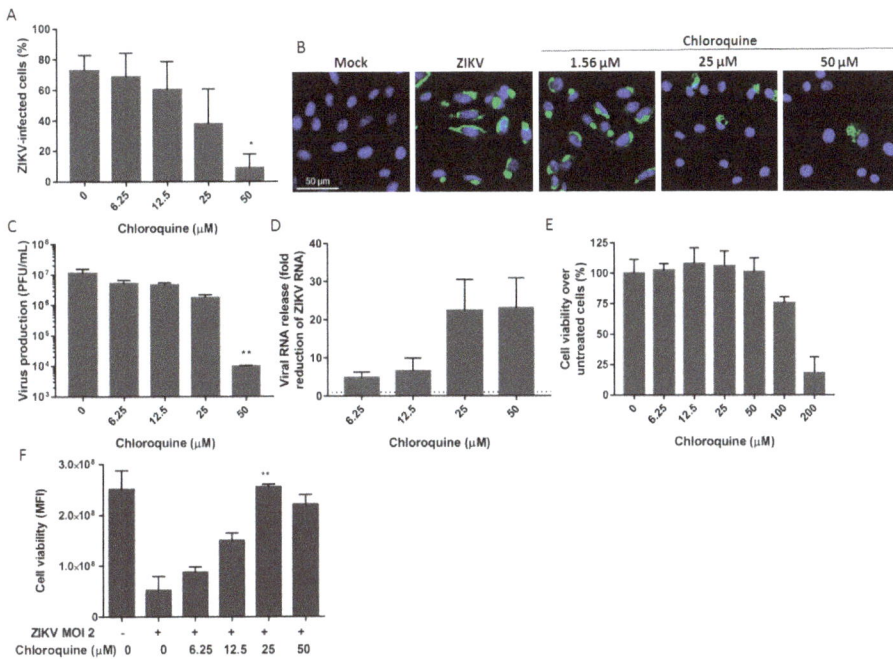

Figure 1. Inhibition of Zika virus (ZIKV) infection by chloroquine in Vero cells. (**A**) Vero cells were infected with ZIKV MR766 at a multiplicity of infection (MOI) of 2, treated with chloroquine for 5 days, and were then stained for the viral envelope protein and analyzed by flow cytometry; (**B**) Vero cells were infected, treated for 2 days, and ZIKV infection was evaluated by immunofluorescence staining with 4G2 antibody (green) and DAPI (blue); Infectious (**C**) or total (**D**) virus particles were quantified on the supernatant 48 h post-infection. Titers are expressed as plaque forming units (PFU) per milliliter. The dashed line represents no reduction in RNA levels; (**E**) Cell viability of uninfected cells treated with increasing concentrations of chloroquine was evaluated by fluorescence measurement at 560/590 nm after viability dye incubation; (**F**) Protection against ZIKV infection was evaluated through cell viability in ZIKV-infected Vero cells treated with chloroquine for 5 days. Data are represented as mean fluorescent intensity (MFI) ± standard deviation (SD) from two to four independent experiments. Statistical analysis was performed with the Kruskal–Wallis test and multiple comparisons with infected and untreated control corrected by Dunn's test (* $p < 0.05$; ** $p < 0,005$).

3.2. Chloroquine Inhibits Infection of Asian ZIKV Strains

Microcephaly cases and neurological disorders have only been associated with infection with strains of ZIKV from the Asian lineage, detected in French Polynesia and in the Americas [6,9,29]. To determine the inhibition spectrum of chloroquine against ZIKV infection, Vero cells were infected with the Brazilian isolate of the Asian lineage (ZIKV BR). Chloroquine decreases the percentage of cells infected with the Brazilian isolate from 70% to 30% and 5% at 12.5 and 25 μM, respectively (Figure 2A,C). Moreover, viral RNA levels in the supernatant of Vero cells were quantified as a direct measurement of ZIKV infection. Treatment with 25 μM chloroquine led to a 16-fold reduction in the level of viral RNA detected in the supernatant (Figure 2B).

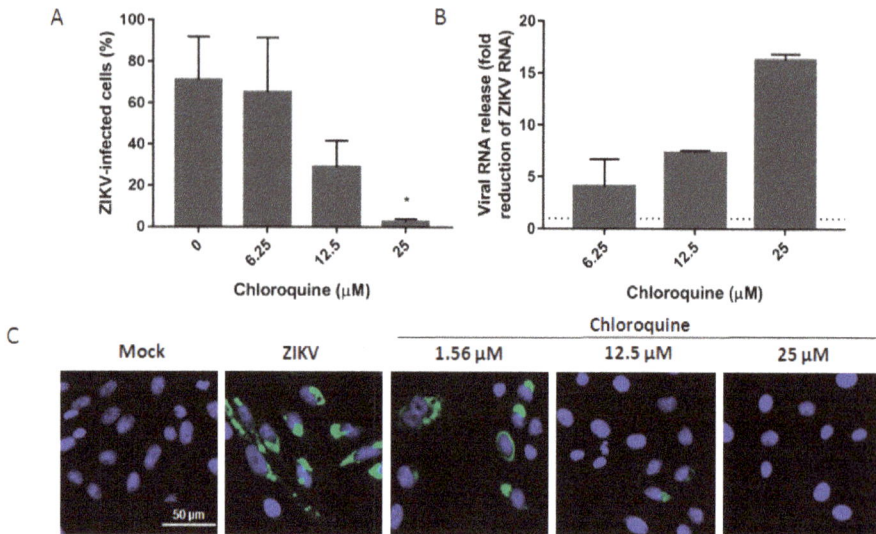

Figure 2. Chloroquine inhibits infection by Asian lineage. (**A**) Vero cells were infected with Brazilian ZIKV strain at an MOI of 2, treated with chloroquine at the indicated concentrations for 5 days, and the frequency of infected cells was evaluated by flow cytometry; (**B**) Vero cells were infected with Brazilian strain at an MOI of 2 and exposed to chloroquine for 48 h. The supernatant was collected and viral RNA was relatively quantified over the untreated infected control (**B**) or infectivity was analyzed by immunofluorescence with 4G2 antibody (**C**). The dashed line represents fold reduction on virus production of 1. Data are represented as mean ± SD from two independent experiments. Statistical significance was assessed by Kruskal–Wallis test and multiple comparisons with infected and untreated control corrected by Dunn's test (* $p < 0.05$).

3.3. Chloroquine Inhibits Early Stages of ZIKV Infection

Inhibition of viral infection mediated by chloroquine can occur in both the early and later stages of the viral replication cycle [30,31]. To evaluate which step of the viral cycle was susceptible to inhibition, chloroquine was added to Vero cells at different time points post-infection with ZIKV MR766. The supernatant was collected 30 h post-infection and virus production was evaluated by relative quantification of viral RNA over the untreated control by RT-qPCR. Virus titers were also determined by plaque assay in Vero cells. Incubation of Vero cells with chloroquine at 0 h post-infection had a greater impact on the production of ZIKV particles, decreasing viral RNA 64-fold over the controls (Figure 3A). The addition of chloroquine from 30 min to 12 h post-infection was able to reduce virus release 9–20 fold over untreated, infected-cells. However, chloroquine added at 24 h post-infection had only a minor effect on viral production (Figure 3A). These results were confirmed by quantification of ZIKV infectious particles released after chloroquine treatment (Figure 3B). These data confirm that chloroquine interferes with the early stages of the ZIKV replication cycle.

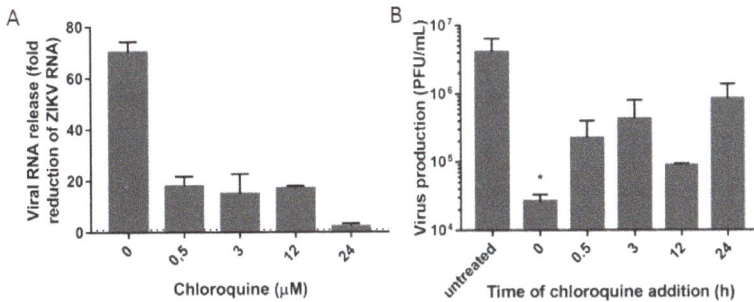

Figure 3. Early stages of infection are inhibited by chloroquine. Vero cells were infected with ZIKV MR766 at an MOI of 10 and 50 μM chloroquine was added at different times post-infection. At 30 h post-infection, the supernatant was collected and viral RNA (**A**) or infectious particles (**B**) were quantified. Viral RNA reduction is represented in fold change ($2^{\Delta Ct}$). The dashed line represents no reduction in RNA levels. Infectious particles are depicted as PFU/mL. Data are represented as mean ± SD from two independent experiments. Statistical analysis was performed with Kruskal–Wallis test and multiple comparisons with infected and untreated control corrected by Dunn's test (* $p < 0.05$).

3.4. Chloroquine Reduces ZIKV Infection in hBMEC, an In Vitro Model of the Blood–Brain Barrier

Considering that ZIKV infects hBMECs [32], we investigated whether chloroquine could inhibit viral infection of these cells. Chloroquine reduced 45% and 50% of the number of ZIKV MR766-infected hBMECs at 25 and 50 μM, respectively (Figure 4A,D). These concentrations are non-cytotoxic (Figure 4B), and protected approximately 80% of hBMECs from ZIKV infection as demonstrated by the increase in cell viability (Figure 4C).

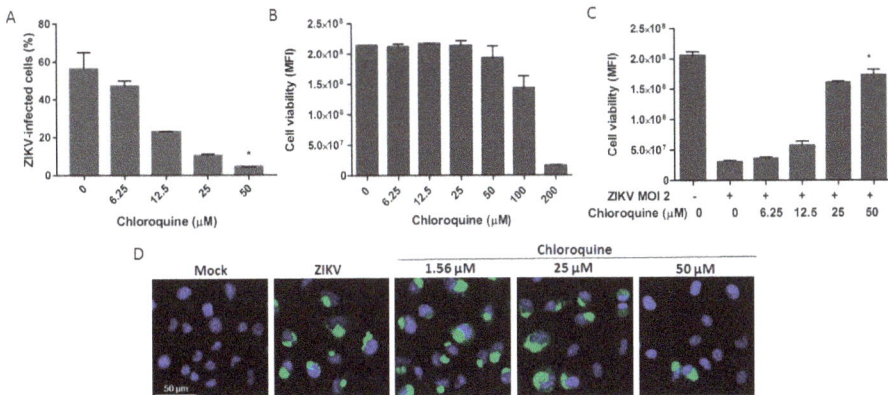

Figure 4. Chloroquine reduces the number of ZIKV-infected human brain microvascular endothelial cells (hBMECs). (**A**) hBMECs were infected with ZIKV MR 766 at an MOI of 2 followed by chloroquine treatment for 5 days. Cells were stained with the 4G2 antibody and analyzed by flow cytometry; (**B**) Uninfected hBMECs were incubated with chloroquine for 5 days and cell viability was analyzed; (**C**) Protection against ZIKV infection was measured through cell viability in chloroquine-treated ZIKV-infected cells; (**D**) Immunofluorescence with 4G2 antibody (green) and DAPI (blue) of ZIKV-infected cells treated with chloroquine for 5 days. Data are represented as mean ± SD from two independent experiments. Statistical significance was assessed by Kruskal–Wallis test and multiple comparisons with infected and untreated control corrected by Dunn's test (* $p < 0.05$).

3.5. Chloroquine Inhibits ZIKV Infection in Human Neural Progenitor Cells

Neural stem cells are key cells in the process of corticogenesis, giving rise to the three main cell types of the central nervous system: neurons, astrocytes, and oligodendrocytes. Depletion of the NSC pool is one of the main mechanisms responsible for primary microcephaly [33]. To evaluate if chloroquine could protect these cells from ZIKV MR766 infection, NSCs were exposed to up to 50 µM chloroquine for 4 days. Chloroquine treatment decreased the number of NSCs infected with ZIKV MR766 by 57%, and, through cell viability assessment protected 70% of NSCs from ZIKV infection, without cytotoxicity effects (Figure 5A–D). Similar results were observed when NSCs were infected with the Brazilian strain (Figure 5E).

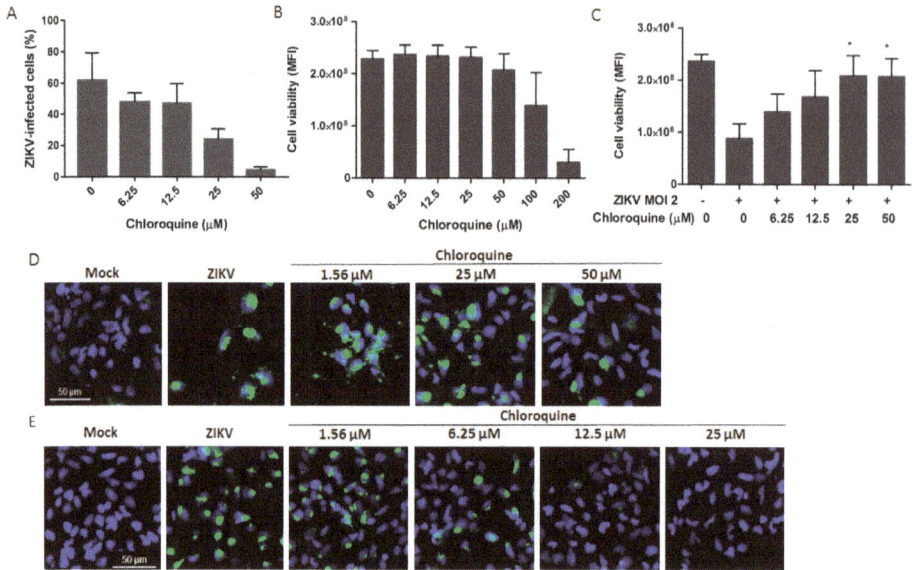

Figure 5. Chloroquine inhibits ZIKV infection in human neural stem cells (hNSCs). hNSCs were infected with ZIKV MR766 at an MOI of 2 and incubated with increasing concentrations of chloroquine for 4 days. (**A**) The frequency of ZIKV-infected cells was analyzed by 4G2 staining and flow cytometry; (**B**) Chloroquine cytotoxicity was assessed by the viability of uninfected hNSCs treated with chloroquine; (**C**) Chloroquine treatment protection from ZIKV infection was evaluated by cell viability measurement at 4 days post-infection. Immunofluorescence NSCs infected with ZIKV MR766 (**D**) or ZIKV BR (**E**) at an MOI of 2 and treated with chloroquine for 4 days with 4G2 antibody (green) and DAPI (blue). Data are represented as mean ± SD from two to three independent experiments. Statistical analysis was performed with Kruskal–Wallis test and multiple comparisons with infected and untreated control corrected by Dunn's test (* $p < 0.05$).

3.6. Chloroquine Inhibits ZIKV Infection in Mouse Neurospheres

Neuroprogenitor-enriched neurospheres, when subjected to differentiation culture conditions, can generate neurons. Our group recently demonstrated that ZIKV infection affected neurosphere size, neurite extension, and neuronal differentiation [34]. As we previously observed, neurospheres infected with ZIKV strain MR766 showed convoluted and misshapen neurites. Neurite extension was evaluated in chloroquine treated cultures by microtubule-associated protein 2 (Map2) staining and phase contrast microscopy and although many neurospheres were severely impacted by the infection, many others displayed the same general characteristics of mock-infected spheres indicating that chloroquine treatment rescued the neurite extension phenotype (Figure 6A–C). Furthermore,

ZIKV infection decreased when neurospheres were treated with 12.5 μM chloroquine, as evaluated by 4G2 staining (Figure 6D–F).

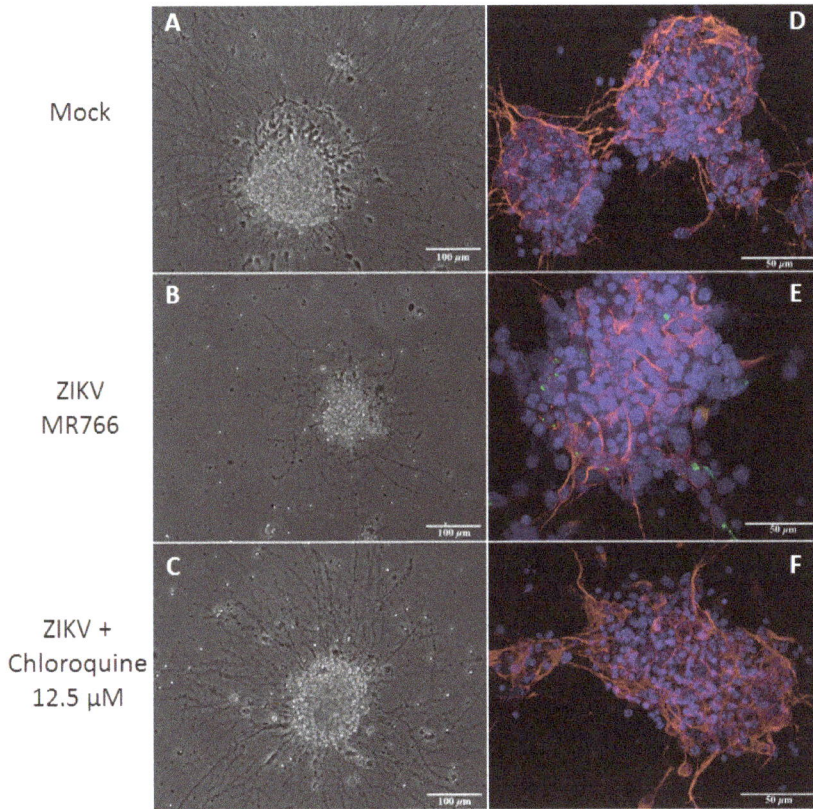

Figure 6. Chloroquine inhibits ZIKV infection in mouse neurospheres. Mouse neurospheres were infected with ZIKV MR766 (2.5×10^5 PFU and were treated with chloroquine for 3 days. Neurospheres were analyzed by phase contrast microscopy (**A–C**), and triple stained for envelope viral protein (green), microtubule-associated protein 2 (Map-2, red), a neuron-specific protein, and DAPI (blue) (**D–F**).

4. Discussion

Chloroquine is known to be a non-specific antiviral agent, but its effect on the Zika virus replication has not been evaluated yet. This is the first report of inhibitory effects of chloroquine on ZIKV replication, which, given the ongoing epidemics, may become interesting both for the scientific knowledge of the virus and for the clinical perspective.

Although Zika virus was first identified in Uganda in 1947, from January 2007 to April 2016, ZIKV transmission has been reported in 64 countries and territories [35]. The Zika virus disease is in general mild, but the recent positive correlation between infection, congenital malformations, and neurological damage in adults has intensified the need for therapeutic approaches. Prophylactic treatments for women intending to get pregnant in epidemic areas and travelers going to affected countries would represent relevant tools to reduce ZIKV transmission and avoid the spread of the disease by travelers. Moreover, a drug that blocks placental transfer of the virus could decrease the chance of vertical transmission in viremic pregnant women as was shown for HIV-infected pregnant women treated with antiretroviral therapy [36].

Here we demonstrated that chloroquine decreases the number of ZIKV-infected cells and protected cells from ZIKV infection as measured by cell viability at non-cytotoxic concentrations (Figures 1, 4 and 5). The EC50 or concentration of chloroquine that protected 50% of the cells from ZIKV infection assessed by cell viability, was 9.82–14.2 μM depending on the cell model and the CC50 ranged from 94.95 to 134.54 μM (Table 1). The values of EC50 obtained for ZIKV MR766 are lower than those obtained for DENV inhibition (~25 μM) and HIV inhibition (100 μM) [20,22]. Furthermore, we observed similar ZIKV inhibitory effects of chloroquine when tested on different ZIKV lineage infections (Figures 2 and 5), supporting the idea that chloroquine could help to manage recent infections caused by Asian ZIKV lineage.

Table 1. Pharmacological parameters of chloroquine in each cell type against ZIKV MR766.

Cell Type	CC50	EC50	TI
Vero	134.54 ± 16.76 μM	9.82 ± 2.79 μM	13.70
hBMEC	116.61 ± 9.70 μM	14.20 ± 0.18 μM	8.21
hNSC	94.95 ± 9.38 μM	12.36 ± 2.76 μM	7.68

CC50: 50% cytotoxicity concentration; EC50: half maximal effective concentration; TI: therapeutic index (CC50/EC50). Data are represented as mean ± SD.

Although chloroquine has shown antiviral activity against a large spectrum of viruses in vitro, few clinical studies have been performed to evaluate chloroquine effects on patients with viral infections. Two clinical trial studies of chloroquine have been conducted to assess chloroquine treatment in patients infected with DENV [37,38]. One of the trials evaluated the benefits of chloroquine treatment for 3 days in patients infected with DENV and showed no reduction in the duration or intensity of DENV viremia or nonstructural 1 protein (NS1) antigenemia clearance [37]. However, a trend towards a reduction in the number of dengue hemorrhagic fever cases was noticed in the chloroquine-treated group [37]. A more recent clinical trial of chloroquine administration to DENV-infected patients, also for 3 days, showed that 60% of the patients in the chloroquine-treated group reported feeling less pain and showed improvement in the performance of daily chores during treatment [38]. Moreover, the symptoms returned after medication withdrawal. However, chloroquine treatment did not reduce the duration and intensity of the fever or duration of the disease [38]. The antiviral effect of chloroquine may be insufficient to produce a decrease in viral load or improvement of the disease progression when chloroquine/hydroxychloroquine is used in monotherapy. However, chloroquine may produce a significant antiflaviviral effect when used in combination therapies, as recently shown in a clinical trial of hydroxychloroquine plus ribavirin and interferon alpha in individuals infected with hepatitis C virus (HCV) [39]. In regard to the potential antiviral therapeutic combinations for Zika, a freshly published screening of drugs already approved for other clinical indications has resulted in the identification of more than 20 candidate drugs [40]. Of note, one of these is mefloquine, a compound related to chloroquine. In terms of safety for pregnant women, however, mefloquine is included in the B category, i.e., a drug for which the animal reproduction studies have failed to demonstrate a risk to the fetus and there are no adequate and well-controlled studies in pregnant women. Be that as it may, the aforementioned study corroborates our results using chloroquine, and provides new anti-ZIKV drugs that could be tested in combination with chloroquine.

Differing from mefloquine administration, the use of chloroquine during pregnancy was thoroughly evaluated and when prophylactic doses of chloroquine were administered for malaria (400 mg/week), no increment in birth defects was observed [41]. Higher concentrations of chloroquine (250 mg to 500 mg/day) were administered to pregnant women with severe diseases, such as lupus or rheumatoid arthritis. In these studies, few cases of abortion and fetal toxicity were observed. However, fetal toxicity or death could not be discarded as direct consequences of the disease itself. In addition, all term deliveries resulted in healthy newborns [19,42].

Chloroquine has been successfully tested in animal models. Twice daily administration of chloroquine (90 mg/kg) has been shown to increase the Balb/c mice survival rate to 80%–90% after infection with Ebola virus (EBOV) [43]. In a C57BL/6 mice model for coronavirus infection, chloroquine (50 mg/kg) protected 100% of 5-day-old suckling mice infected with human coronavirus OC43 (HCoV-OC43) when administered to pregnant mice 1 day prepartum [44]. A survival rate of 70% was observed in Balb/c mice infected with avian influenza H5N1 virus treated with chloroquine at 50 mg/kg/day [45]. These results suggest that chloroquine has the potential to inhibit ZIKV in in vivo mouse studies.

Chloroquine is widely distributed to body tissues as well as its analogue hydroxychloroquine. The concentration of hydroxychloroquine in the brain is 4–30 times higher than in the plasma [46]. The concentration of chloroquine in the plasma reached 10 μM when a daily intake of 500 mg was prescribed to arthritis patients [47]. Chloroquine is able to cross the placental barrier and is supposed to reach similar concentrations in both maternal and fetal plasma [48]. Concentrations of chloroquine, similar to the EC50 values calculated here (Table 1), are achieved in the plasma in current chloroquine administration protocols and might reach the brain.

Different mechanisms for the chloroquine inhibition of viral infection have been described [49–51]. We observed a strong reduction in the release of ZIKV particles when the drug was added at 0 h post-infection (Figure 3), suggesting a higher impact on early stages of infection, possibly during fusion of the envelope protein to the endosome membrane. Chloroquine inhibits acidification of the endosome, consequently inhibiting the low pH-induced conformational changes required for the fusion of the envelope protein of flaviviruses with the endosomal membrane [52]. Chloroquine was also effective in decreasing virus release, although less pronouncedly and not statistically significant, when added after the early stages of virus infection (from 0.5 to 24 h post-infection), suggesting that later stages of the ZIKV replication cycle might also be affected (Figure 3).

ZIKV was detected in the cerebrospinal fluid of ZIKV-infected adult patients that manifested meningoencephalitis, indicating that ZIKV invades the central nervous system through yet unknown mechanisms. Transcytosis through the endothelial cells of the blood brain barrier is a known mechanism of viral access to the central nervous system [53,54]. Here we demonstrated, by different methodologies, that chloroquine protects hBMEC, an immortalized cell line widely used as in vitro model of the blood–brain barrier [25], from ZIKV infection (Figure 4).

Recent studies showed that neural stem cells are highly permissive for ZIKV infection and one of the mechanisms proposed for the cause of microcephaly would be the depletion of the stem cell pool induced by ZIKV [10,11,55]. Our data showed that chloroquine inhibits the infection of human neural stem cells (Figure 5). Using the mouse neurospheres model to study neural stem cell differentiation into neurons, another process that might be disturbed in microcephaly, we observed that chloroquine inhibited the infection of neuronal progenitors and partially protected the ability of these cells to extend neurites (Figure 6). The protective effect of chloroquine on stem cells and committed progenitors is potentially a groundbreaking feature of this compound, as it would be prescribed to women at childbearing age that are traveling to affected countries and women planning pregnancy in endemic areas. This would decrease the chances of infection and thus fetal damage, especially to the developing brain.

Our results suggest that the chloroquine concentrations inhibiting ZIKV replication in vitro may overlap the highest drug concentrations detected in humans [56]. We therefore suggest that the therapeutic potential of chloroquine for Zika be subjected to further study.

Acknowledgments: This work was supported by Conselho Nacional de Desenvolvimento e Pesquisa (CNPq), Fundação de Amparo à Pesquisa do Estado do Rio de Janeiro (FAPERJ), and Departamento de Doenças Sexualmente Transmissíveis, AIDS e Hepatites Virais do Ministério da Saúde do Brasil. The authors acknowledge Manoel Itamar for providing technical support.

Author Contributions: R.D., L.M.H., P.P., A.L.V. designed and performed experiments, prepared figures and/or tables, analyzed the data and wrote the manuscript; P.P.G. designed and performed experiments, analyzed the data and wrote the manuscript; F.L.M. designed and performed experiments, prepared figures and/or tables and analyzed the data; E.C.L., A.A.D., F.J.M.S., M.T.A., E.A.C. performed experiments; M.B., D.H.O., S.R., R.S.d.A., J.E.O., A.S. designed and/or performed experiments, contributed with reagents/materials/analysis tools and wrote the manuscript; L.C. and A.T. designed and performed experiments, prepared figures and/or tables, analyzed the data, contributed with reagents/materials/analysis tools, and wrote the manuscript.

Conflicts of Interest: The authors declare no conflict of interest.

References

1. Dick, G.W.A.; Kitchen, S.F.; Haddow, A.J. Zika virus. I. Isolations and serological specificity. *Trans. R. Soc. Trop. Med. Hyg.* **1952**, *46*, 509–520. [CrossRef]
2. Lanciotti, R.S.; Kosoy, O.L.; Laven, J.J.; Velez, J.O.; Lambert, A.J.; Johnson, A.J.; Stanfield, S.M.; Duffy, M.R. Genetic and serologic properties of Zika virus associated with an epidemic, Yap State, Micronesia, 2007. *Emerg. Infect. Dis.* **2008**, *14*, 1232–1239. [CrossRef] [PubMed]
3. Faye, O.; Freire, C.C.M.; Iamarino, A.; Faye, O.; de Oliveira, J.V.C.; Diallo, M.; Zanotto, P.M.A.; Sall, A.A. Molecular evolution of Zika virus during its emergence in the 20th century. *PLoS Negl. Trop. Dis.* **2014**, *8*, e2636. [CrossRef] [PubMed]
4. Brasil, P.; Pereira, J.P., Jr.; Raja Gabaglia, C.; Damasceno, L.; Wakimoto, M.; Ribeiro Nogueira, R.M.; Carvalho de Sequeira, P.; Machado Siqueira, A.; Abreu de Carvalho, L.M.; Cotrim da Cunha, D.; et al. Zika Virus Infection in Pregnant Women in Rio de Janeiro—Preliminary Report. *N. Engl. J. Med.* **2016**. [CrossRef] [PubMed]
5. Martines, R.B.; Bhatnagar, J.; Keating, M.K.; Silva-Flannery, L.; Muehlenbachs, A.; Gary, J.; Goldsmith, C.; Hale, G.; Ritter, J.; Rollin, D.; et al. Notes from the Field: Evidence of Zika Virus Infection in Brain and Placental Tissues from Two Congenitally Infected Newborns and Two Fetal Losses —Brazil, 2015. *MMWR Morb. Mortal. Wkly. Rep.* **2016**, *65*, 1–2. [CrossRef]
6. Mlakar, J.; Korva, M.; Tul, N.; Popović, M.; Poljšak-Prijatelj, M.; Mraz, J.; Kolenc, M.; Resman Rus, K.; Vesnaver Vipotnik, T.; Fabjan Vodušek, V.; et al. Zika Virus Associated with Microcephaly. *N. Engl. J. Med.* **2016**, *374*, 951–958. [CrossRef] [PubMed]
7. Oliveira Melo, A.S.; Malinger, G.; Ximenes, R.; Szejnfeld, P.O.; Alves Sampaio, S.; Bispo De Filippis, A.M. Zika virus intrauterine infection causes fetal brain abnormality and microcephaly: Tip of the iceberg? *Ultrasound Obstet. Gynecol.* **2016**, *47*, 6–7. [CrossRef] [PubMed]
8. Melo, A.S.; Aguiar, R.S.; Amorim, M.M.; Arruda, M.B.; Melo, F.O.; Ribeiro, S.T.; Batista, A.G.; Ferreira, T.; Dos Santos, M.P.; Sampaio, V.V.; et al. Congenital Zika Virus Infection: Beyond Neonatal Microcephaly. *JAMA Neurol.* **2016**, 1–10. [CrossRef] [PubMed]
9. Calvet, G.; Aguiar, R.S.; Melo, A.S.O.; Sampaio, S.A.; de Filippis, I.; Fabri, A.; Araujo, E.S.M.; de Sequeira, P.C.; de Mendonça, M.C.L.; de Oliveira, L.; et al. Detection and sequencing of Zika virus from amniotic fluid of fetuses with microcephaly in Brazil: A case study. *Lancet Infect. Dis.* **2016**, *16*, 653–660. [CrossRef]
10. Garcez, P.P.; Loiola, E.C.; Madeiro da Costa, R.; Higa, L.M.; Trindade, P.; Delvecchio, R.; Nascimento, J.M.; Brindeiro, R.; Tanuri, A.; Rehen, S.K. Zika virus impairs growth in human neurospheres and brain organoids. *Science* **2016**, *352*, 816–818. [CrossRef] [PubMed]
11. Tang, H.; Hammack, C.; Ogden, S.C.; Jin, P. Zika Virus Infects Human Cortical Neural Progenitors and Attenuates Their Growth. *Stem Cell* **2016**, 1–4. [CrossRef] [PubMed]
12. Duffy, M.R.; Chen, T.-H.; Hancock, W.T.; Powers, A.M.; Kool, J.L.; Lanciotti, R.S.; Pretrick, M.; Marfel, M.; Holzbauer, S.; Dubray, C.; et al. Zika virus outbreak on Yap Island, Federated States of Micronesia. *N. Engl. J. Med.* **2009**, *360*, 2536–2543. [CrossRef] [PubMed]
13. Bearcroft, W.G. Zika virus infection experimentally induced in a human volunteer. *Trans. R. Soc. Trop. Med. Hyg.* **1956**, *50*, 442–448. [CrossRef]
14. Cao-Lormeau, V.-M.; Blake, A.; Mons, S.; Lastère, S.; Roche, C.; Vanhomwegen, J.; Dub, T.; Baudouin, L.; Teissier, A.; Larre, P.; et al. Guillain–Barré Syndrome outbreak associated with Zika virus infection in French Polynesia: A case-control study. *Lancet (Lond. Engl.)* **2016**, *387*, 1531–1539. [CrossRef]
15. Carteaux, G.; Maquart, M.; Bedet, A.; Contou, D.; Brugières, P.; Fourati, S.; Cleret de Langavant, L.; de Broucker, T.; Brun-Buisson, C.; Leparc-Goffart, I.; et al. Zika Virus Associated with Meningoencephalitis. *N. Engl. J. Med.* **2016**, *374*, 1595–1596. [CrossRef] [PubMed]

16. D'Ortenzio, E.; Matheron, S.; de Lamballerie, X.; Hubert, B.; Piorkowski, G.; Maquart, M.; Descamps, D.; Damond, F.; Yazdanpanah, Y.; Leparc-Goffart, I. Evidence of Sexual Transmission of Zika Virus. *N. Engl. J. Med.* **2016**, *374*, 2195–2198. [CrossRef] [PubMed]

17. Deckard, D.T.; Chung, W.M.; Brooks, J.T.; Smith, J.C.; Woldai, S.; Hennessey, M.; Kwit, N.; Mead, P. Male-to-Male Sexual Transmission of Zika Virus - Texas, January 2016. *MMWR Morb. Mortal. Wkly. Rep.* **2016**, *65*, 372–374. [CrossRef] [PubMed]

18. Browning, D. Pharmacology of Chloroquine and Hydroxychloroquine. In *Hydroxychloroquine and Chloroquine Retinopathy*; Springer: New York, NY, USA, 2014; pp. 35–63.

19. Levy, M.; Buskila, D.; Gladman, D.; Urowitz, M.; Koren, G. Pregnancy Outcome Following First Trimester Exposure to Chloroquine. *Am. J. Perinatol.* **1991**, *8*, 174–178. [CrossRef] [PubMed]

20. Tsai, W.P.; Nara, P.L.; Kung, H.F.; Oroszlan, S. Inhibition of human immunodeficiency virus infectivity by chloroquine. *AIDS Res. Hum. Retrovir.* **1990**, *6*, 481–489. [CrossRef] [PubMed]

21. Ooi, E.E.; Chew, J.S.W.; Loh, J.P.; Chua, R.C.S. In vitro inhibition of human influenza A virus replication by chloroquine. *Virol. J.* **2006**, *3*, 39. [CrossRef] [PubMed]

22. Farias, K.J.S.; Machado, P.R.L.; da Fonseca, B.A.L. Chloroquine inhibits dengue virus type 2 replication in Vero cells but not in C6/36 cells. *Sci. World J.* **2013**, *2013*, 282734. [CrossRef] [PubMed]

23. Zhu, Y.; Xu, Q.; Wu, D.; Ren, H.; Zhao, P.; Lao, W.; Wang, Y.; Tao, Q.; Qian, X.; Wei, Y.-H.; et al. Japanese encephalitis virus enters rat neuroblastoma cells via a pH-dependent, dynamin and caveola-mediated endocytosis pathway. *J. Virol.* **2012**, *86*, 13407–13422. [CrossRef] [PubMed]

24. Boonyasuppayakorn, S.; Reichert, E.D.; Manzano, M.; Nagarajan, K.; Padmanabhan, R. Amodiaquine, an antimalarial drug, inhibits dengue virus type 2 replication and infectivity. *Antivir. Res.* **2014**, *106*, 125–134. [CrossRef] [PubMed]

25. Stins, M.F.; Badger, J.; Sik Kim, K. Bacterial invasion and transcytosis in transfected human brain microvascular endothelial cells. *Microb. Pathog.* **2001**, *30*, 19–28. [CrossRef] [PubMed]

26. Paulsen, B.d.S.; Maciel, R.d.M.; Galina, A.; da Silveira, M.S.; Souza, C.d.S.; Drummond, H.; Pozzatto, E.N.; Junior, H.S.; Chicaybam, L.; Massuda, R.; et al. Altered Oxygen Metabolism Associated to Neurogenesis of Induced Pluripotent Stem Cells Derived From a Schizophrenic Patient. *Cell Transplant.* **2012**, *21*, 1547–1559. [CrossRef] [PubMed]

27. Yan, Y.; Shin, S.; Jha, B.S.; Liu, Q.; Sheng, J.; Li, F.; Zhan, M.; Davis, J.; Bharti, K.; Zeng, X.; et al. Efficient and rapid derivation of primitive neural stem cells and generation of brain subtype neurons from human pluripotent stem cells. *Stem Cells Transl. Med.* **2013**, *2*, 862–870. [CrossRef] [PubMed]

28. Donald, C.L.; Brennan, B.; Cumberworth, S.L.; Rezelj, V.V.; Clark, J.J.; Cordeiro, M.T.; Freitas de Oliveira França, R.; Pena, L.J.; Wilkie, G.S.; Da Silva Filipe, A.; et al. Full Genome Sequence and sfRNA Interferon Antagonist Activity of Zika Virus from Recife, Brazil. *PLoS Negl. Trop. Dis.* **2016**, *10*, e0005048. [CrossRef] [PubMed]

29. Cao-Lormeau, V.-M.; Roche, C.; Teissier, A.; Robin, E.; Berry, A.-L.; Mallet, H.-P.; Sall, A.A.; Musso, D. Zika virus, French polynesia, South pacific, 2013. *Emerg. Infect. Dis.* **2014**, *20*, 1085–1086. [CrossRef] [PubMed]

30. Diaz-Griffero, F.; Hoschander, S.A.; Brojatsch, J. Endocytosis is a critical step in entry of subgroup B avian leukosis viruses. *J. Virol.* **2002**, *76*, 12866–12876. [CrossRef] [PubMed]

31. Harley, C.A.; Dasgupta, A.; Wilson, D.W. Characterization of herpes simplex virus-containing organelles by subcellular fractionation: role for organelle acidification in assembly of infectious particles. *J. Virol.* **2001**, *75*, 1236–1251. [CrossRef] [PubMed]

32. Bayer, A.; Lennemann, N.J.; Ouyang, Y.; Bramley, J.C.; Morosky, S.; Marques, E.T.D.A.; Cherry, S.; Sadovsky, Y.; Coyne, C.B. Type III Interferons Produced by Human Placental Trophoblasts Confer Protection against Zika Virus Infection. *Cell Host Microbe* **2016**, *19*, 705–712. [CrossRef] [PubMed]

33. Gilmore, E.C.; Walsh, C.A. Genetic causes of microcephaly and lessons for neuronal development. *Wiley Interdiscip. Rev. Dev. Biol.* **2013**, *2*, 461–478. [CrossRef] [PubMed]

34. Campanati, L.; Higa, L.M.; Delvecchio, R.; Pezzuto, P.; De Filippis, A.M.B.; Aguiar, R.S.; Tanuri, A. The Impact of African and Brazilian ZIKV isolates on neuroprogenitors. *BioRxiv* **2016**, 1–25.

35. WHO. *Zika Virus Microcephaly and Guillain-Barré Syndrome*; WHO: Geneva, Switzerland, 2016.

36. Connor, E.M.; Sperling, R.S.; Gelber, R.; Kiselev, P.; Scott, G.; O'Sullivan, M.J.; VanDyke, R.; Bey, M.; Shearer, W.; Jacobson, R.L.; et al. Reduction of maternal-infant transmission of human immunodeficiency virus type 1 with zidovudine treatment. Pediatric AIDS Clinical Trials Group Protocol 076 Study Group. *N. Engl. J. Med.* **1994**, *331*, 1173–1180. [CrossRef] [PubMed]

37. Tricou, V.; Minh, N.N.; Van, T.P.; Lee, S.J.; Farrar, J.; Wills, B.; Tran, H.T.; Simmons, C.P. A randomized controlled trial of chloroquine for the treatment of dengue in vietnamese adults. *PLoS Negl. Trop. Dis.* **2010**, *4*. [CrossRef] [PubMed]

38. Borges, M.C.; Castro, L.A.; da Fonseca, B.A.L. Chloroquine use improves dengue-related symptoms. *Mem. Inst. Oswaldo Cruz* **2013**, *108*, 596–599. [CrossRef] [PubMed]

39. Helal, G.K.; Gad, M.A.; Abd-Ellah, M.F.; Eid, M.S. Hydroxychloroquine augments early virological response to pegylated interferon plus ribavirin in genotype-4 chronic hepatitis C patients. *J. Med. Virol.* **2016**, *88*, 2170–2178. [CrossRef] [PubMed]

40. Barrows, N.J.; Campos, R.K.; Powell, S.T.; Prasanth, K.R.; Schott-Lerner, G.; Soto-Acosta, R.; Galarza-Muñoz, G.; McGrath, E.L.; Urrabaz-Garza, R.; Gao, J.; et al. A Screen of FDA-Approved Drugs for Inhibitors of Zika Virus Infection. *Cell Host Microbe* **2016**, *20*, 259–270. [CrossRef] [PubMed]

41. Wolfe, M.S.; Cordero, J.F. Safety of chloroquine in chemosuppression of malaria during pregnancy. *Br. Med. J. (Clin. Res. Ed.)* **1985**, *290*, 1466–1467. [CrossRef]

42. Parke, A.L. Antimalarial drugs, systemic lupus erythematosus and pregnancy. *J. Rheumatol.* **1988**, *15*, 607–610. [PubMed]

43. Madrid, P.B.; Chopra, S.; Manger, I.D.; Gilfillan, L.; Keepers, T.R.; Shurtleff, A.C.; Green, C.E.; Iyer, L.V.; Dilks, H.H.; Davey, R.A.; et al. A Systematic Screen of FDA-Approved Drugs for Inhibitors of Biological Threat Agents. *PLoS ONE* **2013**, *8*. [CrossRef] [PubMed]

44. Keyaerts, E.; Li, S.; Vijgen, L.; Rysman, E.; Verbeeck, J.; Van Ranst, M.; Maes, P. Antiviral activity of chloroquine against human coronavirus OC43 infection in newborn mice. *Antimicrob. Agents Chemother.* **2009**, *53*, 3416–3421. [CrossRef] [PubMed]

45. Yan, Y.; Zou, Z.; Sun, Y.; Li, X.; Xu, K.-F.; Wei, Y.; Jin, N.; Jiang, C. Anti-malaria drug chloroquine is highly effective in treating avian influenza A H5N1 virus infection in an animal model. *Cell Res.* **2013**, *23*, 300–302. [CrossRef] [PubMed]

46. Titus, E. Recent developments in the understanding of the pharmacokinetics and mechanism of action of chloroquine. *Ther. Drug Monit.* **1989**, *11*, 369–379. [CrossRef] [PubMed]

47. Mackenzie, A.H. Dose refinements in long-term therapy of rheumatoid arthritis with antimalarials. *Am. J. Med.* **1983**, *75*, 40–45. [CrossRef]

48. Law, I.; Ilett, K.F.; Hackett, L.P.; Page-Sharp, M.; Baiwog, F.; Gomorrai, S.; Mueller, I.; Karunajeewa, H.A.; Davis, T.M.E. Transfer of chloroquine and desethylchloroquine across the placenta and into milk in Melanesian mothers. *Br. J. Clin. Pharmacol.* **2008**, *65*, 674–679. [CrossRef] [PubMed]

49. Gonzalez-Dunia, D.; Cubitt, B.; de la Torre, J.C. Mechanism of Borna disease virus entry into cells. *J. Virol.* **1998**, *72*, 783–788. [PubMed]

50. Ferreira, D.F.; Santo, M.P.; Rebello, M.A.; Rebello, M.C. Weak bases affect late stages of Mayaro virus replication cycle in vertebrate cells. *J. Med. Microbiol.* **2000**, *49*, 313–318. [CrossRef] [PubMed]

51. Savarino, A.; Boelaert, J.R.; Cassone, A.; Majori, G.; Cauda, R. Effects of chloroquine on viral infections: An old drug against today's diseases? *Lancet* **2003**, *3*, 722–727. [CrossRef]

52. Smit, J.M.; Moesker, B.; Rodenhuis-Zybert, I.; Wilschut, J. Flavivirus cell entry and membrane fusion. *Viruses* **2011**, *3*, 160–171. [CrossRef] [PubMed]

53. Dohgu, S.; Ryerse, J.S.; Robinson, S.M.; Banks, W.A. Human immunodeficiency virus-1 uses the mannose-6-phosphate receptor to cross the blood-brain barrier. *PLoS ONE* **2012**, *7*, 1–12. [CrossRef] [PubMed]

54. Suen, W.W.; Prow, N.A.; Hall, R.A.; Bielefeldt-Ohmann, H. Mechanism of west nile virus neuroinvasion: A critical appraisal. *Viruses* **2014**, *6*, 2796–2825. [CrossRef] [PubMed]

55. Qian, X.; Nguyen, H.N.; Song, M.M.; Hadiono, C.; Ogden, S.C.; Hammack, C.; Yao, B.; Hamersky, G.R.; Jacob, F.; Zhong, C.; et al. Brain-Region-Specific Organoids Using Mini-bioreactors for Modeling ZIKV Exposure. *Cell* **2016**, *165*, 1238–1254. [CrossRef] [PubMed]

56. Savarino, A.; Shytaj, I.L. Chloroquine and beyond: exploring anti-rheumatic drugs to reduce immune hyperactivation in HIV/AIDS. *Retrovirology* **2015**, *12*, 51. [CrossRef] [PubMed]

![viruses logo] *viruses*

MDPI

Article

Envelope Protein Mutations L107F and E138K Are Important for Neurovirulence Attenuation for Japanese Encephalitis Virus SA14-14-2 Strain

Jian Yang [1,2,†], Huiqiang Yang [1,†], Zhushi Li [1], Wei Wang [1], Hua Lin [1], Lina Liu [1], Qianzhi Ni [1], Xinyu Liu [3], Xianwu Zeng [1], Yonglin Wu [4] and Yuhua Li [1,3,5,*]

[1] Department of Viral Vaccine, Chengdu Institute of Biological Products Co., Ltd., China National Biotech Group, Chengdu 610023, China; jiany74@163.com (J.Y.); yang-anan@163.com (H.Y.); changdc123@sina.com (Z.L.); suntina926@163.com (W.W.); scciqlh@126.com (H.L.); linaliu@163.com (L.L.); nqz1986@126.com (Q.N.); zengxw64@163.com (X.Z.)
[2] Department of Microbiology and Immunology, North Sichuan Medical College, Nanchong 637007, China
[3] Department of Arbovirus Vaccine, National Institutes for Food and Drug Control, Beijing 100050, China; xinyuliu@hotmail.com
[4] China National Biotech Group, Beijing 100029, China; wuyonglin@sinopharm.com
[5] State Key Laboratory of Biotherapy and Cancer Center, West China Hospital, Sichuan University and Collaborative Innovation Center for Biotherapy, Chengdu 610000, China
[*] Correspondence: liyuhua@nifdc.org.cn; Tel.: +86-10-6709-5986
[†] These authors contributed equally to this work.

Academic Editor: Michael Holbrook
Received: 31 October 2016; Accepted: 16 January 2017; Published: 21 January 2017

Abstract: The attenuated Japanese encephalitis virus (JEV) strain SA14-14-2 has been successfully utilized to prevent JEV infection; however, the attenuation determinants have not been fully elucidated. The envelope (E) protein of the attenuated JEV SA14-14-2 strain differs from that of the virulent parental SA14 strain at eight amino acid positions (E107, E138, E176, E177, E264, E279, E315, and E439). Here, we investigated the SA14-14-2-attenuation determinants by mutating E107, E138, E176, E177, and E279 in SA14-14-2 to their status in the parental virulent strain and tested the replication capacity, neurovirulence, neuroinvasiveness, and mortality associated with the mutated viruses in mice, as compared with those of JEV SA14-14-2 and SA14. Our findings indicated that revertant mutations at the E138 or E107 position significantly increased SA14-14-2 virulence, whereas other revertant mutations exhibited significant increases in neurovirulence only when combined with E138, E107, and other mutations. Revertant mutations at all eight positions in the E protein resulted in the highest degree of SA14-14-2 virulence, although this was still lower than that observed in SA14. These results demonstrated the critical role of the viral E protein in controlling JEV virulence and identified the amino acids at the E107 and E138 positions as the key determinants of SA14-14-2 neurovirulence.

Keywords: attenuation mechanism; Japanese encephalitis virus; SA14-14-2; neuroinvasiveness; neurovirulence

1. Introduction

The Japanese encephalitis virus (JEV) belongs to the *Flavivirus* genus and causes frequent endemic and epidemic infections in Asia, with JEV infection leading to acute encephalitis in humans and resulting in high mortality rates. The wild-type JEV SA14 strain was isolated from mosquitoes in Xi'An, China in 1954, and the attenuated JEV SA14-14-2 strain was obtained by serial passages of the JEV SA14 strain in mouse brain and primary hamster kidney (PHK) cells, followed by purification by plaque screening [1]. The purified SA14-14-2 strain was used to produce the attenuated live Japanese encephalitis (JE) vaccine for humans, with >600 million doses of this vaccine being administered in China and other countries in Southeast Asia, including Korea, Nepal, India, and Thailand, since 1989. The safety and efficacy of this vaccine have been well demonstrated by clinical data [2], and on 10 September 2013, it passed World Health Organization prequalification and was entered into the list of vaccines available for international purchase. As with all attenuated live viral vaccines, its reversion to virulent status remains a concern. This study explored the molecular mechanisms underpinning the attenuated neurovirulence of the live JE vaccine (SA14-14-2) by reverting specific amino acids in the SA14-14-2 envelope (E) protein to their counterparts in the parental virulent strain (SA14) and testing the virulence of the revertant viruses.

Our findings indicated no neurovirulence observed in adult mice inoculated intracerebrally (i.c.) with attenuated JEV SA14-14-2 at 10^6 plaque-forming units (PFUs), as compared with mice inoculated with the parental strain, which caused 100% mortality in mice within 1 week. The marked virulence attenuation of JEV SA14-14-2 is believed to result from specific substitutions at 24 amino acid positions, including eight amino acid mutations in the E protein, throughout the viral genome [3–5], as well as mutations in nonstructural proteins [6]. However, the specific mutations that determine the attenuated SA14-14-2 phenotype remain unknown.

The attenuated yellow fever virus (YFV) 17D strain differs from its parental Asibi strain by 32 amino acid substitutions. Among these, 12 mutations are located in the E protein. Remarkably, as few as one mutation (E303) in the E protein can change the attenuated phenotype of the Asibi strain [7]. The crucial role of amino acid mutations in the E protein, associated with attenuation, was reported in other attenuated viral vaccines, including the chimeric yellow fever-dengue 1 vaccine virus [8]. That study hypothesized that the attenuated phenotype of the JEV SA14-14-2 strain might also be attributed to specific mutations in the E protein. Here, we investigated the roles of five amino acid residues (E107, E138, E176/177, and E279) in the E protein in the attenuated strains, as compared with the virulent parental strain (Table 1), followed by an assessment of the neurovirulence and neuroinvasiveness of these revertants in mice. Our results demonstrated that amino acids at the E138 and E107 positions played key roles in neurovirulence attenuation in the JEV SA14-14-2 strain.

Table 1. Amino acid differences in the viral E protein between Japanese encephalitis virus (JEV) strains SA14, SA14-14-2, and SA14-5-3 [3,4].

Positions in E Protein [a]	Virulent Strain			Attenuated Strain		
	SA14/USA	SA14/CDC	SA14/JAP	SA14-14-2/PHK	SA14-14-2/PDK	SA14-5-3
E107	L	L	L	F	F	F
E138	E	E	E	K	K	K
E176	I	I	I	V	V	V
E177	T	T	T	A	T	T
E264	Q	Q	Q	H	Q	Q
E279	K	K	K	M	M	M
E315	A	V	A	V	V	V
E439	K	R	K	R	R	R

[a] Amino acids that differ between the virulent JEV strain SA14 and the attenuated SA14-14-2 and SA14-5-3 strains are highlighted in bold letters. E177 was studied along with E176 due to their close proximity.

2. Materials and Methods

2.1. Cells, Plasmids, and Viruses

BHK-21 cells (CCL-10; American Type Culture Collection, Manassas, VA, USA) were cultured in an Eagle minimum essential medium (MEM; Gibco; Thermo Fisher Scientific, Waltham, MA, USA) supplemented with 10% heat-inactivated fetal bovine serum. The multiple-cloning site of the low-copy plasmid pACNR was modified to contain the restriction sites *Asc*I, *Kas*I, *Bgl*II, *Bsp*EI,

*Bam*HI, *Bcl*I, *Xba*I, and *Xho*I. The JEV SA14-14-2 strain was generated in PHK cells isolated from 9- to 10-day-old specific pathogen-free (SPF) hamsters at the Chengdu Institute of Biological Products Co., Ltd. (Chengdu, China).

2.2. DNA Cloning

The RNA of the JEV SA14-14-2 strain was extracted using a High Pure viral RNA kit (Roche, Basel, Switzerland), and cDNA was synthesized by reverse transcription (RT) using SuperScript III reverse transcriptase (Invitrogen, Carlsbad, CA, USA). Briefly, 20 ng RNA was mixed with 10 pmol 3'-terminal primers, heated for 5 min at 65 °C, cooled on ice for 1 min, and then incubated with SuperScript III in the recommended buffer for 1 h at 55 °C, followed by heating to 70 °C for 15 min. cDNA amplification was performed with the phusion polymerase (New England Biolabs, Ipswich, MA, USA) using a touchdown polymerase chain reaction (PCR) program: one cycle at 98 °C for 1 min, 10 cycles at 98 °C for 15 s, 58.5 °C to 53.5 °C for 15 s, and 72 °C for 3 min, followed by 20 cycles at an annealing temperature of 53.5 °C and elongation for 10 min at 72 °C. PCR products were purified using a DNA purification kit (Qiagen, Hilden, Germany) and cloned into the pGEM-T easy vector (Promega, Durham, NC, USA). The correct clones were identified by DNA sequencing.

All plasmids were constructed using two-plasmid systems as described previously [9,10]. One plasmid contained the 5' terminal 3.4-kb cDNA and the other contained the 3' terminal 7.6-kb fragment of the SA14-14-2 strain. The first fragment (1–476 nt) contained *Asc*I and *Kas*I restriction sites [11] and was cloned into the low-copy plasmid pACNR. The second fragment, from position 476 to 2654, and the third fragment, from position 2640 to 3446, were inserted into the *Kas*I/*Bgl*II and *Bgl*II/*Bsp*EI sites, respectively, to generate the plasmid pACNR-5'JEV (harboring the 5' terminal 3.4-kb fragment). The fourth fragment, from position 3444 to 5581; the fifth fragment, from position 5575 to 7092; the sixth fragment, from position 7086 to 9136; and the seventh fragment, from position 9130 to 10977, were cloned into the pACNR to create the plasmid pACNR-3'JEV (containing the 3' terminal 7.6-kb fragment). This 7.6-kb fragment of JEV was then inserted into the plasmid pACNR-5'JEV to create the plasmid pACNR-JEV containing the full-length cDNA of JEV SA14-14-2. Mutations in the E protein gene were generated by PCR-based site-directed mutagenesis, and all plasmids were sequenced to verify the engineered mutations.

2.3. In Vitro RNA Transcription, Transfection, and Viral Recovery

The pACNR-JEV plasmid was linearized by restriction digest using *Xho*I and used as a template for in vitro transcription. The RNA used for transfection was synthesized using the RiboMAX large-scale RNA production system Sp6 kit (Promega) in the presence of Ribo m7G cap analog (Promega). Reaction products were treated with DNase I (RQ1 RNase-free DNase; Promega), followed by purification with the RNeasy mini kit (Qiagen). BHK-21 cells were washed twice with cold phosphate-buffered saline, then 4×10^6 cells in 200 µL were mixed with the synthesized RNA in vitro (1 µg) and pulsed at 140 V for 25 ms using a Gene Pulser II apparatus (Bio-Rad, Hercules, CA, USA). Transfected BHK-21 cells were cultivated at 37 °C in a 5% CO_2 incubator, and the viruses were harvested at day 5 post-transfection upon observation of the cytopathic effect. The harvested viruses were passaged two additional times in BHK-21 cells, titrated for the plaque assay, and stored at −80 °C until further use.

2.4. Nucleotide Sequencing of the Revertant Viruses

Briefly, viral RNA was extracted from the recovered viruses using the High Pure viral RNA kit (Roche). cDNA from position 468 to 2667 containing the prM/E protein gene was synthesized by RT, followed by the amplification of the prM/E fragment using the phusion polymerase (New England Biolabs). The PCR products were purified using the QIAquick gel extraction kit (Qiagen) and sequenced to determine the consensus sequence (Invitrogen).

2.5. Growth Analysis of Revertants and Control Viruses

BHK-21 cells were infected with the revertants or control viruses at a multiplicity of infection of 0.5. After 1 h of absorption at 37 °C, viral inocula were removed, and 20 mL MEM containing 2% inactivated newborn calf serum was added. Culture supernatant (1 mL) was collected every 24-h post-infection for 96 h. Titers of the collected viruses were determined as described for the plaque assay.

2.6. Mouse Experiments

To assess and compare neurovirulence, groups (n = 4) of 3-week-old SPF Kunming mice were inoculated with 0.03 mL of 10-fold dilutions of the revertants or the control viruses by the i.c. route, and inoculated mice were monitored for 14 days. All moribund mice were euthanized, and the median lethal dose (LD_{50}) was determined by the Reed and Muench calculation. Neurovirulence results for each virus were recorded as LD_{50} ($\log_{10}PFU$; the viral dose capable of inducing 50% mortality). Neuroinvasiveness was measured by inoculating 3-week-old SPF Kunming mice with 0.1 mL of 10-fold dilutions of the revertants or the control viruses by the subcutaneous (s.c.) route, and the neuroinvasive results were also recorded as LD_{50} ($\log_{10}PFU$). The average survival time (AST) was determined by inoculating 0.03 mL of viruses containing equal plaque titers (5.18 $\log_{10}PFU$) in another group of mice (n = 6) by the i.c. route. Mice in a moribund condition were euthanized and scored as deaths.

2.7. Statistical Analysis

Statistical analysis of the AST was performed using analysis of variance, and a $p < 0.05$ was considered statistically significant. All analyses were performed using SPSS software version 17.0 (SPSS, Inc., Chicago, IL, USA).

2.8. Ethical Approval

The experimental protocols involving mice were approved by the Experimental Animal Welfare and Ethical Committee of the National Institutes for Food and Drug Control, China.

3. Results

3.1. Construction of Infectious JEV Full-Length cDNA Clones Containing Specific Reverse Mutations in the E Protein

All pACNR-JEV plasmids containing specific mutations were verified by sequencing, and the viruses used for testing were amplified by three passages in the BHK-21 cells. The E protein-coding region of each virus was sequenced an additional time, with the results confirming that the sequences of all engineered plasmids and revertant viruses were correct and that no new mutations had been introduced.

3.2. Growth Analysis of Revertants and Control Viruses

One mechanism of viral attenuation involves crippled viral replication [12]; therefore, the effects of reverse mutations on JEV replication were measured by infecting BHK-21 cells, followed by a determination of the production of revertants and control viruses at different time intervals following infection. Growth-curve results showed that all viruses exhibited similar replication capacities, although the SA14 virus replicated at a modestly faster rate, with 5.7 $\log_{10}PFU/mL$ at 24-h post-infection, which was higher than the other viruses tested. However, the peak SA14 titer was not the highest among all viruses, which was likely due to the highest SA14 titer not being collected at the denoted time points (Figure 1).

Additionally, analysis of the plaque sizes of all viruses revealed that those of SA14 (2–3 mm) were larger than those of SA14-14-2 (1–2 mm) and the other viruses (1–2 mm). We observed no significant difference in plaque size between SA14-14-2 and all the revertant viruses, except for that of rJEV4 (E279) (0.5–1 mm), which was significantly smaller than those of the other viruses (Figure 2).

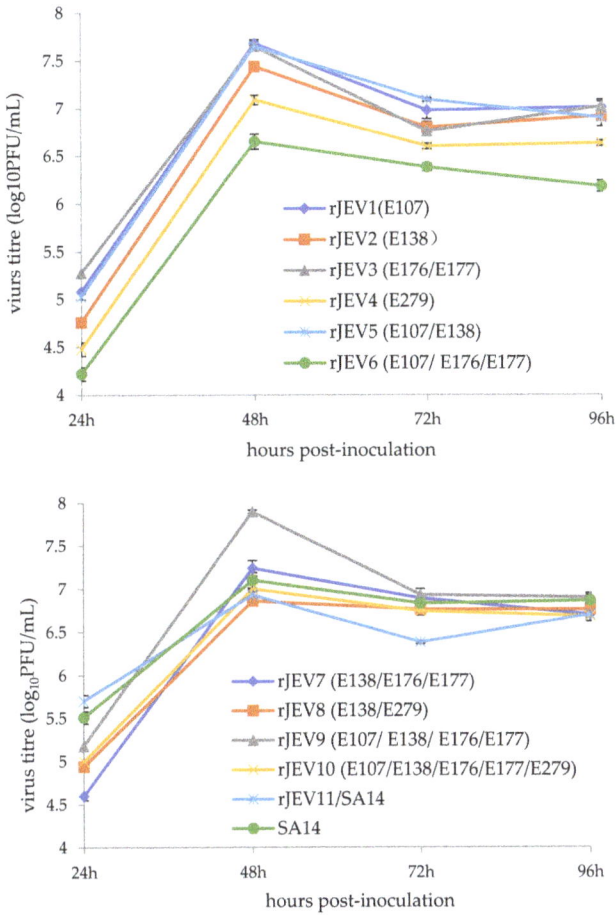

Figure 1. Growth curves of the revertants and the control viruses.

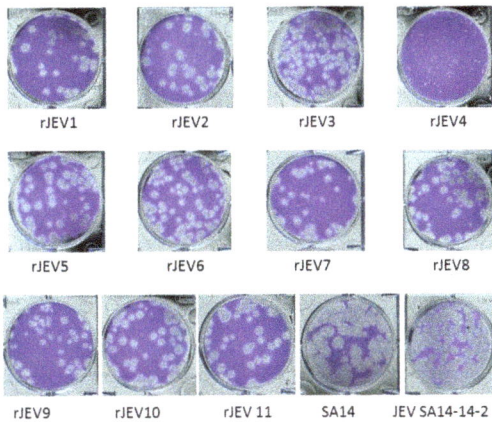

Figure 2. Plaque sizes of the revertants and the control viruses.

3.3. Mutation at Residue E138 in Combination with E107 Is Critical to the Attenuated Neurovirulence of JEV SA14-14-2

To determine the amino acids in the E protein that attenuate JEV SA14-14-2 neurovirulence, we measured the LD_{50} ($log_{10}PFU$) values of all the revertant viruses (Table 2), with low LD_{50} ($log_{10}PFU$) values indicating high degrees of neurovirulence. Among revertants containing a single amino acid substitution, rJEV1 (E107) and rJEV2 (E138) exhibited lower LD_{50} ($log_{10}PFU$) values as compared with that of the SA14-14-2 virus, whereas the revertant viruses rJEV3 (E176/E177) and rJEV4 (E279) exhibited similar LD_{50} ($log_{10}PFU$) values to that of SA14-14-2. Among these four revertants, rJEV2 (E138) exhibited the lowest LD_{50} ($log_{10}PFU$) value, indicating the highest degree of neurovirulence, followed by rJEV1 (E107). The reverse mutation of E138 in combination with E107 significantly decreased the LD_{50} ($log_{10}PFU$) value as compared with those of rJEV1 (E107) and rJEV2 (E138), and the LD_{50} ($log_{10}PFU$) value of rJEV10 (E107, E138, E176/177, and E279; 1.43) was slightly lower than that of rJEV9 (E107, E138, and E176/177; 1.99). The rJEV11/SA14 virus, wherein the E protein of SA14-14-2 was replaced with the E protein of wild-type SA14, exhibited the lowest LD_{50} ($log_{10}PFU$) value (0.66) among all the revertants, although it was still higher than that of the virulent wild-type SA14 virus (-0.92). These findings suggested, that among the five tested amino acid residues, E138 and E107 played the most important roles in the attenuation of SA14-14-2 neurovirulence.

Table 2. Neurovirulence of the mutated viruses in 3-week-old mice inoculated by the i.c. route.

Viruses	LD_{50} ($log_{10}PFU$) *
rJEV (SA14-14-2)	$\geq$6.48
rJEV1 (E107)	3.97
rJEV2 (E138)	2.89
rJEV3 (E176/E177)	$\geq$6.43
rJEV4 (E279) [†]	$\geq$6.24
rJEV5 (E107/E138)	1.70
rJEV6 (E107/E176/E177)	5.69
rJEV7 (E138/E176/E177)	3.64
rJEV8 (E138/E279)	2.82
rJEV9 (E107/E138/E176/E177)	1.99
rJEV10 (E107/E138/E176/E177/E279)	1.43
rJEV11/SA14	0.66
SA14	-0.92

* LD_{50} ($log_{10}PFU$) represents the plaque titers that cause death in 50% of tested mice; [†] Virus rJEV4 did not cause neurovirulence.

3.4. Reverse Mutations in the E Protein Increased the Mortality and Decreased the AST of I.C.-Inoculated Mice

The virulence phenotype of the revertant viruses was further evaluated by determining the mortality and AST of mice inoculated by the i.c. route with 5.18 $log_{10}PFU$ revertant virus (Table 3). Mice inoculated with rJEV2, rJEV5, rJEV7, rJEV8, rJEV9, rJEV10, rJEV11, or SA14 exhibited 100% mortality, whereas rJEV1 or rJEV3 inoculation resulted in 83.3% and 16.7% mortality, respectively, and SA14-14-2 or rJEV4 (E279) inoculation resulted in 0% mortality. These results showed that the E138 and E107 residues were more important than E279 and E176/177 at effecting SA14-14-2 virulence. The wild-type SA14 group exhibited the shortest AST (4 days), followed by the rJEV11 group (E107/E138/E176/177/E264/E279/E315/E439), with an AST of 5 days ($p \leq 0.05$, compared with SA14). The AST of mice inoculated with rJEV2 (E138) exhibited an AST of 6 days, and the ASTs of the rJEV1 (E107) and the rJEV6 (E107/E176/177) groups were 6.6 and 9 days, respectively ($p \leq 0.05$, comparing rJEV1 to rJEV6). The rJEV3 group (E176/177) exhibited the longest AST of 11 days. These results demonstrated that the E138 residue was a greater determinant of attenuated SA14-14-2 virulence, as compared with other residues in the E protein.

Table 3. AST and mortality of mice inoculated with the viruses by the i.c. route.

Viruses	No. of Dead Mice/ Total No. of Mice (%)	AST (day) Mean ± SD
rJEV (SA14-14-2)	0/6 (0)	-
rJEV1 (E107)	5/6 (83.3%)	6.6 ± 0.9 [$]
rJEV2 (E138)	6/6 (100%)	6 ± 0 [#,$]
rJEV3 (E176/E177)	1/6 (16.7%)	11 ± 0
rJEV4 (E279)	0/6 (0)	-
rJEV5 (E107/E138)	6/6 (100%)	6 ± 0 [#]
rJEV6 (E107/E176/E177)	3/6 (50%)	9 ± 0
rJEV7 (E138/E176/E177)	6/6 (100%)	6 ± 0 [#]
rJEV8 (E138/E279)	6/6 (100%)	6 ± 0 [#]
rJEV9 (E107/E138/E176/E177)	6/6 (100%)	6 ± 0 [#]
rJEV10 (E107/E138/E176/E177/E279)	6/6 (100%)	6 ± 0 [#]
rJEV11/SA14	6/6 (100%)	5 ± 0 [*]
SA14	6/6 (100%)	4 ± 0 [*]

[#] $p = 1$, compared with each other; [*] $p \leq 0.05$, compared with each other; [$] $p \leq 0.05$, compared with each other.

3.5. Effects of Specific Reverse Mutations on JEV SA14-14-2 Neuroinvasiveness

The neuroinvasiveness of all the revertants was tested using the same protocol as that used to test neurovirulence, except that the mice were inoculated by the s.c. route (Table 4). The LD_{50} ($\log_{10}$PFU) values of all the revertants containing single amino acid substitutions were similar to that of the attenuated SA14-14-2 strain, whereas the other revertants showed lower LD_{50} ($\log_{10}$PFU) values than that of SA14-14-2. The LD_{50} ($\log_{10}$PFU) values of the mice inoculated with rJEV5 (E107 and E138), rJEV7 (E138 and E176/177), or rJEV8 (E138 and E279) were ≥ 6.54, 5.76, and 6.01, respectively, suggesting that the E107 revertant mutation combined with E138 did not show the same synergistic effect as observed in the neurovirulence test. The LD_{50} ($\log_{10}$PFU) value of mice inoculated with rJEV10 (E107/E138/E176/E177/E279) was slightly higher than that of rJEV9 (E107/E138/E176/E177)-inoculated mice, and rJEV9 (E107/E138/E176/E177)-, rJEV10 (E107/E138/E176/E177/E279)-, and rJEV11/SA14-inoculated mice exhibited low LD_{50} ($\log_{10}$PFU) values of 5.40, 5.53, and 3.17, respectively. Furthermore, the LD_{50} ($\log_{10}$PFU) value of mice inoculated with rJEV11 (rJEV11/SA14) was higher than that of mice infected with virulent SA14, indicating that other regions in the JEV genome also contributed to the neuroinvasive phenotype.

Table 4. Neuroinvasiveness of the revertant viruses in 3-week-old mice inoculated by the s.c. route.

Inocula	LD_{50} ($\log_{10}$PFU)
rJEV (SA14-14-2)	≥6.14
rJEV1 (E107)	≥7.32
rJEV2 (E138)	≥6.20
rJEV3 (E176/E177)	≥6.93
rJEV4 (E279)	≥6.74
rJEV5 (E107/E138)	≥6.54
rJEV6 (E107/E176/E177)	≥6.71
rJEV7 (E138/E176/E177)	5.76
rJEV8 (E138/E279)	6.01
rJEV9 (E107/E138/E176/E177)	5.40
rJEV10 (E107/E138/E176/E177/E279)	5.53
rJEV11/SA14	3.17
SA14	1.86

4. Discussion

Reverse genetics is a powerful tool for studying the replication, virulence attenuation, and gene functions of positive-strand RNA viruses. The key step in this strategy involves constructing stable cDNA clones containing the full-length viral sequence. However, constructing the full-length cDNA clone of the JEV SA14-14-2 strain was hindered by the instability and toxicity of some gene products in *Escherichia coli* [13–15]. Previous studies utilized two strategies to overcome these hurdles. One was an in vitro ligation approach [16] and the second involved using low-copy plasmids, such as those containing artificial bacterial chromosomes, to stabilize the full-length cDNA of JEV [17]. Here, a different low-copy plasmid (pACNR) was employed to stably maintain the full-length cDNA of the infectious JEV. To generate a marker in the recombinant viruses, a silent mutation was inserted at nucleotide 473 (from A to C) that also created a new restriction site (*Kas*I) for DNA cloning. This genetic marker allowed confirmation that the recovered viruses were derived from the infectious cDNA. Furthermore, this cDNA cloning system previously enabled the mechanistic study of the virulence attenuation of *Flaviviruses* and the development of other *Flavivirus* vaccines [18].

Several studies reported nucleotide changes potentially underlying the attenuated phenotype of the JEV SA14-14-2 strain through comparisons with its parental strain SA14 [3,4,19]. Major nucleotide changes in *Flavivirus* E proteins responsible for viral neurovirulence were also revealed by comparing the JEV AT31 strain with its attenuated derivative [20] and between YFV (the Asibi train) and its attenuated 17D strain [21]. The results of mouse-infection studies showed that single substitutions at amino acid positions E107, E138, E176/177, or E279 differentially modulated viral virulence [22]. Our results showed that inoculation with the rJEV2 (E138) mutation increased SA14-14-2 neurovirulence to the highest level, followed by that of the rJEV1 (E107) mutation, whereas the single revertant mutation of E279 had no effect on neurovirulence. A synergistic virulence effect was observed when the E138 revertant mutation was combined with E107, but not E176/177 or E279, whereas infection with rJEV5 (E138/E107) exhibited the lowest LD_{50} ($\log_{10}$PFU) value (1.70). These results were consistent with observations that after five passages in the suckling mouse brain, the revertant mutations at E138 and E107, from JEV SA14-14-2 to those of the parental SA14 strain, increased the neurovirulence of the resulting virus [22].

Residue E107 is located within a highly-conserved hairpin motif spanning amino acids 98 to 111 in domain II [23]. This region contains a fusogenic peptide according to studies of the tick-borne encephalitis virus, the Murray Valley encephalitis virus, and the dengue type 2 virus [24,25]. Mutations in close proximity to this region alter the fusion properties of the E protein in cell culture and are associated with changes in the neurovirulence of the dengue virus and the tick-borne encephalitis virus [26,27]. Residue E138 is located in the 'hinge' region at the interface of domains I and II of the E protein, and mutation at this position alters E protein conformation and function. Previous studies of *Flaviviruses* indicated that mutations within this region modulate viral virulence in mice [28–32], thereby supporting the results of this study.

The effect of the E176/177 cluster on viral neurovirulence is interesting. In contrast to other reverse mutations that increased SA14-14-2 virulence, single substitutions in the E176/177 cluster elevated viral virulence to a lesser extent than other substitutions; however, when combined with mutations at E107 or E138, E176/177 mutations significantly decreased viral virulence. Additionally, the virulence of rJEV6 (E107/E176/E177) was lower than that observed for rJEV1 (E107) (5.69 vs. 3.97), the virulence of rJEV7 (E138/E176/E177) was lower than that observed for rJEV2 (E138) (3.64 vs. 2.89), and the virulence of rJEV9 (E107/E138/E176/E177) was lower than that of rJEV5 (E107/E138) (1.99 vs. 1.77). Therefore, we concluded that the E176/177 mutations significantly neutralized the function of E107 and E138.

Residue E279 is located in the hinge region of the E protein, suggesting a possible regulatory role in E protein function, similar to that of E138. Previous studies showed that reverse mutation of residue E279 from methionine to lysine significantly increased neurovirulence [33]. Additionally, mutations in close proximity to E279 in the Murray Valley encephalitis virus impair hemagglutination and fusion properties of the E protein and reduce neuroinvasiveness in mice [28,34]. By contrast, reverse mutation of E279 in this study did not affect SA14-14-2 neurovirulence. The neutral effect of the E279 mutation might be explained by the decreased ability of the virus harboring the mutation to infect host cells, given that inoculation with rJEV4 (E279) resulted in that formation of the smallest plaques among all tested viruses, including SA14-14-2.

A previous study reported that the molecular determinants associated with the prM-E region of the attenuated JE SA14-14-2 virus are insufficient to confer an attenuated phenotype upon the JE Nakayama virus [6]. This suggested a role for determinants located in the 5′ untranslated region and/or the capsid protein of the JE SA 14-14-2 viral genome in influencing the virulent properties of the JE Nakayama virus in mouse models. Here, we observed that the revertant rJEV11 virus, having the same eight amino acids in the E protein as the parental virulent SA14 strain, exhibited significantly lower neurovirulence and neuroinvasiveness in mice as compared with JEV SA14 (Tables 2 and 4). These data demonstrated that mutations in viral proteins (including nonstructural protein) other than the E protein in JEV SA14-14-2 may also contribute to attenuated neurovirulence and neuroinvasiveness.

A previous report utilized a chimeric YFV/JEV SA14-14-2 virus to characterize the attenuation mechanism [35]. This chimeric virus contained the backbone of YFV and the prM and E protein sequences from the attenuated JEV SA14-14-2 strain. Consistent with our findings, Arroyo et al. [35] reported the importance of the E138 amino acid together with amino acid residues at other positions in neurovirulence attenuation; however, other results from that study differed from our findings. Arroyo et al. [35] reported that inoculation with variants harboring single reverse mutations of E107, E138, and E176/177 did not cause sickness or death in mice and that the single reverse mutation of E279 caused death in only 13% of mice. By contrast, we observed that inoculation with each of the three single reverse mutations (E107, E138, and E176/177) resulted in sickness or death in some of the mice and isolation of the revertant viruses in the brain. Additionally, the single reverse mutation of E279 did not cause sickness or death in mice. These discrepancies might be explained by the different inoculation doses used between the two studies, given that our inoculated mice became sick or died only when inoculated with >4.0 $\log_{10}$PFU of the revertants rJEV1 (E107), rJEV3 (E176/E177), and rJEV6 (E107/E176/E177), whereas the previous study used inoculation doses via i.c. of <4.0 $\log_{10}$PFU (10,000 PFU) [35]. Additionally, the results of that study did not suggest an important role for the single mutation of E107 in JEV attenuation, whereas our results provided a more detailed account of the roles of E176/177 and E279 in JEV attenuation, in combination with E138. Furthermore, Arroyo et al. [35] reported that the single substitution at E176/177 greatly enhanced virulence in combination with other mutations, whereas we observed decreased virulence associated with this mutation in combination with others. One explanation for these discrepancies might be use of a chimeric virus with YFV as the backbone in the previous study [35], whereas the JEV SA14 virus was used in this study to investigate the attenuation mechanism.

Yun et al. [36] showed that the passage of the JEV SA14-14-2 strain in the mouse brain was selected for mutations at the E244 position, which drastically altered the viral phenotype [36]. In this study, this amino acid position was not tested because the E244 position in the SA14 viral population harbors two different amino acids in the mouse brain (Figure 3), with glutamic acid at this position in wild-type SA14/USA and glycine at this position in the SA14/CDC and SA14/JAP strains (Table 1 and Figure 3).

The mechanisms associated with viral attenuation are complicated and involve membrane fusion [24,25], replication capacity [12], and heparin-binding activity [37]. In addition to the SA-14 E protein, our findings suggested that other regions of the viral genome likely also contribute to the attenuated phenotype. This was supported by our results showing that the virulence of the revertant rJEV11/SA14 strain remained lower than that of the parental SA14 strain, despite substitution with the intact E protein from wild-type SA14.

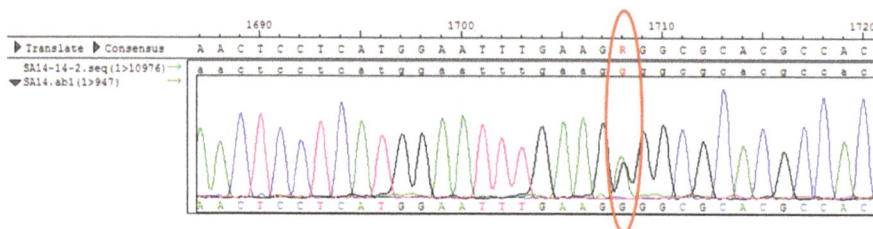

Figure 3. An equal proportion of two different amino acids exists at position E244 in the SA14 virus.

5. Conclusions

In summary, this study demonstrated that amino acids at positions E107, E138, E176/177, and E279 differentially contributed to virulence attenuation in the SA14-14-2 virus. Our findings indicated that the E138 position played the most important role in sustaining neurovirulence, but not the attenuated neuroinvasive phenotype associated with JEV SA14-14-2. Additionally, the role of the E107 position in attenuating virulence was revealed by its synergistic effect with the E138 position, although E107 alone also contributed to virulence attenuation. Compared with the E107 and E138 positions, E176/177 and E279 exhibited relatively minor roles in virulence attenuation. These results identified the key residues in the E protein involved in regulating attenuated JEV SA14-14-2 virulence, thereby elucidating the molecular mechanisms of JEV attenuation. The data presented in this study supported JEV vaccine guidelines stating that the stability of the E protein sequence should be used as the main safety indicator for the attenuated live JE vaccine (SA14-14-2 strain).

Acknowledgments: We would like to thank all members of the Animal Biotechnological Center for their contributions to this study. This study was supported by the national key projects of "863" High Technology, China (No. 2012AA02A401) and the Chinese Mega Project of Science Research for major new drug innovation and developmental research for quality control of the JE live attenuated vaccine and polio vaccine (Grant No. 2014ZX09304316-003). The funders had no role in study design, data collection and analysis, decision to publish, or manuscript preparation.

Author Contributions: Y.L. and H.Y. conceived and designed the experiments. J.Y., H.Y., Z.L., W.W., H.L., L.L., X.L., and Q.N. performed the experiments. X.Z. and Y.W. analyzed the data. J.Y. and H.Y. wrote the paper. J.Y. and H.Y. contributed equally to this work.

Conflicts of Interest: The authors declare no conflict of interest.

References

1. Wills, M.R.; Sil, B.K.; Cao, J.X.; Yu, Y.X.; Barrett, A.D. Antigenic characterization of the live attenuated Japanese encephalitis vaccine virus SA14-14-2: A comparison with isolates of the virus covering a wide geographic area. *Vaccine* **1992**, *10*, 861–872. [CrossRef]
2. Liu, Z.L.; Hennessy, S.; Strom, B.L.; Tsai, T.F.; Wan, C.M.; Tang, S.C.; Xiang, C.F.; Bilker, W.B.; Pan, X.P.; Yao, Y.J.; et al. Short-term safety of live attenuated Japanese encephalitis vaccine (SA14-14-2): Results of a randomized trial with 26,239 subjects. *J. Infect. Dis.* **1997**, *176*, 1366–1369. [CrossRef] [PubMed]
3. Ni, H.; Burns, N.J.; Chang, G.J.; Zhang, M.J.; Wills, M.R.; Trent, D.W.; Sanders, P.G.; Barrett, A.D. Comparison of nucleotide and deduced amino acid sequence of the 5' non-coding region and structural protein genes of the wild-type Japanese encephalitis virus strain SA14 and its attenuated vaccine derivatives. *J. Gen. Virol.* **1994**, *75*, 1505–1510. [CrossRef] [PubMed]
4. Ni, H.; Chang, G.J.; Xie, H.; Trent, D.W.; Barrett, A.D. Molecular basis of attenuation of neurovirulence of wild-type Japanese encephalitis virus strain SA14. *J. Gen. Virol.* **1995**, *76*, 409–413. [CrossRef] [PubMed]
5. Yu, Y. Phenotypic and genotypic characteristics of Japanese encephalitis attenuated live vaccine virus SA14-14-2 and their stabilities. *Vaccine* **2010**, *28*, 3635–3641. [CrossRef] [PubMed]
6. Chambers, T.J.; Droll, D.A.; Jiang, X.; Wold, W.S.; Nickells, J.A. JE Nakayama/JE SA14-14-2 virus structural region intertypic viruses: Biological properties in the mouse model of neuroinvasive disease. *Virology* **2007**, *366*, 51–61. [CrossRef] [PubMed]

7. Jennings, A.D.; Gibson, C.A.; Miller, B.R.; Mathews, J.H.; Mitchell, C.J.; Roehrig, J.T.; Wood, D.J.; Taffs, F.; Sil, B.K.; Whitby, S.N.; et al. Analysis of a yellow fever virus isolated from a fatal case of vaccine-associated human encephalitis. *J. Infect. Dis.* **1994**, *169*, 512–518. [CrossRef] [PubMed]

8. Taffs, R.E.; Chumakov, K.M.; Rezapkin, G.V.; Lu, Z.; Douthitt, M.; Dragunsky, E.M.; Levenbook, I.S. Genetic stability and mutant selection in Sabin 2 strain of oral poliovirus vaccine grown under different cell culture conditions. *Virology* **1995**, *209*, 366–373. [CrossRef] [PubMed]

9. Chambers, T.J.; Liang, Y.; Droll, D.A.; Schlesinger, J.J.; Davidson, A.D.; Wright, P.J.; Jiang, X. Yellow fever virus/dengue-2 virus and yellow fever virus/dengue-4 virus chimeras: Biological characterization, immunogenicity, and protection against dengue encephalitis in the mouse model. *J. Virol.* **2003**, *77*, 3655–3668. [CrossRef] [PubMed]

10. Chambers, T.J.; Nestorowicz, A.; Mason, P.W.; Rice, C.M. Yellow fever/Japanese encephalitis chimeric viruses: Construction and biological properties. *J. Virol.* **1999**, *73*, 3095–3101. [PubMed]

11. Chambers, T.J.; Jiang, X.; Droll, D.A.; Liang, Y.; Wold, W.S.; Nickells, J. Chimeric Japanese encephalitis virus/dengue 2 virus infectious clone: Biological properties, immunogenicity, and protection against dengue encephalitis in mice. *J. Gen. Virol.* **2006**, *87*, 3131–3140. [CrossRef] [PubMed]

12. Muylaert, I.R.; Chambers, T.J.; Galler, R.; Rice, C.M. Mutagenesis of the N-linked glycosylation sites of the yellow fever virus NS1 protein: Effects on virus replication and mouse neurovirulence. *Virology* **1996**, *222*, 159–168. [CrossRef] [PubMed]

13. Mishin, V.P.; Cominelli, F.; Yamshchikov, V.F. A "minimal" approach in design of flavivirus infectious DNA. *Virus Res.* **2001**, *81*, 113–123. [CrossRef]

14. Ruggli, N.; Rice, C.M. Functional cDNA clones of the Flaviviridae: Strategies and applications. *Adv. Virus Res.* **1999**, *53*, 183–207. [PubMed]

15. Zhang, F.; Huang, Q.; Ma, W.; Jiang, S.; Fan, Y.; Zhang, H. Amplification and cloning of the full-length genome of Japanese encephalitis virus by a novel long RT-PCR protocol in a cosmid vector. *J. Virol. Methods* **2001**, *96*, 171–182. [CrossRef]

16. Sumiyoshi, H.; Hoke, C.H.; Trent, D.W. Infectious Japanese encephalitis virus RNA can be synthesized from in vitro-ligated cDNA templates. *J. Virol.* **1992**, *66*, 5425–5431. [PubMed]

17. Yun, S.I.; Kim, S.Y.; Rice, C.M.; Lee, Y.M. Development and application of a reverse genetics system for Japanese encephalitis virus. *J. Virol.* **2003**, *77*, 6450–6465. [CrossRef] [PubMed]

18. Lai, C.J.; Monath, T.P. Chimeric flaviviruses: Novel vaccines against dengue fever, tick-borne encephalitis, and Japanese encephalitis. *Adv. Virus Res.* **2003**, *61*, 469–509. [PubMed]

19. Nitayaphan, S.; Grant, J.A.; Chang, G.J.; Trent, D.W. Nucleotide sequence of the virulent SA-14 strain of Japanese encephalitis virus and its attenuated vaccine derivative, SA-14-14-2. *Virology* **1990**, *177*, 541–552. [CrossRef]

20. Zhao, Z.; Date, T.; Li, Y.; Kato, T.; Miyamoto, M.; Yasui, K.; Wakita, T. Characterization of the E-138 (Glu/Lys) mutation in Japanese encephalitis virus by using a stable, full-length, infectious cDNA clone. *J. Gen. Virol.* **2005**, *86*, 2209–2220. [CrossRef] [PubMed]

21. Galler, R.; Freire, M.S.; Jabor, A.V.; Mann, G.F. The yellow fever 17D vaccine virus: Molecular basis of viral attenuation and its use as an expression vector. *Braz. J. Med. Biol. Res.* **1997**, *30*, 157–168. [CrossRef] [PubMed]

22. Wu, Y.L.; Liu, J.; Yang, H.Q.; Zhao, Y.; Wang, W.; Mu, J.C.; Huang, Y.X.; Liu, R.; Sun, Y.; Yu, Y.X.; et al. Genetic property of attenuated Japanese encephalitis virus strain SA14-14-2 after subculture in suckling mouse brain. *Chin. J. Biol.* **2007**, *20*, 19–21.

23. Kolaskar, A.S.; Kulkarni-Kale, U. Prediction of three-dimensional structure and mapping of conformational epitopes of envelope glycoprotein of Japanese encephalitis virus. *Virology* **1999**, *261*, 31–42. [CrossRef] [PubMed]

24. Roehrig, J.T.; Hunt, A.R.; Johnson, A.J.; Hawkes, R.A. Synthetic peptides derived from the deduced amino acid sequence of the E-glycoprotein of Murray Valley encephalitis virus elicit antiviral antibody. *Virology* **1989**, *171*, 49–60. [CrossRef]

25. Roehrig, J.T.; Johnson, A.J.; Hunt, A.R.; Bolin, R.A.; Chu, M.C. Antibodies to dengue 2 virus E-glycoprotein synthetic peptides identify antigenic conformation. *Virology* **1990**, *177*, 668–675. [CrossRef]

26. Despres, P.; Frenkiel, M.P.; Deubel, V. Differences between cell membrane fusion activities of two dengue type-1 isolates reflect modifications of viral structure. *Virology* **1993**, *196*, 209–219. [CrossRef] [PubMed]

27. Rey, F.A.; Heinz, F.X.; Mandl, C.; Kunz, C.; Harrison, S.C. The envelope glycoprotein from tick-borne encephalitis virus at 2 A resolution. *Nature* **1995**, *375*, 291–298. [CrossRef] [PubMed]

28. Chen, L.K.; Lin, Y.L.; Liao, C.L.; Lin, C.G.; Huang, Y.L.; Yeh, C.T.; Lai, S.C.; Jan, J.T.; Chin, C. Generation and characterization of organ-tropism mutants of Japanese encephalitis virus in vivo and in vitro. *Virology* **1996**, *223*, 79–88. [CrossRef] [PubMed]

29. Gualano, R.C.; Pryor, M.J.; Cauchi, M.R.; Wright, P.J.; Davidson, A.D. Identification of a major determinant of mouse neurovirulence of dengue virus type 2 using stably cloned genomic-length cDNA. *J. Gen. Virol.* **1998**, *79*, 437–446. [CrossRef] [PubMed]

30. Hasegawa, H.; Yoshida, M.; Shiosaka, T.; Fujita, S.; Kobayashi, Y. Mutations in the envelope protein of Japanese encephalitis virus affect entry into cultured cells and virulence in mice. *Virology* **1992**, *191*, 158–165. [CrossRef]

31. McMinn, P.C.; Marshall, I.D.; Dalgarno, L. Neurovirulence and neuroinvasiveness of Murray Valley encephalitis virus mutants selected by passage in a monkey kidney cell line. *J. Gen. Virol.* **1995**, *76*, 865–872. [CrossRef] [PubMed]

32. Sumiyoshi, H.; Tignor, G.H.; Shope, R.E. Characterization of a highly attenuated Japanese encephalitis virus generated from molecularly cloned cDNA. *J. Infect. Dis.* **1995**, *171*, 1144–1151. [CrossRef] [PubMed]

33. Monath, T.P.; Arroyo, J.; Levenbook, I.; Zhang, Z.X.; Catalan, J.; Draper, K.; Guirakhoo, F. Single mutation in the flavivirus envelope protein hinge region increases neurovirulence for mice and monkeys but decreases viscerotropism for monkeys: Relevance to development and safety testing of live, attenuated vaccines. *J. Virol.* **2002**, *76*, 1932–1943. [CrossRef] [PubMed]

34. McMinn, P.C.; Weir, R.C.; Dalgarno, L. A mouse-attenuated envelope protein variant of Murray Valley encephalitis virus with altered fusion activity. *J. Gen. Virol.* **1996**, *77*, 2085–2088. [CrossRef] [PubMed]

35. Arroyo, J.; Guirakhoo, F.; Fenner, S.; Zhang, Z.X.; Monath, T.P.; Chambers, T.J. Molecular basis for attenuation of neurovirulence of a yellow fever virus/Japanese encephalitis virus chimera vaccine (ChimeriVax-JE). *J. Virol.* **2001**, *75*, 934–942. [CrossRef] [PubMed]

36. Yun, S.I.; Song, B.H.; Kim, J.K.; Yun, G.N.; Lee, E.Y.; Li, L.; Kuhn, R.J.; Rossmann, M.G.; Morrey, J.D.; Lee, Y.M. A molecularly cloned, live-attenuated Japanese encephalitis vaccine SA14–14–2 virus: A conserved single amino acid in the ij hairpin of the viral E glycoprotein determines neurovirulence in mice. *PLoS Pathog.* **2014**, *10*, e1004290. [CrossRef] [PubMed]

37. Silva, L.A.; Khomandiak, S.; Ashbrook, A.W.; Weller, R.; Heise, M.T.; Morrison, T.E.; Dermody, T.S. A single-amino-acid polymorphism in Chikungunya virus E2 glycoprotein influences glycosaminoglycan utilization. *J. Virol.* **2014**, *88*, 2385–2397. [CrossRef] [PubMed]

viruses

MDPI

Article

Mx Is Not Responsible for the Antiviral Activity of Interferon-α against Japanese Encephalitis Virus

Jing Zhou [1], Shi-Qi Wang [1], Jian-Chao Wei [2], Xiao-Min Zhang [1], Zhi-Can Gao [1], Ke Liu [2], Zhi-Yong Ma [2], Pu-Yan Chen [1] and Bin Zhou [1,*]

[1] Key Laboratory of Animal Diseases Diagnosis and Immunology, Ministry of Agriculture, College of Veterinary Medicine, Nanjing Agricultural University, Nanjing 210095, China; 2015107081@njau.edu.cn (J.Z.); 15150560620@163.com (S.-Q.W.); xiaomin107228@126.com (X.-M.Z.); 2014107082@njau.edu.cn (Z.-C.G.); puyanchennj@163.com (P.-Y.C.)
[2] Shanghai Veterinary Research Institute, Chinese Academy of Agricultural Science, Shanghai 200241, China; weijianchao@shvri.ac.cn (J.-C.W.); liuke@shvri.ac.cn (K.L.); zhiyongma@shvri.ac.cn (Z.-Y.M.)
*   Correspondence: zhoubin@njau.edu.cn; Tel./Fax: +86-25-8439-6028

Academic Editor: Michael R. Holbrook
Received: 6 November 2016; Accepted: 28 December 2016; Published: 10 January 2017

Abstract: Mx proteins are interferon (IFN)-induced dynamin-like GTPases that are present in all vertebrates and inhibit the replication of myriad viruses. However, the role Mx proteins play in IFN-mediated suppression of Japanese encephalitis virus (JEV) infection is unknown. In this study, we set out to investigate the effects of Mx1 and Mx2 expression on the interferon-α (IFNα) restriction of JEV replication. To evaluate whether the inhibitory activity of IFNα on JEV is dependent on Mx1 or Mx2, we knocked down Mx1 or Mx2 with siRNA in IFNα-treated PK-15 cells and BHK-21 cells, then challenged them with JEV; the production of progeny virus was assessed by plaque assay, RT-qPCR, and Western blotting. Our results demonstrated that depletion of Mx1 or Mx2 did not affect JEV restriction imposed by IFNα, although these two proteins were knocked down 66% and 79%, respectively. Accordingly, expression of exogenous Mx1 or Mx2 did not change the inhibitory activity of IFNα to JEV. In addition, even though virus-induced membranes were damaged by Brefeldin A (BFA), overexpressing porcine Mx1 or Mx2 did not inhibit JEV proliferation. We found that BFA inhibited JEV replication, not maturation, suggesting that BFA could be developed into a novel antiviral reagent. Collectively, our findings demonstrate that IFNα inhibits JEV infection by Mx-independent pathways.

Keywords: Mx1; Mx2; interferon-α (IFNα); Japanese encephalitis virus (JEV); antivirus; Brefeldin A (BFA)

1. Introduction

Japanese encephalitis virus (JEV)—a member of the genus *Flavivirus* within the family Flaviviridae—causes serious epidemics in tropical and subtropical areas with a high mortality rate of approximately 25% in humans, and is a serious public health problem in southern and eastern Asia [1,2]. It is well known that JEV infects boars and sows, which are the major amplifying hosts of JEV in nature. The treatment of JEV infection in pigs is important for controlling the prevalence of JEV in humans and economic losses in pig production. Even though two kinds of vaccines—the attenuated vaccine (SA14-14-2) and the inactivated vaccines (mouse brain-derived and Vero cell culture-derived)—are widely used to vaccinate human and pigs, JE is widespread in the south, southeast, and the east regions of Asia, with epidemics breaking out every few years [3,4]. Therefore, it is necessary to develop new strategies against JEV.

Type I interferons (IFNs, including IFN-α) mediate a wide range of biological activities, including antiviral activity, cell growth, differentiation, apoptosis, and immune response [5]. Type I IFNs bind a heterodimeric transmembrane receptor termed the IFN-α receptor to activate interferon-stimulated gene factor 3 (ISGF3) via the JAK-STAT signaling pathway and induce the coordinated upregulation of hundreds of interferon-stimulated genes (ISGs) that orchestrate an antiviral state in the cells [6]. Of these ISGs, Mx (myxovirus-resistant), PKR (Double-stranded RNA-dependent protein kinase), and OAS (2′,5′-oligoadenylate synthetases) are the three major mediators of innate antiviral mechanism induced in the host cells, and have been studied extensively. Recently, it has been shown that porcine IFN-α inhibits JEV replication [7]. Furthermore, transient overexpression of OAS isoforms inhibits JEV replication [8]. However, whether the inhibitory activity of type I IFNs on JEV is mediated by Mx proteins is largely unknown.

Mx proteins are interferon-induced dynamin-like GTPases that are present in all vertebrates [9–11]. These proteins have a broad range of antiviral activities against various viruses [12], such as vesicular stomatitis virus (VSV) [13,14], influenza virus [15,16], classic swine fever virus (CSFV) [17], foot mouth disease virus (FMDV) [18], and bovine viral diarrhea virus (BVDV) [19]. Mx proteins consist of an N-terminal globular GTPase domain, a connecting bundle signaling element, and the C-terminal stalk that mediates oligomerization and antiviral specificity [20]. It is well known that the dynamin-like GTPase activity—including GTP binding and GTP hydrolysis—is required for Mx to function [5,10,21]. Human MxB—which previously had not been ascribed an antiviral function—was recently found to be a suppressor of human immunodeficiency virus type 1 (HIV-1) [22,23]. Based on the nucleotide and amino acid sequences, porcine Mx1 (poMx1) has 78% homology with human MxA (huMxA) and is located in the cytoplasm of target cells, suggesting that they share similar antiviral activities against some RNA viruses. Our previous study showed that a commercial recombinant human interferon-α (huIFNα) was used to characterize the antiviral effect on JEV replication in BHK-21 cells. In this study, we sought to investigate the roles of Mx1 and Mx2 during the inhibition of JEV infection, overexpression, and knockdown of Mx1 and Mx2 were performed to determine the antiviral activities of Mx. Our findings indicate that Mx protein does not contribute to the antiviral effect of IFNα against *Flavivirus*.

2. Materials and Methods

2.1. Cells, Virus, and Interferon

The baby hamster kidney (BHK-21) cells were maintained in Dulbecco's modified essential medium (DMEM, GIBCO, Invitrogen, Carlsbad, CA, USA) supplemented with 10% fetal bovine serum (FBS) (GIBCO, Invitrogen), 0.2% NaHCO$_3$, 100 μg/mL streptomycin, and 100 IU/mL penicillin (GIBCO, Invitrogen) at 37 °C with 5% CO$_2$. Porcine kidney (PK-15) cells were grown in RPMI 1640 (GIBCO, Invitrogen) supplemented with 10% FBS, 100 μg/mL streptomycin, and 100 U/mL penicillin. JEV virulent strain NJ2008 (GenBank: GQ918133) used in this study was described previously [24]. JEV attenuated vaccine strain (SA14-14-2) was purchased from Wuhan Keqian Biology Co., Ltd. (Wuhan, China). CSFV virulent strain Shimen (GenBank: AF092448) was obtained from the National Institute of Veterinary Drug Control (Beijing, China). The commercial human interferon α-1b (huIFNα) was purchased from Shenzhen Kexing Biotech Co., Ltd. (Shenzhen, China).

2.2. Virus Infection and Titration

Viral infection and titration were performed as previously described [25]. Briefly, cells were adsorbed with virus at the indicated multiplicity of infection (MOI) for 1 h at 37 °C, washed to remove nonadherent virus, then incubated at 37 °C. The culture supernatants were collected, and virus titers were determined by plaque-forming assay in BHK-21 cells.

2.3. Antibodies

Mouse anti-JEV NS1 (2B8), NS5 (1G6), and E (2A5) mAbs were kindly provided by Professor Shengbo Cao (Huazhong Agricultural University, Wuhan, China). Mouse anti-Mx1 mAb (ab79609), rabbit anti-Mx2 antibody (ab196833) and rabbit anti-Viperin antibody (ab121042) were purchased from Abcam (Cambridge, UK). Rabbit anti-green fluorescent protein (GFP) antibody was purchased from Sigma-Aldrich (St. Louis, MO, USA). Goat anti-mouse immunoglobulin G (IgG) (Alexa Fluor-568) was purchased from Thermo Fisher (Cambridge, MA, USA). Rabbit anti-β-actin mAb (13E5), goat anti-rabbit IgG-HRP (sc-2004) and goat anti-mouse IgG-HRP (sc-2005) were purchased from Santa Cruz Biotechnology (Santa Cruz, CA, USA). Stat-1 α/β rabbit polyclonal antibody was purchased from Beyotime Biotech Co., Ltd. (Nanjing, China).

2.4. Plasmids

pEGFP-poMx1 expressing wild-type poMx1 fused to green fluorescent protein was reported previously [17]. pcDNA3.0-poMx1 expressing wild-type poMx1 fused to HA tag was constructed from pEGFP-poMx1 using a pair of primers (pcDNA3.0-poMx1-F/R). pcDNA3.0-poTMx1 expressing a nuclear form of wild-type poMx1 [26] fused to HA tag was constructed from pEGFP-poMx1 using a pair of primers (pcDNA3.0-poTMx1-F/R). pcDNA3.0-poMx1(ΔL4) expressing a poMx1 variant (deletion of residues 534 to 573) [27] fused to HA tag was constructed by directed mutagenesis (Vazyme Biotech Co., Ltd., Nanjing, China) based on pEGFP-poMx1. Porcine Mx2 cDNA (GenBank: AB258432) was commercially synthesized by Nanjing Genscript Corporation (Nanjing, China) and cloned in-frame with EGFP into the vector pEGFP-C1 at Hind III and Xho I sites using a pair of primers (pEGFP-poMx2-F/R). pcDNA3.0-poMx2 was constructed from pEGFP-poMx2 using a pair of primers (pcDNA3.0-poMx2-F/R). Human MxA (GenBank: P20591) and mouse Mx1 (GenBank: P09922) cDNA were kindly provided by Dr. Song Gao (Sun Yat-sen University Cancer Center, Guangzhou, China) and cloned in-frame with EGFP into the vector pEGFP-C1 at Xho I and BamH I sites using two pairs of primers (pEGFP-huMxA-F/R, pEGFP-mmMx1-F/R). All of the primer pairs are listed in tab:viruses-09-00005-t001. The corresponding genes were verified by sequencing.

Table 1. Primers used in this study.

Primer	Sequence (5' → 3')
pEGFP-poMx2-F	TGACAAGCTTACCATGCCTAAACCCCGCATGTCG
pEGFP-poMx2-R	TGACCTCGAGTTACCCCTGTAATGACTGAGC
pcDNA3.0-poMx2-F	TGACAAGCTTACCATGCCTAAACCCCGCATGTCG
pcDNA3.0-poMx2-R	TGACCTCGAGTTAAGCGTAGTCTGGGACGTCGTATGGGTAC CCCTGTAATGACTGAGC
pEGFP-huMxA-F	TGACCTCGAGCTACCATGGTTGTTTCCGAAGTGGACATC
pEGFP-huMxA-R	TGACGGATCCACCGGGGAACTGGGCAAGCCGGCG
pEGFP-mmMx1-F	TGACCTCGAGCTACCATGGATTCTGTGAATAATCTGTGC
pEGFP-mmMx1-R	TGACGGATCCATCGGAGAATTTGGCAAGCTTCTG
pcDNA3.0-poMx1-F	TGACAAGCTTACCATGGTTTATTCCAGCTGTG
pcDNA3.0-poMx1-R	TGACCTCGAGTTAAGCGTAGTCTGGGACGTCGTATGGGTAGC CTGGGAACTTGGCGA
pcDNA3.0-poTMx1-F	TGACAAGCTTACCATGGACAAGGAGTTCCTGGAGGCTCCTAA GAAGAAGAGAAAGGTTGAGTTCAGAATTGTTTATTCC
	AACTGTGAAAGTAAAGAACCTGATTCAGTT
pcDNA3.0-poTMx1-R	TGACCTCGAGTTAAGCGTAGTCTGGGACGTCGTATGGGTAGCC TGGGAACTTGGCGA

2.5. Immunofluorescence Assay

Cells grown on glass coverslips were infected with JEV at an MOI of 0.05. At 24 hpi, cells were washed with PBS, and fixed with 4% paraformaldehyde in PBS. Cells were then permeabilized with

0.2% Triton X-100, washed again, then reacted with either anti-JEV E, NS1, or NS5 mAbs diluted 1:500. After washing, the coverslips were reacted with goat anti-mouse IgG (Alexa Fluor-568). After washing, cells were visualized by confocal microscopy (Leica Sp5 AOBS confocal system) with a 63 _HCX PL Apo 1.4 oil immersion objective.

2.6. Western Blot Analysis

Cells were washed three times with PBS and lysed in cold lysis buffer (1% Triton X-100, 1 mM PMSF in PBS) for 10 min. The lysates were clarified by centrifugation at 12,000× g for 10 min. Total cell extracts were separated by SDS-PAGE, transferred to nitrocellulose membranes, and then probed with the indicated antibodies (anti-JEV E, NS1, or NS5 mAbs), followed by goat anti-mouse IgG-HRP conjugate antibody or goat anti-rabbit IgG-HRP conjugate antibody. β-actin was used as a loading control.

2.7. Brefeldin A (BFA) Treatment

BHK-21 cells (1.2 × 10^6) were transfected with 2 µg pEGFP-poMx1 or pEGFP-poMx2. Twenty-four hours post transfection, cells were infected with JEV at an MOI of 0.05. At 12 hpi, BFA (5 µg/mL) was added to the culture medium, and incubation continued for an additional 12 h as described previously [28]. The effect of poMx1 or poMx2 in these cells was analyzed by Western blot analysis and plaque assay. In order to establish the parameters for BFA treatment, the following experiments were conducted. (i) the cytotoxic effect of BFA on BHK-21 cells was established by viability assay, as described previously [29]. Briefly, sub-confluent cell cultures grown in 96-well plates were incubated with various concentrations (0–8, 10, and 15 µg/mL) of BFA for 24 h. An MTS-based viability assay (CellTiter 96 aqueous nonradioactive cell proliferation assay from Promega (Madison, WI, USA) was conducted as recommended by the manufacturer; (ii) the dose-dependent activity of BFA was characterized, 1.2 × 10^6 BHK-21 cells were seeded into six-well plates and infected with JEV NJ2008 at an MOI of 0.05. After virus adsorption and washing, cells were maintained in medium containing BFA at various concentrations (2.5, 5, 7.5, and 10 µg/mL) or an equivalent volume of the DMSO carrier. At 24 hpi, cell supernatants were used to determine the levels of infectious virus by plaque assay. Whole cell-culture lysates were used to determine the viral protein levels by Western blot analysis; (iii) to assess the antiviral activity of BFA over time, 1.2 × 10^6 BHK-21 cells were seeded into six-well plates and infected with JEV strain NJ2008 at an MOI of 0.05. After virus adsorption and washing, cells were maintained in medium containing 5 µg/mL BFA or an equivalent volume of the DMSO carrier. At 13, 15, 18, and 24 hpi, cell supernatants were used to determine the levels of infectious virus by plaque assay, and whole cell-culture lysates were used to determine the viral protein levels by Western blot analysis; (iv) the antiviral activity of BFA was assessed by immunofluorescence. Briefly, the JEV-infected cells were maintained in medium containing 5 µg/mL BFA or an equivalent volume of the DMSO carrier. At 24 hpi, cells were fixed and reacted with anti-JEV E or NS5 mAbs. After washing, cells were visualized by confocal microscopy (Leica Sp5 AOBS confocal system) with a 63_HCX PL Apo 1.4 oil immersion objective.

2.8. Knockdown Experiments

siRNA experiments were carried out in six-well plates containing BHK-21 cells or PK-15 cells starting at 2.5 × 10^5 cells/well. siRNAs for siMx1 (sc-45260), siMx2 (sc-45261), and the negative-control siRNA (sc-37007) (Santa Cruz Biotechnology) were transfected at a concentration of 100 nM into cells using Lipofectamine 3000 (Invitrogen) according to the manufacturer's instructions. Six hours after transfection, the medium was aspirated, fresh complete medium containing 200 ng/mL huIFNα was added, and incubation continued for an additional 12 h [30]. Cells were washed with PBS, and infected with JEV at an MOI of 0.05. Cell supernatants were used to determine the levels of infectious virus by plaque assay. Whole cell-culture lysates were used to determine viral protein levels by Western blot analysis.

2.9. RT-qPCR

Viral RNA was extracted from each sample using TRIzol reagent (Invitrogen). JEV and CSFV RNA loads were measured using RT-qPCR assay, as described previously [7,31].

2.10. Statistical Analysis

All data were presented as means ± standard deviation (S.D.) as indicated. Student's *t*-test was used to compare the data from pairs of treated or untreated groups. Statistical significance was indicated as ns ($p > 0.05$), * ($p < 0.05$), and ** ($p < 0.01$). All statistical analyses and calculations were performed using GraphPad Prism 5 (GraphPad Software Inc., La Jolla, CA, USA).

3. Results

3.1. Mx Proteins Were Not Detectable in JEV-Infected Cells

Flaviviruses have evolved specific strategies to avoid and/or attenuate induction of IFN and its effector responses. In JEV, the N-terminal 83 residues of NS5 inhibit JAK-STAT signaling through a protein-tyrosine phosphatase-dependent mechanism, resulting in suppressed expression of a wide variety of interferon-stimulated genes [32]. Here, the innate immune response of the host cells against JEV infection was determined by Western blot analysis using the antibodies against ISG proteins Mx1, Mx2, and Viperin. As shown in Figure 1, endogenous Mx1, Mx2, and Viperin proteins were produced in huIFNα-treated BHK-21 and PK-15 cells (lane 3); in contrast, no endogenous Mx1, Mx2, or Viperin proteins were detectable in cells infected with virulent (NJ2008) or attenuated JEV (SA14-14-2). STAT1 expression was observed after viral infection, suggesting that the IFN-induced JAK-STAT signaling was not blocked. These data are consistent with previous studies [32] demonstrating that JEV NS5 expression hijacks STAT1 protein and blocks its nuclear translocation, causing loss of endogenous interferon-induced proteins.

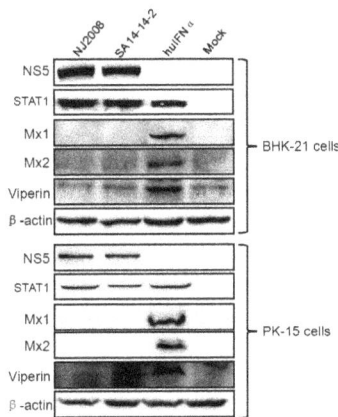

Figure 1. Japanese encephalitis virus (JEV) inhibits the production of interferon-stimulated gene (ISG) proteins. Cells were infected with JEV NJ2008 or SA14-14-2 strain at a multiplicity of infection (MOI) of 0.05. Alternatively, cells were treated with 200 ng/mL human interferon-α (huIFNα) for 12 h at 37 °C. After these treatments, ISG proteins in cell lysates were probed with the indicated antibodies using Western blot analysis.

3.2. Exogenous Mx Proteins Have No Anti-JEV Activity in Infected Cells

Previous reports have shown that exogenous Mx proteins inhibit the replication of a wide range of viruses [12]. In this study, poMx1 and poMx2 were over-expressed in BHK-21 or PK-15 cells, and

JEV replication was assessed. First, BHK-21 cells were transfected with various concentrations of pEGFP-poMx1 and infected with JEV at an MOI of 0.05. At 24 hpi, the effect of GFP-poMx1 on JEV replication was analyzed by Western blot analysis and plaque assay. Although the expression level of the GFP-poMx1 fusion protein increased with increasing construct concentration, the level of viral proteins E, NS1, and NS5 were comparable in all transfected cells, and roughly equal amounts of progeny virus were produced (Figure 2A). The Mx protein family is highly conserved, and similar results were obtained with JEV-infected cells over-expressing pEGFP-huMxA or pEGFP-mmMx1. As shown in Figure 2B by plaque assay and Western blot analysis, viral titers and NS5 protein expression were roughly equal for each construct-transfected sample. These data demonstrate that JEV replication is not inhibited by over-expression of GFP-poMx1, huMxA, or mmMx1. Similar experiments were performed to evaluate Mx2 antiviral activity. BHK-21 cells were transfected with the pEGFP-poMx2 construct and infected with JEV at MOI 0.001, 0.01, and 0.1. As shown in Figure 2C, JEV replication in pEGFP-poMx2-transfected cells was the same as that in pEGFP-C1-transfected cells, suggesting that GFP-poMx2 has no direct anti-JEV activity. Immunofluorescence assays were performed to assess JEV replication in pEGFP-poMx1- or pEGFP-poMx2-transfected cells. As shown in Figure 3, red fluorescence—indicating viral proteins—was observed in GFP-poMx1 and GFP-poMx2-positive cells (indicted by green fluorescence). These data demonstrate that exogenous porcine Mx1 and Mx2, human MxA, and mouse Mx1 proteins fused to GFP had no obvious anti-JEV activity.

To address the possible influence of GFP on the function of Mx proteins, we constructed another set of plasmids as follows: pcDNA3.0-poMx1, pcDNA3.0-poMx2, pcDNA3.0-poTMx1 and pcDNA3.0-poMx1(ΔL4). PK-15 cells were transfected with these constructs and infected with JEV at an MOI of 0.05. At 12 and 24 hpi, JEV replication was analyzed by Western blot analysis, RT-qPCR, and plaque assay. As shown in Figure 4, at 12 and 24 hpi, JEV RNA levels were roughly equal among the cells overexpressing the different isoforms of porcine Mx, suggested that none of the isoforms affected JEV replication. Plaque assay data showed that viral titers in cells overexpressing Mx proteins were the same as that in the control cells. However, as a positive control, we saw that CSFV replication was inhibited in cells overexpressing poMx1, poMx2, and huMxA, but not mmMx1. This is consistent with previous studies [17,33]. Overall, exogenous poMx1 or poMx2 had no demonstrable anti-JEV activity.

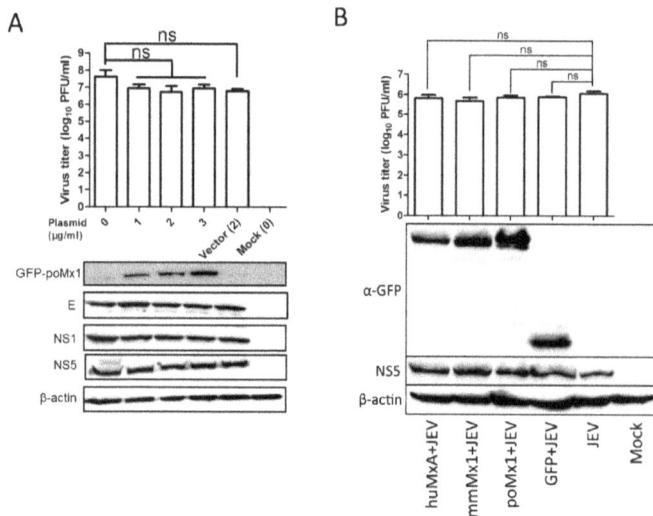

Figure 2. *Cont.*

C

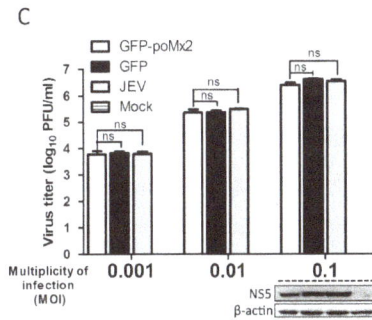

Figure 2. Exogenous Mx proteins fused to green fluorescent protein (GFP) have no anti-JEV activity. Three separate experiments were performed to assess the anti-JEV activity of the fused Mx1 and Mx2 protein. Cells were transfected with the constructs and then infected with JEV. At 24 hpi, lysates of the harvested cell culture were used to determine viral protein levels by Western blot analysis, the cell supernatants were used to determine the levels of infectious virus by plaque assay. (**A**) Cells transfected with the various concentrations of pEGFP-poMx1 and then infected with JEV at an MOI of 0.05; (**B**) Cells transfected with pEGFP-huMxA (human MxA), pEGFP-mmMx1, pEGFP-poMx1 (porcine Mx1), or pEGFP-C1 then infected with JEV at an MOI of 0.05; (**C**) Cells transfected with pEGFP-poMx2 or pEGFP-C1 and then infected with JEV at an MOI of 0.001, 0.01, and 0.1. All data are presented as means ± standard deviation (S.D.) as indicated. Statistical significance is indicated as ns ($p > 0.05$).

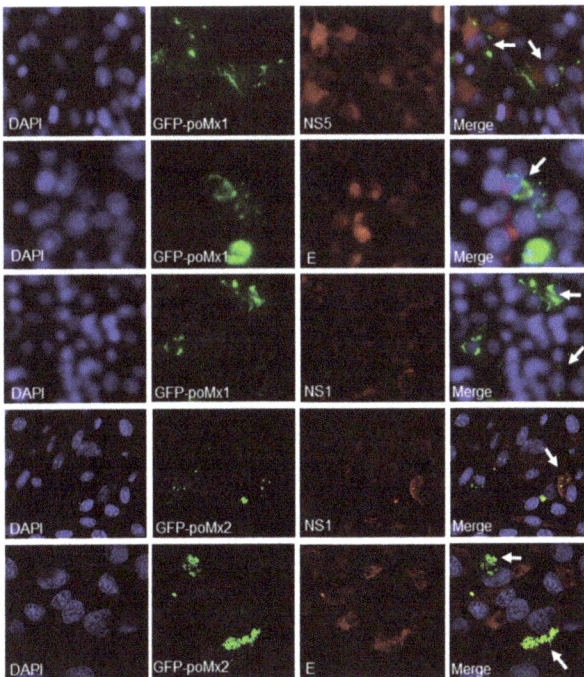

Figure 3. JEV replication in pEGFP-poMx1 or pEGFP-poMx2-transfected BHK-21 cells. Transfected cells grown on glass coverslips were infected with JEV at an MOI of 0.05. At 24 hpi, cells were washed with PBS and subjected to immunofluorescence assay. The cell nucleus was counterstained with DAPI (blue). White arrows indicate cells with high GFP-poMx1 or GFP-poMx2 expression levels along with high JEV viral proteins (NS1, NS5, or E).

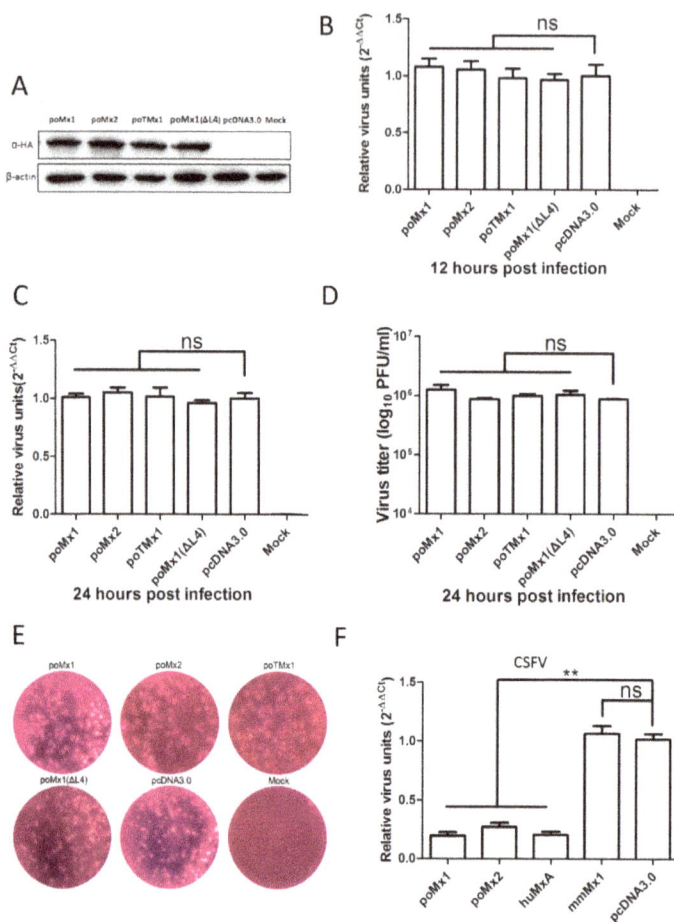

Figure 4. The different isoforms of porcine Mx proteins have no anti-JEV activity. (**A**) Western blot analysis of the lysates of PK-15 cells transfected with the indicated constructs. Overexpression of the different porcine Mx proteins was determined using anti-HA Flag mouse monoclonal antibody; (**B,C**) PK-15 cells were transfected with these constructs and infected with JEV at an MOI of 0.05. At 12 and 24 hpi, JEV replication was analyzed by RT-qPCR; (**D,E**) At 24 hpi, JEV replication was analyzed by plaque assay; (**F**) Classic swine fever virus (CSFV) replication (positive control) was inhibited by some exogenous Mx proteins. All data are presented as means ± standard deviation (S.D.) as indicated. Statistical significance is indicated as ns ($p > 0.05$) and ** ($p < 0.01$).

3.3. Mx1 or Mx2 Depletion Did Not Affect the Antiviral Activity of IFN

To determine whether Mx expression is necessary to inhibit JEV replication, endogenous Mx1 or Mx2 was knocked down in interferon-treated BHK-21 cells and PK-15 cells prior to virus infection. Cells were transfected for 6 h with the commercial Mx1, Mx2 siRNA, or negative control siRNA (siCtrl), then treated with 200 ng/mL of huIFNα for 12 h. Mx1 expression was reduced by 66% in cells transfected with Mx1 siRNA compared to the negative control (Figure 5A). Subsequently, these siRNA-transfected-huIFN-treated cells were infected with JEV at an MOI of 0.05, and viral protein expression levels and virus titers were assessed. NS5 levels were roughly equally suppressed in siCtrl- and siMx1-transfected cells and non-transfected control cells, and significantly suppressed compared to untreated controls. Likewise, virus titers in these samples were roughly equal and significantly reduced compared to untreated controls (Figure 5B). These data demonstrate that knockdown of endogenous Mx1 does not impair the antiviral ability of IFN. Mx2 expression was reduced by 79% in cells transfected with Mx2 siRNA compared to the negative control (Figure 5C), and results similar to those described above were observed in huIFNα-treated cells with Mx2 knockdown. NS5 levels were suppressed in siCtrl- and siMx2-transfected cells similar to that in huIFNα-treated cells transfected with siCtrl and siMx1, and significantly suppressed compared to untreated controls. Virus titers in these samples were roughly equal and significantly reduced compared to untreated controls (Figure 5D). These data demonstrate that knockdown of endogenous Mx2 does not impair the antiviral ability of IFN.

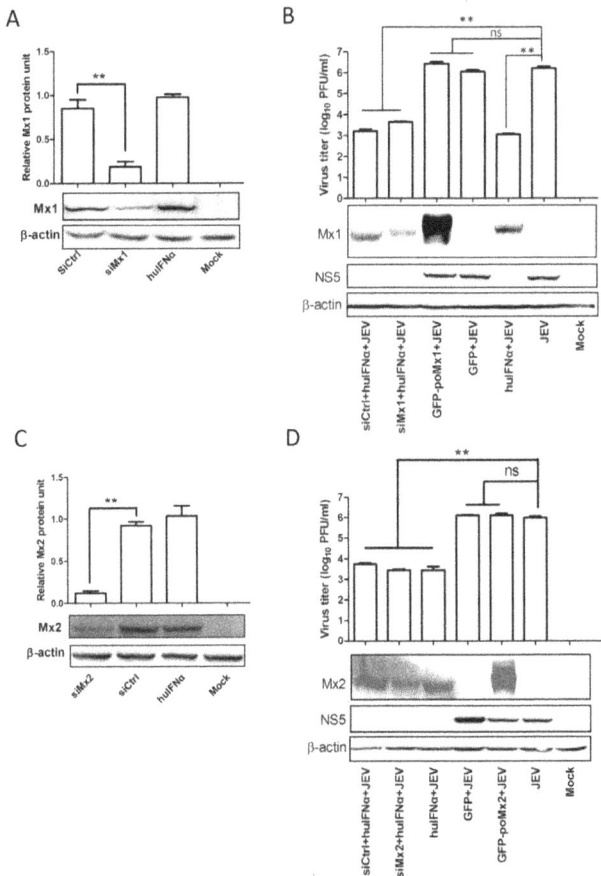

Figure 5. IFNα inhibits JEV replication in Mx-knockdown BHK-21 cells. Cells were transfected with siRNA targeting Mx1, Mx2, and a siRNA control (siCtrl). Six hours after transfection cells were treated with huIFNα for 12 h, then infected with JEV at an MOI of 0.05. At 24 hpi, cell supernatants were used to determine the levels of infectious virus by plaque assay, and the cell culture lysates were used to determine the viral protein levels by Western blot analysis. Percent knockdown of (**A**) Mx1 and (**C**) Mx2. JEV replication determined by plaque assay and Western blot analysis in (**B**) siMx1- or (**D**) siMx2-transfected and control cells. Quantification of the blotted proteins was performed using Image J software. All data are presented as means ± standard deviation (S.D.) as indicated. Statistical significance is indicated as ns ($p > 0.05$) and ** ($p < 0.01$).

Similar siRNA experiments were performed in PK-15 cells. After the successful knockdown of endogenous porcine Mx1 or Mx2 as described above, the levels of NS5 (Figure 6A) and virus titers (Figure 6B) were significantly reduced in huIFNα-treated cells compared to the controls, which suggests that IFN can effectively inhibit JEV replication by an Mx-independent pathway.

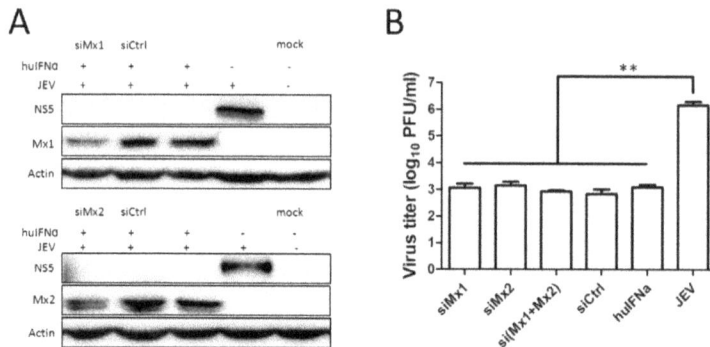

Figure 6. IFNα inhibits JEV replication in Mx-knockdown PK-15 cells. Mx1 and Mx2 knockdown in PK-15 cells was performed as described above. At 24 hpi, lysates of cell culture were used to determine the viral protein levels by Western blot analysis (**A**), and cell supernatants were used to determine the levels of infectious virus by plaque assay (**B**). All data are presented as means ± standard deviation (S.D.) as indicated. Statistical significance is indicated as ** ($p < 0.01$).

3.4. Mx Does Not Inhibit JEV Replication in Cells Treated with BFA

BFA is a Golgi apparatus-disrupting agent which prevents the development of virus-induced membranes when added before the end of the latent period of *Flavivirus* infection [34]. A previous report showed that West Nile Virus (WNV) replication was significantly reduced in BFA-treated Vero cells overexpressing huMxA, compared to BFA-treated Vero cells, suggesting that WNV-induced membranes may provide partial protection against huMxA [28]. Here, we performed a series of experiments to examine whether a similar mechanism exists in JEV infection. Initially, we confirmed that BFA treatment disrupted the Golgi apparatus by visualizing giantin (Golgi marker) in BFA-treated and untreated cells using the anti-giantin antibody. Subsequently, GFP-poMx1 (or poMx2)-overexpressing BHK-21 cells and mock-transfected cells were infected with JEV at an MOI of 0.05. At 12 hpi, BFA (5 μg/mL) was added to the culture medium for an additional 12 h, as described previously. As shown in Figure 7A,C, in the absence of BFA, GFP-poMx1- (or poMx2)-overexpressing cells, and mock-transfected cells showed comparable levels of NS5 expression, as expected. However, in the presence of BFA, NS5 and E expression was reduced significantly, with no statistical difference between GFP-poMx1- (or poMx2) transfected cells and mock-transfected cells. Virus titers were also decreased significantly (Figure 7C,D). These data indicated that BFA—not Mx—inhibits JFV replication. Collectively, these results demonstrated that poMx1 or poMx2 inhibit JEV replication in the presence of BFA, suggesting that JEV is resistant to IFN-induced Mx protein via an unknown pathway.

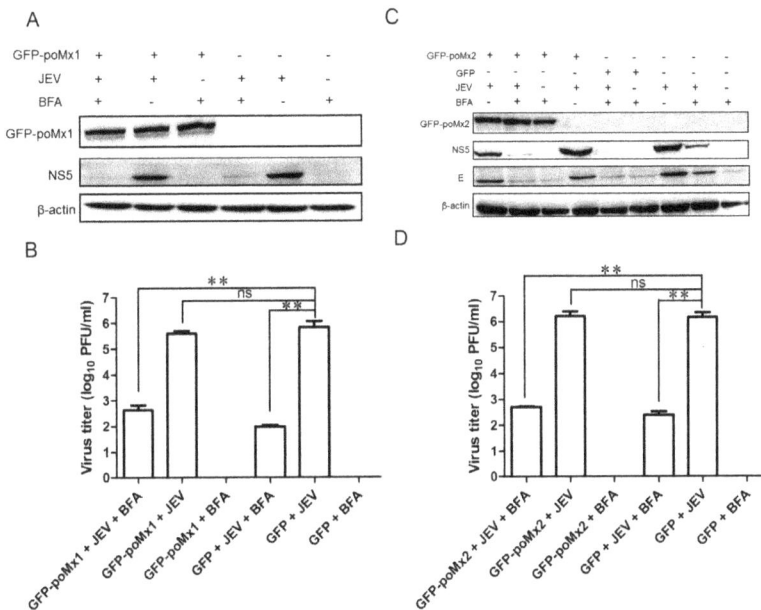

Figure 7. Mx1 or Mx2 overexpression has no impact on JEV replication in cells treated with Brefeldin A (BFA). GFP-poMx1 (or poMx2)-overexpressing BHK-21 cells and control cells were infected with JEV at an MOI of 0.05. At 12 hpi, BFA (5 µg/mL) was added to the culture medium for an additional 12 h. The lysates of cell culture were used to determine the viral protein levels by Western blot analysis, and cell supernatants were used to determine the levels of infectious virus by plaque assay. Western blot analysis (**A**) and virus titer (**B**) of infectious virus of GFP-poMx1-overexpressing infected cells treated with or without BFA. Alternatively, Western blot analysis (**C**) and virus titer (**D**) of infectious virus of the GFP-poMx2-overexpressing infected cells treated with or without BFA. All data are presented as means ± standard deviation (S.D.) as indicated. Statistical significance is indicated as ns ($p > 0.05$) and ** ($p < 0.01$).

3.5. BFA Effectively Inhibits JEV Replication

To further explore the antiviral activity of BFA, we performed a series of experiments as follows. Cells were treated with 0 to 15 µg/mL BFA for 24 h, and the cytotoxic effect was evaluated to ensure the sub-toxic doses of BFA. As shown in Figure 8A, cells tolerated up to 10 µg/mL BFA. The cell viability was reduced only slightly in the presence of 15 µg/mL BFA. To test the effects of BFA on JEV production, cells were infected with JEV and then treated with various concentrations of BFA. At 24 h post treatment, the viral protein levels of E, NS1, and NS5 in BFA-treated cells were significantly reduced compared to that in DMSO-treated cells, which is consistent with the decreasing of virus titer. The results showed that virus titer was reduced by about 209-fold, suggesting that BFA up to 2.5 µg/mL inhibits JEV replication in a dose-independent manner (Figure 8B). To test whether BFA inhibits JEV replication in a time-dependent manner, cells were infected with JEV at an MOI of 0.05 and then treated with 5 µg/mL BFA. At 13, 15, 18, and 24 hpi, the viral protein levels in lysed cells were determined by Western blot analysis, and the amount of infectious virus in cell supernatants was determined by plaque assay. The results showed that NS5 protein level in the BFA-treated cells reduced significantly by 18 hpi when compared to untreated cells (Figure 8C). Plaque numbers were in accord with the results above, and virus titers at 18 and 24 hpi were reduced by 20-fold and 8175-fold, respectively (Figure 8D), suggesting that BFA strongly inhibits JEV replication in a time-dependent manner. The inhibitory effect of BFA was analyzed by immunofluorescence assay. The results showed that viral protein levels were significantly decreased at 24 hpi (as indicated by red fluorescence) compared to untreated cells (Figure 8E), suggesting that JEV replication was strongly inhibited by BFA.

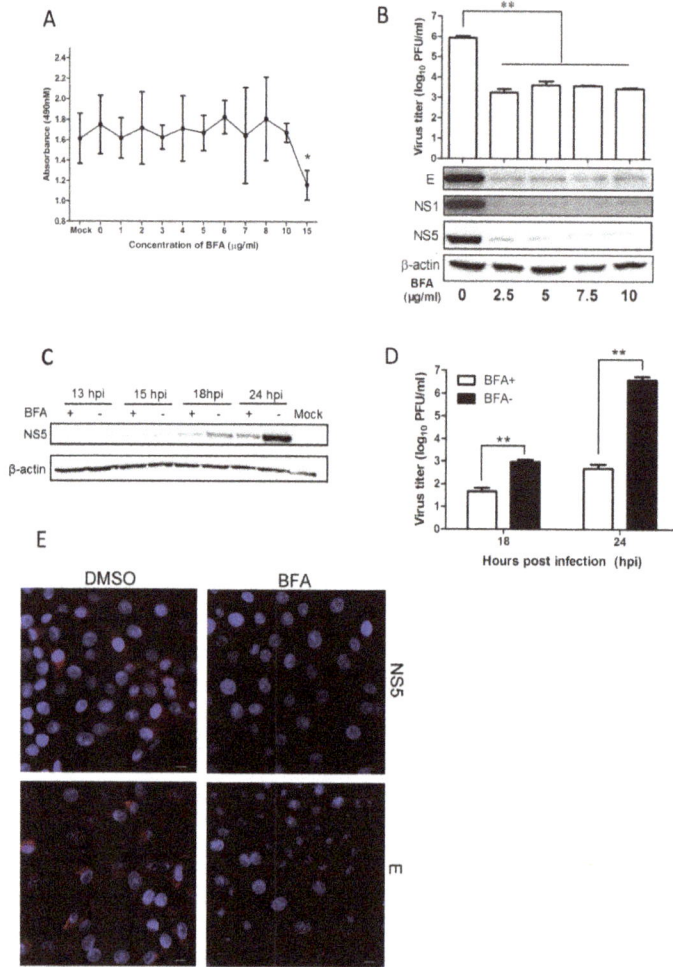

Figure 8. Antiviral activity of BFA. (**A**) Cytotoxic effect of BFA. BHK-21 cells at 80% confluence in 24-well plates were treated with various concentrations of BFA for 24 h. After treatment, a cell proliferation reagent was added to each well, and 2 h later, the absorbance at 490 nm was recorded. (**B–E**) BHK-21 cells were seeded into six-well plates and infected with JEV at an MOI of 0.05. After virus adsorption and washing, cells were maintained in medium containing BFA at various concentrations or an equivalent volume of DMSO. At 24 hpi, cell culture lysates were used to determine viral protein levels by Western blot analysis, cell supernatants were used to determine levels of infectious virus by plaque assay (**B**); Infected cells were maintained in medium containing BFA at 5 μg/mL or an equivalent volume of DMSO. At 13, 15, 18, and 24 hpi, cell culture lysates were used to determine viral protein levels by Western blot analysis (**C**); Cell supernatants were used to determine levels of infectious virus by plaque assay (**D**); The inhibitory effect of BFA was detected using anti-JEV NS5 or E mAbs by confocal microscopy (**E**). JEV was strained with red fluorescence, and nucleus was strained with DAPI. All data are presented as means ± standard deviation (S.D.) as indicated. Statistical significance is indicated as ** ($p < 0.01$).

4. Discussion

Type I interferon (IFN) is abundantly produced in virus-infected cells soon after infection, as well as a myriad additional virus-initiated modulatory effects, including induction of cellular inhibitors or repressors of transcription, and activation of IFN-I stimulated genes (ISG) and proteins, all in order to antagonize an antiviral host response [35]. Thus, IFNs have been used as antiviral agents in the treatment of several pathogens, including Flaviviridae [36]. Previous reports have shown that IFNα has activity against JEV in PK-15 cells [7,8] and BHK-21 cells [37]. In addition, IFNα and IFNγ can both efficiently prevent WNV infection, though IFNα demonstrated the greater antiviral efficacy [38]. Furthermore, the ISGs induced by interferons inhibit WNV and DENV replication at different stages through different mechanisms [39–41]. Although type I IFNs—IFN-α/β—are important innate immune regulators for resisting viral infections, it has been demonstrated that flaviviruses produce effective immune modulatory proteins and utilize multiple immune evasion mechanisms that limit host immune responses and advance viral replication. Previous reports have shown that JEV and DEV NS5 protein is an IFN antagonist and that it may play a role in blocking IFN-stimulated JAK-STAT signaling via activation of PTPs during JEV infection, resulting in suppression of the expression of a wide variety of ISGs which can establish antiviral, anti-proliferative, and/or immune-regulatory states in host cells [32,42]. In this study, BHK-21 cells and PK-15 cells were infected with JEV virulent strain NJ2008 or attenuated vaccine strain SA14-14-2. Figure 1 showed that JEV-infected cells did not produce endogenous Mx1, Mx2, or Viperin above background levels, suggesting that the production of IFN-induced ISGs were suppressed by JEV. Our findings are consistent with the recent report that a low-level induction of IFNα mRNA expression was observed after JEV infection [8].

ISGs are the key antiviral factors in the tug-of-war between interferons and JEV, and the different ISGs have been demonstrated to be induced by different stimuli [43]. To date, previous reports have shown that ISG15, Viperin, and OAS have anti-JEV activities in the respective manners. Overexpression of ISG15 significantly reduced the JEV-induced cytopathic effect and inhibited JEV replication by activating the expression of STAT1-dependent genes including IRF-3, IFN-β, IL-8, PKR, and OAS before and post-JEV infection [44]. In addition, although the antiviral activities of porcine OAS1, OAS2, and OSAL against JEV were demonstrated in PK-15 cells, their antiviral mechanisms need further investigation [8]. Overexpression of Viperin significantly decreased the production of JEV in the presence of the proteasome inhibitor MG132 that sustained Viperin levels [45]. Mx proteins—the IFN-induced GTPase—are key components of the antiviral state induced by interferons in many species. Our previous work showed that overexpression of porcine Mx1 could inhibit CSFV replication in vitro and in vivo [13,17], as well as VSV replication [31]. Here, we explore the role of Mx proteins during IFNα-inhibited JEV infection. Unexpectedly, even though IFNα effectively blocks JEV infection, Mx proteins play no apparent role inhibiting JEV replication. Overexpression of Mx isoforms including porcine Mx1 and Mx2, human Mx1, and mouse Mx1 did not inhibit JEV replication as determined by Western blot analysis, plaque assay, and immunofluorescence assay. Furthermore, we found that huIFNα still inhibited JEV replication where Mx1 or Mx2 was knocked down using RNA interference, suggesting that Mx is not a critical factor in the pathway whereby IFN inhibits JEV infection. Previous studies of the overexpression of MxA in Vero cells showed WNV$_{KUN}$ replication, maturation, and secretion was uninhibited [28]. However, retargeting MxA expression from cytoplasmic inclusions to the endoplasmic reticulum during WNV$_{KUN}$ replication did significantly hamper the formation and spread of infectious WNV$_{KUN}$ virions [46]. We speculated that JEV may be resistant to Mx protein through a similar mechanism.

BFA has been widely used to study membrane trafficking and protein processing in eukaryotic cells [47,48]. After the treatment of mammalian cells with BFA, ER to Golgi transport is rapidly inhibited. Previous reports have shown that BFA acts in a variety of ways as an antiviral, such as arresting the maturation and egress of herpes simplex virus particles during infection [49] and inhibiting *Pestivirus* release from infected cells without affecting its assembly and infectivity [50]. In addition, short-term (1 h) BFA treatment inhibits VSV gene expression, while long-term (12 h) treatment blocks VSV entry [51]. BFA completely inhibits poliovirus RNA synthesis by preventing the formation of secretory vesicles [52,53]. This is the first report that BFA harbors anti-JEV activity. We found that BFA treatment at low concentration effectively inhibits JEV proliferation in a dose-independent and time-dependent manner. Because overexpression of human MxA inhibits WNV replication in the presence of BFA [28], we hypothesized that BFA targets a similar mechanism in JEV-infected cells. Unexpectedly, we found overexpression of porcine Mx1 or Mx2 did not inhibit JEV replication in the presence of BFA, clearly, understanding its antiviral mechanisms against JEV at the cellular level needs further study. Our data suggests that it would be feasible to develop BFA as a potential reagent against JEV infection.

Taken together, the data from overexpression and knockdown of Mx proteins have indicated that IFNα inhibits JEV replication by Mx-independent pathway. Moreover, JEV-induced membranes did not provide any protection against Mx protein in the presence of BFA. That BFA can inhibit JEV replication suggests that BFA could be developed into an antiviral reagent.

Acknowledgments: This work was supported by a grant from the National Natural Science Foundation of China (31572554), Agro-Scientific Research in the Public Interest (201203082), and by the Priority Academic Program Development of Jiangsu Higher Education Institutions (PAPD). Also, we thank Elizabeth Wills from Cornell University and Kui Yang from Louisiana State University for critical reading and editing of the manuscript.

Author Contributions: B.Z. and X.-M.Z. designed the experiments. J.Z., S.-Q.W. and Z.-C.G. performed experiments. B.Z., Z.-Y.M., and P.-Y.C. analyzed data. K.L. and J.-C.W. contributed reagents and materials. B.Z. wrote the paper.

Conflicts of Interest: The authors declare no conflict of interest.

References

1. Van den Hurk, A.F.; Ritchie, S.A.; Mackenzie, J.S. Ecology and geographical expansion of Japanese encephalitis virus. *Annu. Rev. Entomol.* **2009**, *54*, 17–35. [CrossRef] [PubMed]
2. Wang, H.Y.; Takasaki, T.; Fu, S.H.; Sun, X.H.; Zhang, H.L.; Wang, Z.X.; Hao, Z.Y.; Zhang, J.K.; Tang, Q.; Kotaki, A.; et al. Molecular epidemiological analysis of Japanese encephalitis virus in China. *J. Gen. Virol.* **2007**, *88*, 885–894. [CrossRef] [PubMed]
3. Pang, R.; He, D.N.; Zhou, B.; Liu, K.; Zhao, J.; Zhang, X.M.; Chen, P.Y. In vitro inhibition of Japanese encephalitis virus replication by capsid-targeted virus inactivation. *Antivir. Res.* **2013**, *97*, 369–375. [CrossRef] [PubMed]
4. Verma, R. Japanese encephalitis vaccine: Need of the hour in endemic states of India. *Hum. Vaccines Immunother.* **2012**, *8*, 491–493. [CrossRef] [PubMed]
5. Doughty, L.; Nguyen, K.; Durbin, J.; Biron, C. A role for IFN-alpha beta in virus infection-induced sensitization to endotoxin. *J. Immunol.* **2001**, *166*, 2658–2664. [CrossRef] [PubMed]
6. Takeuchi, O.; Akira, S. Innate immunity to virus infection. *Immunol. Rev.* **2009**, *227*, 75–86. [CrossRef] [PubMed]
7. Liu, K.; Liao, X.; Zhou, B.; Yao, H.; Fan, S.; Chen, P.; Miao, D. Porcine alpha interferon inhibit Japanese encephalitis virus replication by different ISGs in vitro. *Res. Vet. Sci.* **2013**, *95*, 950–956. [CrossRef] [PubMed]
8. Zheng, S.; Zhu, D.; Lian, X.; Liu, W.; Cao, R.; Chen, P. Porcine 2′,5′-oligoadenylate synthetases inhibit Japanese encephalitis virus replication in vitro. *J. Med. Virol.* **2016**, *88*, 760–768. [CrossRef] [PubMed]
9. Charleston, B.; Stewart, H.J. An interferon-induced Mx protein: cDNA sequence and high-level expression in the endometrium of pregnant sheep. *Gene* **1993**, *137*, 327–331. [CrossRef]
10. Haller, O.; Kochs, G. Interferon-induced Mx proteins: Dynamin-like GTPases with antiviral activity. *Traffic* **2002**, *3*, 710–717. [CrossRef] [PubMed]
11. Haller, O.; Stertz, S.; Kochs, G. The Mx GTPase family of interferon-induced antiviral proteins. *Microbes Infect.* **2007**, *9*, 1636–1643. [CrossRef] [PubMed]
12. Verhelst, J.; Hulpiau, P.; Saelens, X. Mx proteins: Antiviral gatekeepers that restrain the uninvited. *Microbiol. Mol. Biol. Rev.* **2013**, *77*, 551–566. [CrossRef] [PubMed]
13. Zhang, X.; Jing, J.; Li, W.; Liu, K.; Shi, B.; Xu, Q.; Ma, Z.; Zhou, B.; Chen, P. Porcine Mx1 fused to HIV Tat protein transduction domain (PTD) inhibits classical swine fever virus infection in vitro and in vivo. *BMC Vet. Res.* **2015**, *11*. [CrossRef] [PubMed]

14. Asano, A.; Ko, J.H.; Morozumi, T.; Hamashima, N.; Watanabe, T. Polymorphisms and the antiviral property of porcine Mx1 protein. *J. Vet. Med. Sci.* **2002**, *64*, 1085–1089. [CrossRef] [PubMed]

15. Nakajima, E.; Morozumi, T.; Tsukamoto, K.; Watanabe, T.; Plastow, G.; Mitsuhashi, T. A naturally occurring variant of porcine Mx1 associated with increased susceptibility to influenza virus in vitro. *Biochem. Genet.* **2007**, *45*, 11–24. [CrossRef] [PubMed]

16. Palm, M.; Leroy, M.; Thomas, A.; Linden, A.; Desmecht, D. Differential anti-influenza activity among allelic variants at the *Sus scrofa Mx1* locus. *J. Interferon Cytokine Res.* **2007**, *27*, 147–155. [CrossRef] [PubMed]

17. He, D.N.; Zhang, X.M.; Liu, K.; Pang, R.; Zhao, J.; Zhou, B.; Chen, P.Y. In vitro inhibition of the replication of classical swine fever virus by porcine Mx1 protein. *Antivir. Res.* **2014**, *104*, 128–135. [CrossRef] [PubMed]

18. Yuan, B.; Fang, H.; Shen, C.; Zheng, C. Expression of porcine Mx1 with FMDV IRES enhances the antiviral activity against foot-and-mouth disease virus in PK-15 cells. *Arch. Virol.* **2015**, *160*, 1989–1999. [CrossRef] [PubMed]

19. Shi, H.; Fu, Q.; Ren, Y.; Wang, D.; Qiao, J.; Wang, P.; Zhang, H.; Chen, C. Both foot-and-mouth disease virus and bovine viral diarrhea virus replication are inhibited by Mx1 protein originated from porcine. *Anim. Biotechnol.* **2015**, *26*, 73–79. [CrossRef] [PubMed]

20. Patzina, C.; Haller, O.; Kochs, G. Structural requirements for the antiviral activity of the human MxA protein against Thogoto and influenza A virus. *J. Biol. Chem.* **2014**, *289*, 6020–6027. [CrossRef] [PubMed]

21. Haller, O.; Gao, S.; von der Malsburg, A.; Daumke, O.; Kochs, G. Dynamin-like MxA GTPase: Structural insights into oligomerization and implications for antiviral activity. *J. Biol. Chem.* **2010**, *285*, 28419–28424. [CrossRef] [PubMed]

22. Goujon, C.; Moncorge, O.; Bauby, H.; Doyle, T.; Ward, C.C.; Schaller, T.; Hue, S.; Barclay, W.S.; Schulz, R.; Malim, M.H. Human Mx2 is an interferon-induced post-entry inhibitor of HIV-1 infection. *Nature* **2013**, *502*, 559–562. [CrossRef] [PubMed]

23. Kane, M.; Yadav, S.S.; Bitzegeio, J.; Kutluay, S.B.; Zang, T.; Wilson, S.J.; Schoggins, J.W.; Rice, C.M.; Yamashita, M.; Hatziioannou, T.; et al. Mx2 is an interferon-induced inhibitor of HIV-1 infection. *Nature* **2013**, *502*, 563–566. [CrossRef] [PubMed]

24. Zhang, Y.; Chen, P.; Cao, R.; Gu, J. Mutation of putative N-linked glycosylation sites in Japanese encephalitis virus premembrane and envelope proteins enhances humoral immunity in BALB/C mice after DNA vaccination. *Virol. J.* **2011**, *8*. [CrossRef] [PubMed]

25. Wu, S.F.; Lee, C.J.; Liao, C.L.; Dwek, R.A.; Zitzmann, N.; Lin, Y.L. Antiviral effects of an iminosugar derivative on flavivirus infections. *J. Virol.* **2002**, *76*, 3596–3604. [CrossRef] [PubMed]

26. Ponten, A.; Sick, C.; Weeber, M.; Haller, O.; Kochs, G. Dominant-negative mutants of human MxA protein: Domains in the carboxy-terminal moiety are important for oligomerization and antiviral activity. *J. Virol.* **1997**, *71*, 2591–2599. [PubMed]

27. Von der Malsburg, A.; Abutbul-Ionita, I.; Haller, O.; Kochs, G.; Danino, D. Stalk domain of the dynamin-like MxA GTPase protein mediates membrane binding and liposome tubulation via the unstructured L4 loop. *J. Biol. Chem.* **2011**, *286*, 37858–37865. [CrossRef] [PubMed]

28. Hoenen, A.; Liu, W.; Kochs, G.; Khromykh, A.A.; Mackenzie, J.M. West Nile virus-induced cytoplasmic membrane structures provide partial protection against the interferon-induced antiviral MxA protein. *J. Gen. Virol.* **2007**, *88*, 3013–3017. [CrossRef] [PubMed]

29. Zhou, B.; Yang, K.; Wills, E.; Tang, L.; Baines, J.D. A mutation in the DNA polymerase accessory factor of herpes simplex virus 1 restores viral DNA replication in the presence of raltegravir. *J. Virol.* **2014**, *88*, 11121–11129. [CrossRef] [PubMed]

30. Wang, H.; Bai, J.; Fan, B.; Li, Y.; Zhang, Q.; Jiang, P. The Interferon-Induced Mx2 inhibits porcine reproductive and respiratory syndrome virus replication. *J. Interferon Cytokine Res.* **2016**, *36*, 129–139. [CrossRef] [PubMed]

31. Zhang, X.M.; He, D.N.; Zhou, B.; Pang, R.; Liu, K.; Zhao, J.; Chen, P.Y. In vitro inhibition of vesicular stomatitis virus replication by purified porcine Mx1 protein fused to HIV-1 Tat protein transduction domain (PTD). *Antivir. Res.* **2013**, *99*, 149–157. [CrossRef] [PubMed]

32. Lin, R.J.; Chang, B.L.; Yu, H.P.; Liao, C.L.; Lin, Y.L. Blocking of interferon-induced JAK-STAT signaling by Japanese encephalitis virus NS5 through a protein tyrosine phosphatase-mediated mechanism. *J. Virol.* **2006**, *80*, 5908–5918. [CrossRef] [PubMed]

33. Zhao, Y.; Pang, D.; Wang, T.; Yang, X.; Wu, R.; Ren, L.; Yuan, T.; Huang, Y.; Ouyang, H. Human MxA protein inhibits the replication of classical swine fever virus. *Virus Res.* **2011**, *156*, 151–155. [CrossRef] [PubMed]

34. Mackenzie, J.M.; Jones, M.K.; Westaway, E.G. Markers for trans-Golgi membranes and the intermediate compartment localize to induced membranes with distinct replication functions in flavivirus-infected cells. *J. Virol.* **1999**, *73*, 9555–9567. [PubMed]

35. Theofilopoulos, A.N.; Baccala, R.; Beutler, B.; Kono, D.H. Type I interferons (alpha/beta) in immunity and autoimmunity. *Annu. Rev. Immunol.* **2005**, *23*, 307–336. [CrossRef] [PubMed]

36. Reid, E.; Juleff, N.; Windsor, M.; Gubbins, S.; Roberts, L. Type I and III IFNs produced by plasmacytoid dendritic cells in response to a member of the Flaviviridae suppress cellular immune responses. *J. Immunol.* **2016**, *196*, 4214–4226. [CrossRef] [PubMed]

37. Lin, Y.L.; Huang, Y.L.; Ma, S.H.; Yeh, C.T.; Chiou, S.Y.; Chen, L.K.; Liao, C.L. Inhibition of Japanese encephalitis virus infection by nitric oxide: Antiviral effect of nitric oxide on RNA virus replication. *J. Virol.* **1997**, *71*, 5227–5235. [PubMed]

38. Ma, D.; Jiang, D.; Qing, M.; Weidner, J.M.; Qu, X.; Guo, H.; Chang, J.; Gu, B.; Shi, P.Y.; Block, T.M.; et al. Antiviral effect of interferon lambda against West Nile virus. *Antivir. Res.* **2009**, *83*, 53–60. [CrossRef] [PubMed]

39. Lazear, H.M.; Pinto, A.K.; Vogt, M.R.; Gale, M., Jr.; Diamond, M.S. Beta interferon controls West Nile virus infection and pathogenesis in mice. *J. Virol.* **2011**, *85*, 7186–7194. [CrossRef] [PubMed]

40. Pulit-Penaloza, J.A.; Scherbik, S.V.; Brinton, M.A. Type 1 IFN-independent activation of a subset of interferon stimulated genes in West Nile virus Eg101-infected mouse cells. *Virology* **2012**, *425*, 82–94. [CrossRef] [PubMed]

41. Samuel, M.A.; Diamond, M.S. Alpha/beta interferon protects against lethal West Nile virus infection by restricting cellular tropism and enhancing neuronal survival. *J. Virol.* **2005**, *79*, 13350–13361. [CrossRef] [PubMed]

42. Ashour, J.; Laurent-Rolle, M.; Shi, P.Y.; Garcia-Sastre, A. NS5 of dengue virus mediates STAT2 binding and degradation. *J. Virol.* **2009**, *83*, 5408–5418. [CrossRef] [PubMed]

43. Sadler, A.J.; Williams, B.R. Interferon-inducible antiviral effectors. *Nat. Rev. Immunol.* **2008**, *8*, 559–568. [CrossRef] [PubMed]

44. Hsiao, N.W.; Chen, J.W.; Yang, T.C.; Orloff, G.M.; Wu, Y.Y.; Lai, C.H.; Lan, Y.C.; Lin, C.W. ISG15 over-expression inhibits replication of the Japanese encephalitis virus in human medulloblastoma cells. *Antivir. Res.* **2010**, *85*, 504–511. [CrossRef] [PubMed]

45. Chan, Y.L.; Chang, T.H.; Liao, C.L.; Lin, Y.L. The cellular antiviral protein viperin is attenuated by proteasome-mediated protein degradation in Japanese encephalitis virus-infected cells. *J. Virol.* **2008**, *82*, 10455–10464. [CrossRef] [PubMed]

46. Hoenen, A.; Gillespie, L.; Morgan, G.; van der Heide, P.; Khromykh, A.; Mackenzie, J. The West Nile virus assembly process evades the conserved antiviral mechanism of the interferon-induced MxA protein. *Virology* **2014**, *448*, 104–116. [CrossRef] [PubMed]

47. Nuchtern, J.G.; Bonifacino, J.S.; Biddison, W.E.; Klausner, R.D. Brefeldin A implicates egress from endoplasmic reticulum in class I restricted antigen presentation. *Nature* **1989**, *339*, 223–226. [CrossRef] [PubMed]

48. Yewdell, J.W.; Bennink, J.R. Brefeldin A specifically inhibits presentation of protein antigens to cytotoxic T lymphocytes. *Science* **1989**, *244*, 1072–1075. [CrossRef] [PubMed]

49. Cheung, P.; Banfield, B.W.; Tufaro, F. Brefeldin A arrests the maturation and egress of herpes simplex virus particles during infection. *J. Virol.* **1991**, *65*, 1893–1904. [PubMed]

50. Macovei, A.; Zitzmann, N.; Lazar, C.; Dwek, R.A.; Branza-Nichita, N. Brefeldin A inhibits pestivirus release from infected cells, without affecting its assembly and infectivity. *Biochem. Biophys. Res. Commun.* **2006**, *346*, 1083–1090. [CrossRef] [PubMed]

51. Cureton, D.K.; Burdeinick-Kerr, R.; Whelan, S.P. Genetic inactivation of COPI coatomer separately inhibits vesicular stomatitis virus entry and gene expression. *J. Virol.* **2012**, *86*, 655–666. [CrossRef] [PubMed]

52. Cuconati, A.; Molla, A.; Wimmer, E. Brefeldin A inhibits cell-free, de novo synthesis of poliovirus. *J. Virol.* **1998**, *72*, 6456–6464. [PubMed]

53. Maynell, L.A.; Kirkegaard, K.; Klymkowsky, M.W. Inhibition of poliovirus RNA synthesis by brefeldin A. *J. Virol.* **1992**, *66*, 1985–1994. [PubMed]

viruses

MDPI

Article

Equine Immunoglobulin and Equine Neutralizing F(ab′)₂ Protect Mice from West Nile Virus Infection

Jiannan Cui [1,2,†], Yongkun Zhao [2,3,†], Hualei Wang [2,3,†], Boning Qiu [4], Zengguo Cao [2], Qian Li [2], Yanbo Zhang [2], Feihu Yan [2], Hongli Jin [4], Tiecheng Wang [2,3], Weiyang Sun [2,3], Na Feng [2,3], Yuwei Gao [2,3], Jing Sun [5], Yanqun Wang [5], Stanley Perlman [6], Jincun Zhao [5,*], Songtao Yang [2,3,*] and Xianzhu Xia [2,3,*]

1   College of Veterinary Medicine, Northeast Agricultural University, Harbin 150030, China; cjn052015@163.com
2   Institute of Military Veterinary, Academy of Military Medical Sciences, Changchun 130122, China; zhaoyongkun1976@126.com (Y.Z.); wangh25@hotmail.com (H.W.); CC_xiaocao9ye@163.com (Z.C.); linglong662222@163.com (Q.L.); m13894809942@163.com (Y.Z.); yanfeihu62041@126.com (F.Y.); wgcha@163.com (T.W.); sunweiyang1987@163.com (W.S.); fengna0308@126.com (N.F.); gaoyuwei@gmail.com (Y.G.)
3   Jiangsu Co-innovation Center for Prevention and Control of Important Animal Infectious Disease and Zoonoses, Yangzhou 225009, China
4   College of Veterinary Medicine, Jilin University, Changchun 130062, China; zerg1230@163.com (B.Q.); jin8616771@163.com (H.J.)
5   State Key Laboratory of Respiratory Diseases, Guangzhou Institute of Respiratory Disease, The First Affiliated Hospital of Guangzhou Medical University, Guangzhou 510120, China; sj-ji@163.com (J.S.); wangyanqun2008@163.com (Y.W.)
6   Department of Microbiology, University of Iowa, Iowa City, IA 52242, USA; stanley-perlman@uiowa.edu
*   Correspondence: zhaojincun@gird.cn (J.Z.); yst62041@163.com (S.Y.); xiaxzh@cae.cn (X.X.); Tel.: +86-20-8306-2867 (J.Z.); +86-431-8698-5928 (S.Y.); +86-431-8698-5808 (X.X.); Fax: +86-431-8698-5869 (S.Y.); +86-431-8698-5888 (X.X.)
†   These authors contributed equally to this paper.

Academic Editor: Michael Holbrook
Received: 8 October 2016; Accepted: 13 December 2016; Published: 18 December 2016

Abstract: West Nile virus (WNV) is prevalent in Africa, Europe, the Middle East, West Asia, and North America, and causes epidemic encephalitis. To date, no effective therapy for WNV infection has been developed; therefore, there is urgent need to find an efficient method to prevent WNV disease. In this study, we prepared and evaluated the protective efficacy of immune serum IgG and pepsin-digested F(ab′)₂ fragments from horses immunized with the WNV virus-like particles (VLP) expressing the WNV M and E proteins. Immune equine F(ab′)₂ fragments and immune horse sera efficiently neutralized WNV infection in tissue culture. The passive transfer of equine immune antibodies significantly accelerated the virus clearance in the spleens and brains of WNV infected mice, and reduced mortality. Thus, equine immunoglobulin or equine neutralizing F(ab′)₂ passive immunotherapy is a potential strategy for the prophylactic or therapeutic treatment of patients infected with WNV.

Keywords: West Nile virus; equine immunoglobulin; F(ab′)₂ fragments; mice

1. Introduction

In recent years, the incidences of West Nile virus (WNV) infection from Culex to humans have increased greatly due to global travel, which raises the public health concern that the 1999–2002 WNV pandemic in North America could recur [1,2]. Some vaccines and treatments have been developed, including imino sugar derivatives [3], DNA vaccines [4], interferon treatment, and human monoclonal antibodies [5]. The clinical evaluation of hE16 and other mAb therapeutics for flavivirus infections should monitor for the selection or emergence of resistant variants, especially in immunocompromised patients who are a likely target population for the treatment of severe WNV infections [6]. The use of a large dose of ribavirin inhibited WNV replication in vitro [7]; however, it was less efficient in vivo—the mortality of the hamsters infected with WNV increased after the ribavirin injection [8]. The WNV RNA replication process is inhibited in vitro by mycophenolate (another candidate inhibitor) [9], which showed no protective effect in vivo. Another clinical study indicated that interferon alpha (IFN-α) could be protective against WNV infection [10]; however, only limited evidence has shown that IFN-α could cure patients with serious WNV encephalitis.

Several WNV vaccines have been recently developed, including inactivated, subunit, and live attenuated vaccines [11]. However, vaccine immunizations might not protect patients who have already developed WNV neuroinvasion. As of now, there are no licensed vaccines or therapeutics for human use. In the study of Engle and Diamond, they found that immune mouse antibodies cannot be directly used in humans due to the heterogeneity of intact mouse IgG, which has the risk of inducing an allergic response in humans; human IgGs are not readily available in large quantities [12]. Passive immunotherapy has been proven useful for many infectious diseases. The administration of polyclonal immunoglobulins from hyperimmune animals or humans has controlled hepatitis B virus (HBV) [13] and human immunodeficiency virus (HIV) [14]. The equine antibodies—especially the F(ab')2 fragments with reduced heterogeneity—are polyclonal, easy to purify in large quantities, and relatively inexpensive and successfully used to prevent Rabies virus infection [15] and snake venom in humans [16,17]. They can serve as a useful preparedness for WNV pandemic.

Here, we prepared serum IgG from horses immunized with WNV virus-like particles (VLP). To avoid potential allergic responses, pepsin was used to digest the antibodies and to generate the F(ab')2 fragments. The protective efficacy of the serum IgG and F(ab')2 fragments against WNV was evaluated in vitro and in vivo. The serum IgG and F(ab')2 fragments neutralized WNV infection in tissue culture, as determined by a plaque reduction assay. Furthermore, prophylactic and therapeutic treatment of mice with purified IgG or F(ab')2 fragments significantly decreased the viral loads in the spleens and brains of WNV-infected mice, and reduced mortality. These results indicate that equine immunoglobulin and equine neutralizing F(ab')2 passive immunotherapy could be potentially useful for the treatment of patients infected with WNV.

2. Materials and Methods

2.1. Antigen Preparation

Sf9 cells were infected with recombinant baculoviruses co-expressing the WNV structural proteins M and E at a multiplicity of infection (MOI) of 0.5. The culture supernatant was harvested 72 h post infection, and was centrifuged at 5000 rpm for 20 min to remove the cell debris. The supernatant was ultra-centrifuged at 38,000 rpm for 1.5 h at 4 °C. The VLP pellets were resuspended in phosphate-buffered saline (PBS) and loaded on a 10%–30%–50% discontinuous sucrose gradient. After 1.5 h of ultra-centrifugation at 30,000 rpm at 4 °C, the WNV-VLP-containing bands between the 10%–30% sucrose were collected.

2.2. Inoculation of Horses

Two 4–6-year-old healthy brown horses (300–350 kg in weight) that had no detectable antibodies against WNV were provided by the Military Hongshan Stud Farm (Changchun, China), and were intramuscularly multi-point injected in the submandibular region and backside with 0.5, 1.5, 2.0, 3.0, and 5.0 mg of WNV-VLP with Freund's adjuvant (complete/incomplete) (Sigma, St. Louis, MO, USA) on days 0, 7, 14, 28, and 42 (5 times), respectively. The sera were collected from the jugular vein 2 weeks after each immunization, and were stored at −20 °C for further analysis.

2.3. WNV-Specific Antibody Measurement

WNV-specific antibodies in the serum were measured by an indirect enzyme-linked immunosorbent assay (ELISA) using purified WNV EDIII (structural domain III of the E protein). The 96-well microtiter plates (Corning Costar, cat# 42592, Corning, NY, USA) used for ELISA were EIA/RIA 1 × 8 Stripwell plates. The featured polystyrene 96-well microtiter plates are flat bottomed and certified for high capacity binding of proteins. Briefly, 96-well microtiter plates were pre-coated with 100 μL of purified EDIII antigen diluted in 0.02 mol/L carbonate sodium buffer (pH 9.6) at a final concentration of 10 μg/mL, and were incubated at 4 °C overnight. After blocking with skim milk for 2 h at 37 °C, 100 μL of two-fold serially diluted serum samples were added to the wells, which were then incubated at 37 °C for 1.5 h. The plate was washed three times with PBS containing 0.5% Tween-20 (PBST), and then 100 μL of an HRP (horseradish peroxidase)-labeled goat antibody against horse IgG (Bios, China) was added (diluted 1:20,000), and the plate was incubated at 37 °C for 1 h. After washing with PBST, 100 μL of the substrate 3,3',3,5'-tetramethyl benzidine (TMB) (Sigma) was added to each well and incubated for 30 min. The reaction was stopped by adding 50 μL of 0.5 M H_2SO_4. The optical density (OD) values were measured using a 96-well ELISA plate reader at a wavelength of 450 nm (Bio-Rad, Los Angeles, CA, USA). The titer results were indicated as the reciprocal of the serum maximum dilution multiple, as the ratio of positive OD value to negative OD value was more than 2.

2.4. Immunoglobulin Purification

Horse anti-WNV serum was diluted with equal volumes of saline, and 1/2 volume of a saturated ammonium sulfate solution was then added. The solution was mixed gently at room temperature for 30 min and then centrifuged at 5000 rpm for 20 min. The precipitates were dissolved in saline before 1/3 volume of ammonium sulfate was added. After incubation at room temperature for 30 min, the solution was centrifuged at 5000 rpm for 20 min. The precipitates were dissolved in saline and dialyzed overnight at 4 °C to remove the salt.

2.5. Preparation of F(ab')₂ Fragments

The pH of the horse anti-WNV serum was adjusted to 3.3 with 1 mol/L HCl. Pepsin was activated by a NaAc solution (pH 3.3) and then added to the diluted horse antiserum and digested at 30 °C for 2.5 h. The reaction was stopped by adjusting the pH to 7.2 with 1 mol/L NaOH. Then, the solution was applied to a Protein-A column, followed by a Protein-G column. The purified protein was subjected to sodium dodecyl sulfate polyacrylamide gel electrophoresis (SDS-PAGE) followed by Coomassie blue staining. The F(ab')₂ purity was analyzed with a thin slice scan, and the protein was stored at 4 °C until use.

2.6. Mice, Virus, and Cells

Mavs$^{-/-}$ mice on a C57BL/6 background were initially obtained as a generous gift from S. Akira (Osaka University, Osaka, Japan) [18–20]. C57BL/6mice were purchased from Charles River Laboratories. All mice were maintained in the specific-pathogen-free Animal Care Facility at the University of Iowa. All protocols were approved by the University of Iowa Institutional Animal Care and Use Committee. The WNV strain TX 2002-HC (WNV-TX) was propagated as previously described [18]. The working stocks of WNV-TX were generated by a single round of amplification in Vero cells. The titers of virus stocks were determined by a standard plaque assay using Vero cells, as previously described [21]. The Vero cells were maintained in Dulbecco's modified Eagle medium (DMEM) supplemented with 10% fetal bovine serum (FBS). All work with WNV was conducted in the University of Iowa bio-safety level 3 (BSL 3) laboratory.

2.7. WNV Plaque Reduction Neutralization Assay

The serum samples, purified IgG, or F(ab')$_2$ were serially diluted in DMEM and mixed 1:1 with 80 PFU (Plaque Forming Unit) WNV. After incubation at 37 °C for 1 h, the mixture was added to Vero cells, and the plates were incubated at 37 °C in 5% CO$_2$ for an additional 1 h. After absorption, the cells were overlaid with 1.2% agarose. After further incubation for 3 days, the agarose plugs were removed. The plaques were visualized by 0.1% crystal violet (contains 2% ethanol and 20% methanol) staining.

2.8. Mouse Challenge and Antibody Treatment

Eight to 12-week-old wild type C57BL/6 mice or *Mavs*$^{-/-}$ mice were treated with 500 µg of purified horse immune IgG or purified horse immune F(ab')$_2$ via the intraperitoneal (i.p.) route one day before and/or after subcutaneous (s.c.) footpad infection with 100 PFU WNV. The mice were monitored for morbidity and mortality daily. The spleens and brains were harvested at the indicated time points and titered using Vero cells.

3. Results

3.1. Immunization and Evaluation of Equine Antibodies

WNV-VLP were produced and purified as previously described [22], and 4–6-year-old healthy horses were intramuscularly immunized with WNV-VLP with Freund's adjuvant and boosted every 2 weeks for 4 cycles. Serial serum samples were collected. The IgG titers against WNV EDIII were measured by ELISA (Figure 1). The EDIII-specific IgG titers in the serum were all above 1:20,480 after five immunizations (Figure 1).

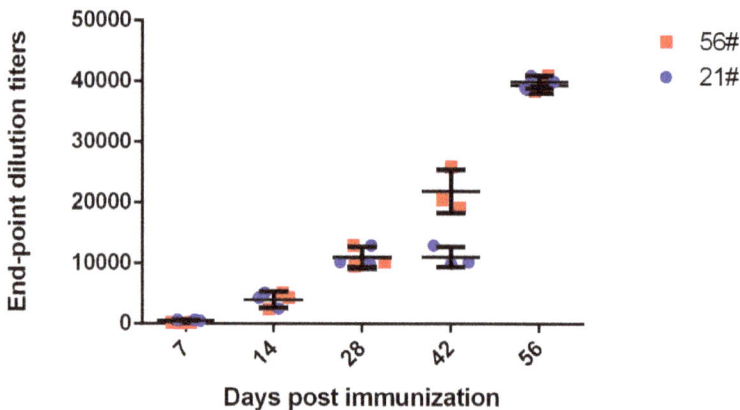

Figure 1. ELISA titers of West Nile virus (WNV) EDIII (structural domain III of the E protein)-specific antibody in immunized horse sera. Horses (*n* = 2, labeled as 21# and 56#) were intramuscularly immunized with WNV-VLP (virus-like particles) and boosted every two weeks for an additional four cycles. The sera were collected 2 weeks after each immunization.

3.2. Generation and Purification of IgG and F(ab')$_2$

Immunoglobulins were precipitated using saturated ammonium sulfate and digested by pepsin. The F(ab')$_2$ fragments were purified by Protein-A/G chromatography. The integrity of IgG and F(ab')$_2$ fragments was evaluated using SDS-PAGE (Figure 2A). The purity of the F(ab')$_2$ fragments after Protein-A/G chromatography was higher than 93.5% (Figure 2B).

A

B

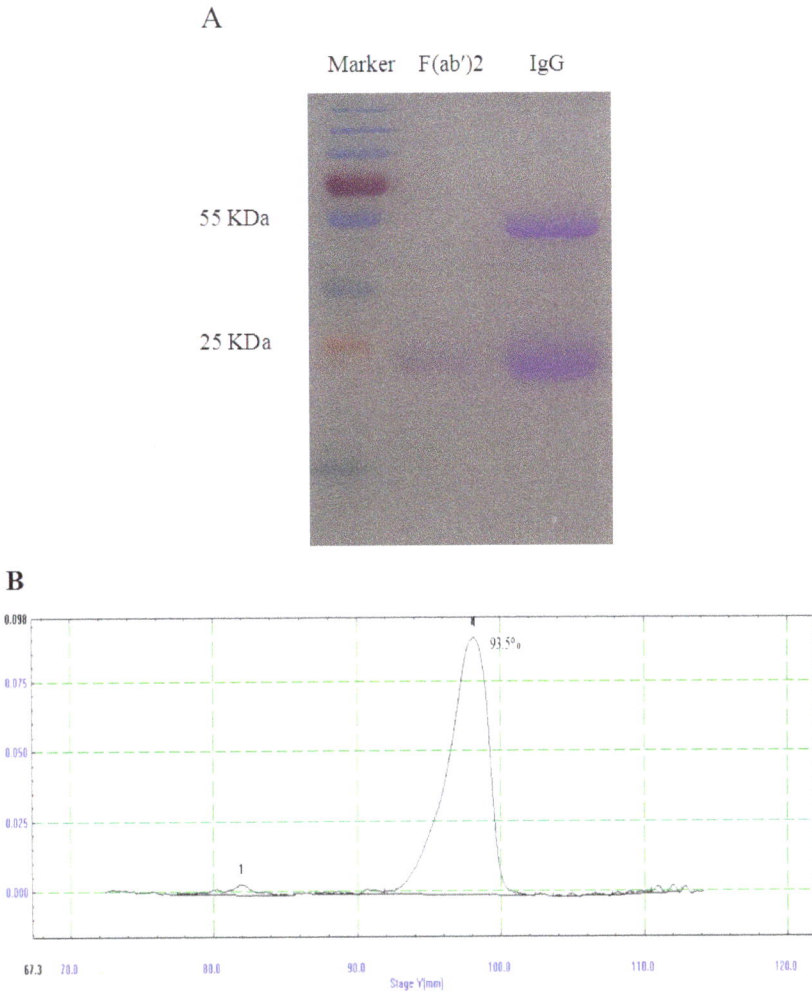

Figure 2. Generation and purification of IgG and F(ab′)₂. A saturated solution of ammonium sulfate was added to the serum to precipitate the IgG, and F(ab′)₂ was generated by digesting the IgG with pepsin followed by Protein-A/G chromatography. (**A**) SDS-PAGE and Coomassie blue staining of the purified IgG and F(ab′)₂; (**B**) The purity of the F(ab′)₂ fragment was determined by a thin slice scan.

3.3. Equine Antibodies Neutralized WNV in Cell Culture

Using a plaque reduction-neutralizing assay, we confirmed that the immune sera efficiently neutralized WNV infection in vitro, with half effect maximal dilutions of $10^{-4.3}$ for the 21–25 serum and $10^{-4.6}$ for the 56–25 serum (Figure 3A,B). Furthermore, we found that the purified equine IgG and F(ab′)₂ also neutralized WNV infection with half effective maximal concentrations (EC50) of 4.1 µg/mL and 16.5 µg/mL for the IgG and F(ab′)₂ fragments, respectively (Figure 3C,D). Of note, the EC50 of the F(ab′)₂ fragments was lower than that of the IgG in vitro. These results suggest that equine antibody products exhibit highly potent neutralizing activity against WNV in tissue culture.

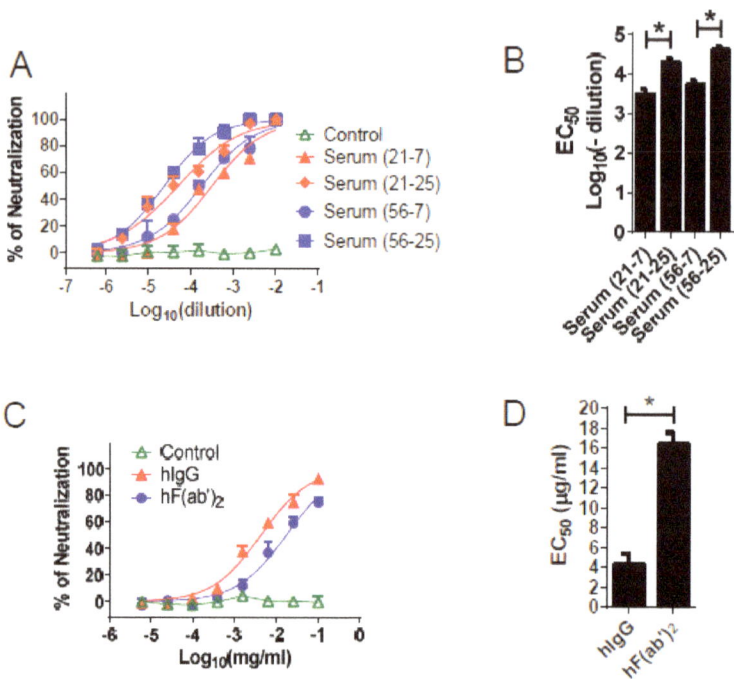

Figure 3. Neutralization assay for WNV using immune horse serum, purified IgG, and F(ab′)$_2$ in vitro. The "21" and "56" are the horse ID numbers. The "21–7", "56–7" and the "21–25", "56–25" represent the sera that were collected after the fourth or fifth immunization, respectively. (**A,B**) Serum or (**C,D**) antibody samples from "56–25" were serially diluted in Dulbecco's modified Eagle medium (DMEM) and mixed 1:1 with 80 PFU (Plaque Forming Unit) WNV. After a 1-h incubation at 37 °C, the mixture was added to Vero cells for an additional 1 h. After absorption, the cells were then overlaid with 1.2% agarose. After a further incubation for 3 days, the agarose plugs were removed. The plaques were visualized by 0.1% crystal violet staining; (**B**) The dilutions; and (**D**) Concentrations for 50% of the maximal neutralizing effect are shown. * $p < 0.05$. Normal horse IgG was used as the control.

3.4. Passive Transfer of Equine Antibodies Protected WNV-Infected Mice

To validate the protective effect of the equine antibodies in vivo, we transferred purified horse immune IgG or purified horse immune F(ab′)$_2$ intraperitoneally into mice one day before or after infection, or on both days. After WNV infection, all wild type (WT) mice receiving the immune IgG or F(ab′)$_2$ survived, whereas 30%–40% of WT mice receiving the control horse IgG died, regardless of whether the antibodies were administered before or after the infection. The mitochondrial antiviral-signaling protein (MAVS) is a key adaptor molecule in the RIG (Retinoic Acid-inducible Gene)-I-like receptor RNA-sensing pathway [23]; *Mavs*$^{-/-}$ mice were highly susceptible and uniformly succumbed to WNV infection, with death occurring by 11 days (Figure 4C–E) [18]. However, 50% of the *Mavs*$^{-/-}$ mice that received the horse IgG one day before the infection survived. In contrast, in the in vitro neutralizing assays (Figure 3C,D), the mice that received the F(ab′)$_2$ fragments showed delayed mortality; however, 100% of these mice died within 15 days (Figure 4C). An immune antibody transfer one day after the infection was not protective in the *Mavs*$^{-/-}$ mice (Figure 4D). Furthermore, to determine whether two doses of the equine antibody treatment could enhance the prophylactic function in *Mavs*$^{-/-}$ mice, we treated mice both one day before and one day after WNV infection. All *Mavs*$^{-/-}$ mice from the control group died, whereas 90% of mice that received the horse immune

IgG survived; the F(ab')$_2$ treatment protected only 30% of the mice. Next, we asked whether a passive transfer of the equine immune antibodies could decrease the WNV load in vivo. WT and *Mavs*$^{-/-}$ mice were infected with WNV at the indicated time points, and the spleens and brains were harvested for titers. As expected, antibody treatment decreased the virus load in both the WT and *Mavs*$^{-/-}$ mice (Figure 5).

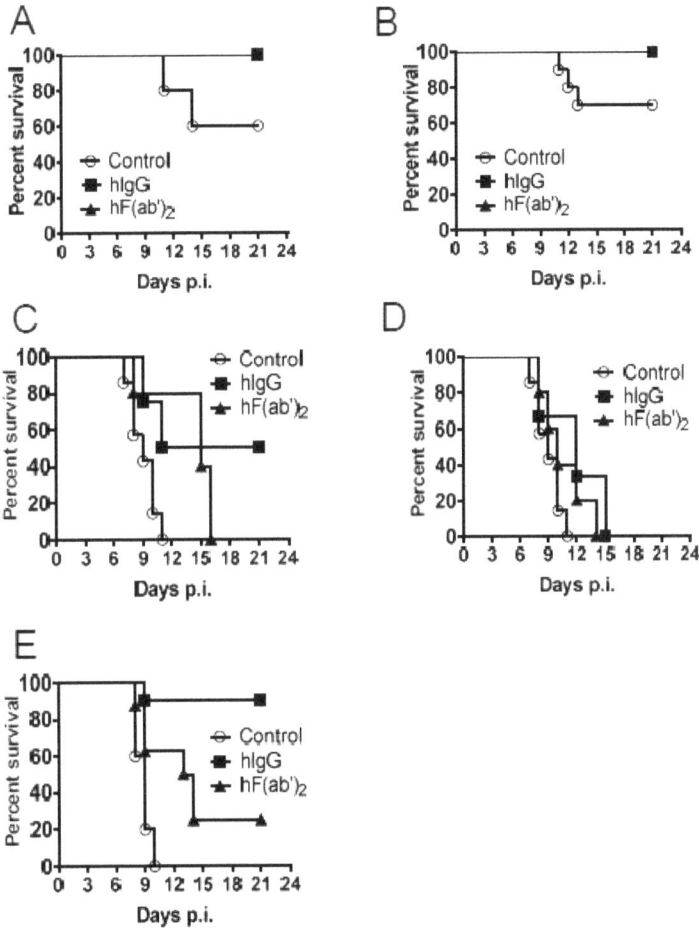

Figure 4. Immune purified IgG and F(ab')$_2$ transfer protected WNV infected mice. Eight to twelve-week-old wild type (WT) C57BL/6 mice (**A,B**) or *Mavs*$^{-/-}$ mice (**C–E**) were infected with 100 PFU WNV in the footpad (s.c.), and 500 µg of the purified horse immune IgG (hIgG) or the purified horse immune F(ab')$_2$ (hF(ab')$_2$) was injected (i.p.) into infected mice one day before (**A,C**) or after infection (**B,D**), or on both days (**E**). The mice were monitored for morbidity and mortality daily. The mice that received normal horse IgG were used as the controls. (**A**) $n = 10$, Control; $n = 5$, hIgG; $n = 5$, hF(ab')$_2$; (**B**) $n = 10$, Control; $n = 5$, hIgG; $n = 5$, hF(ab')$_2$; (**C**) $n = 7$, Control; $n = 8$, hIgG; $n = 5$, hF(ab')$_2$; (**D**) $n = 7$, Control; $n = 6$, hIgG; $n = 5$, hF(ab')$_2$; (**E**) $n = 5$, Control; $n = 10$, hIgG; $n = 8$, hF(ab')$_2$. For Figure 4C, $p = 0.009$ when Control group compared to hIgG group. $p = 0.0123$ when Control group compared to F(ab')$_2$ group. For Figure 4E, $p = 0.0009$ when Control group compared to hIgG group. $p = 0.0291$ when Control group compared to F(ab')$_2$ group.

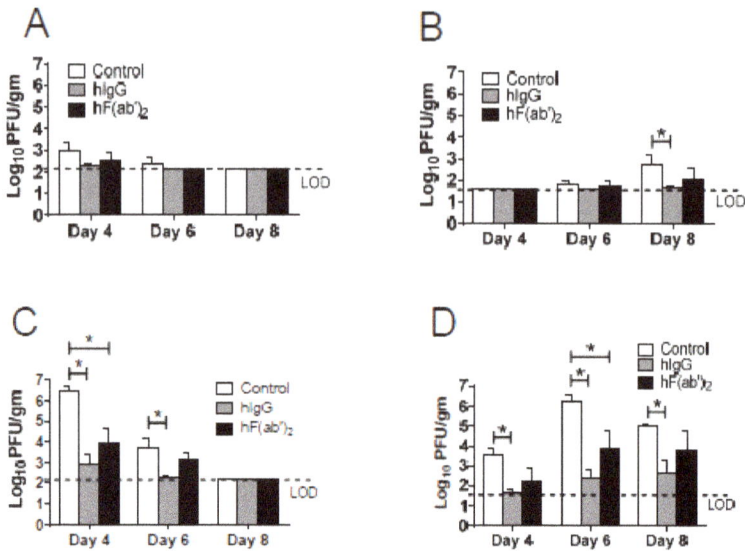

Figure 5. Immune purified IgG and F(ab')$_2$ transfer reduced the WNV load in the mice. Eight to twelve-week-old WT (**A,B**) and *Mavs*$^{-/-}$ mice; (**C,D**) were infected with 100 PFU WNV in the footpad (s.c.), and 500 µg of the purified horse immune IgG (hIgG) or the purified horse immune F(ab')$_2$ (hF(ab')$_2$) was transferred (i.p.) one day before and one day after infection. To obtain the virus titers, the spleens (**A,C**) and brains (**B,D**) were homogenized at the indicated time points and titered using Vero cells. The titers are expressed as PFU/g tissue. n = 3–4 mice/group/time point. * $p < 0.05$. The mice that received normal horse IgG were used as the controls.

4. Discussion

Previous studies have shown that the WNV is spread by air transportation or the migration of birds [24]. Currently, WNV is circulating in not only North America but also Russia, India, and Australia [24]. No effective treatment or licensed vaccine exists that has been proven for human use.

In this study, we demonstrated that EDIII-specific IgG titers in the serum were up to 1:40,960 after five immunizations, and we also verified that the purity of the F(ab')2 fragments after Protein-A/G chromatography was higher than 93.5%. Then, we found that the equine antibody products exhibit highly potent neutralizing activity against WNV in tissue culture. Importantly, we confirmed that equine anti-WNV immunoglobulin and neutralizing F(ab')2 decreased the virus load and protected mice from a lethal WNV infection.

We used C57BL/6 mice and *Mavs*$^{-/-}$ mice on C57BL/6 background for this study. C57BL/6 mice are immunocompetent and are broadly used in WNV studies with a mortality rate of around 20%–40%. *Mavs*$^{-/-}$ mice are immunocompromised with the deficiency of MAVS gene which is important for viral RNA sensing. These mice were uniformly succumbed to WNV infection. These models perfectly mimic WNV infection in humans. In human infections, most of infected patients with intact immune system get mild disease, while the immunocompromised patients and the elderly tend to develop severe neuroinvasion.

WNV-VLP were used to immunize healthy horses and generate the equine anti-WNV IgG and F(ab')2. The results showed that it is technically feasible to rapidly generate large amounts of equine neutralizing antibodies against WNV. The F(ab')$_2$ fragments help alleviate immune hypersensitivity, making the antibodies safer for human use. More importantly, the equine anti-WNV antibodies are polyclonal and can recognize more antigen determinants on WNV glycoproteins and membranes than a monoclonal antibody, which could potentially avoid antibody escape.

Highly purified equine F(ab′)2 fragments have been shown to be safe in humans and have been used to treat many diseases for decades. Equine F(ab′)2 fragments surmount the bottleneck of antibody yield, possess broader antigenic coverage, and avoid the emergence of escape mutants [25–27]. Of note, equine antibodies are relatively economical and are readily available. Although convalescent plasma from patients has shown some protective efficacy, the neutralizing antibody titers in patients are generally low, and the limited number of survivors makes this approach impractical [28–30].

The quality of the neutralization antibody can influence the treatment efficacy, which may depend on the quantity and quality of the F(ab′)2 fragments. According to our assays, the purity of the F(ab′)2 fragments was 93.5%, which met the requirement of the bio-preparation in China. The results of this study show that F(ab′)2 fragments are not as efficient as IgG, both in vitro and in vivo, and a possible factor contributing to this outcome is that F(ab′)2 has a shorter half-life than IgG.

In summary, by immunizing healthy horses with WNV-VLP, we successfully developed the first equine IgG-derived F(ab′)2 fragments that could neutralize WNV both in vitro and vivo. Both prophylactic and therapeutic treatments with equine anti-WNV immunoglobulin or neutralizing F(ab′)2 decreased the viral loads in the spleens and brains, and protected mice from a lethal WNV infection. Therefore, horses immunized with WNV-VLP could serve as a useful initial source for developing protective F(ab′)2 fragments, which may be useful for preparedness and serve as a strategic reserve for a potential WNV epidemic and other emergent pathogens.

5. Conclusions

We successfully developed the first equine IgG-derived F(ab′)$_2$ fragments that could neutralize WNV both in vitro and vivo. Both prophylactic and therapeutic treatments with equine anti-WNV immunoglobulin or neutralizing F(ab′)$_2$ decreased the viral loads in the spleens and brains, and protected mice from a lethal WNV infection. These results indicate that equine immunoglobulin and equine neutralizing F(ab′)$_2$ passive immunotherapy could be potentially useful for the treatment of patients infected with WNV.

Acknowledgments: This work was supported in part by the Study on the Emergency Treatment and Agents for the Outburst of New Mutation Pathogen Project Grant (2016YFC1200902). This work was also supported by the Thousand Talents Plan Award of China 2015 (J.Z.) and the Municipal Healthcare Joint-Innovation Major Project of Guangzhou (201604020011) (J.Z.).

Author Contributions: Jiannan Cui, Yongkun Zhao, Jincun Zhao, Songtao Yang, and Xianzhu Xia conceived and designed the experiments. Jiannan Cui and Yongkun Zhao performed the experiments. Jiannan Cui, Yongkun Zhao, and Hualei Wang analyzed the data. Boning Qiu, Zengguo Cao, Qian Li, Yanbo Zhang, Feihu Yan, Hongli Jin, Tiecheng Wang, Weiyang Sun, Na Feng, Yuwei Gao, Jing Sun, Yanqun Wang, and Stanley Perlman contributed reagents/materials/analysis tools. Jiannan Cui wrote the paper. Xianzhu Xia, Songtao Yang, and Jincun Zhao requested financial support. All authors read and approved the manuscript.

Conflicts of Interest: The authors declare no conflict of interest.

References

1. Malone, R.W.; Homan, J.; Callahan, M.V.; Glasspool-Malone, J.; Damodaran, L.; Schneider, A.B.; Zimler, R.; Talton, J.; Cobb, R.R.; Ruzic, I.; et al. Zika virus: Medical countermeasure development challenges. *PLoS Negl. Trop. Dis.* **2016**. [CrossRef] [PubMed]

2. Suthar, M.S.; Diamond, M.S.; Gale, M., Jr. West Nile virus infection and immunity. *Nat. Rev. Microbiol.* **2013**, *11*, 115–128. [CrossRef] [PubMed]

3. Chang, J.; Wang, L.; Ma, D.; Qu, X.; Guo, H.; Xu, X.; Mason, P.M.; Bourne, N.; Moriarty, R.; Gu, B.; et al. 2009 Novel imino sugar derivatives demonstrate potent antiviral activity against flaviviruses. *Antimicrob. Agents Chemother.* **2009**, *53*, 1501–1508. [CrossRef] [PubMed]

4. Chang, D.C.; Liu, W.J.; Anraku, I.; Clark, D.C.; Pollitt, C.C.; Suhrbier, A.; Hall, R.A.; Khromykh, A.A. Single-round infectious partious enhance immunogenicity of a DNA vaccine against West Nile virus. *Nat. Biotechnol.* **2008**, *26*, 571–577. [CrossRef] [PubMed]

5. Vogt, M.R.; Moesker, B.; Goudsmit, J.; Jongeneelen, M.; Austin, S.K.; Oliphant, T.; Nelson, S.; Pierson, T.C.; Wilschut, J.; Throsby, M.; et al. Human monoclonal antibodies against West Nile virus induced by natural infection neutralize at a postattachment step. *J. Virol.* **2009**, *83*, 6494–6507. [CrossRef] [PubMed]

6. Zhang, S.; Vogt, M.R.; Oliphant, T.; Engle, M.; Bovshik, E.I.; Diamond, M.S.; Beasley, D.W. The development of resistance to passive therapy by a potently neutralizing humanized West Nile virus monoclonal antibody. *J. Infect. Dis.* **2009**, *200*, 202–205. [CrossRef] [PubMed]

7. Day, C.W.; Smee, D.F.; Julander, J.G.; Yamshchikov, V.F.; Sidwell, R.W.; Morrey, J.D. Error-prone replication of West Nile virus caused by ribavirin. *Antivir. Res.* **2005**, *67*, 38–45. [CrossRef] [PubMed]

8. Morrey, J.D.; Day, C.W.; JuLander, J.G.; Blatt, L.M.; Smee, D.F.; Sidwell, R.W. Effect of interferon-alpha and interferon-inducers on West Nile virus in mouse and hamster animal models. *Antivir. Chem. Chemother.* **2004**, *15*, 101–109. [CrossRef] [PubMed]

9. Ng, C.Y.; Gu, F.; PHong, W.Y.; Chen, Y.L.; Lim, S.P.; Davidson, A.; Vasudevan, S.G. Construction and characterization of a stable subgenomic dengue virus type 2 replicon system for antiviral compound and siRNA testing. *Antivir. Res.* **2007**, *76*, 222–231. [CrossRef] [PubMed]

10. Lewis, M.; Amsden, J.R. Successful treatment of West Nile virus infection after approximately 3 weeks into the disease course. *Pharmacotherapy* **2007**, *27*, 455–458. [CrossRef] [PubMed]

11. Diamond, M.S. Progresson the development of therapeutics against West. *Nile Virus. Antivir. Res.* **2009**, *83*, 215–216.

12. Engle, M.J.; Diamond, M.S. Antibody prophylaxis and therapy against West Nile virus infection in wild-type and immunodeficient mice. *J. Virol.* **2003**, *77*, 12941–12949. [CrossRef] [PubMed]

13. Dahmen, U.; Dirsch, O.; Li, J.; Fiedle, M.; Lu, M.; Rispeter, K.; Picucci, M.; Broelsch, C.E.; Roggendorf, M. Adoptive transfer of immunity: A new strategy to interfere with severe hepatitis virus reinfection after woodchuck liver transplantation. *Transplant* **2004**, *77*, 965–972. [CrossRef]

14. Watt, G.; Kantipong, P.; Jongsakul, K.; de Souza, M.; Burnouf, T. Passive transfer of scrub typhus plasma to patients with AIDS: A descriptive clinical study. *QJM* **2001**, *94*, 599–607. [CrossRef] [PubMed]

15. Lang, J.; Attanath, P.; Quiambao, B.; Singhasivanon, V.; Chanthavanich, P.; Montalban, C.; Lutsch, C.; Pepin-Covatta, S.; le Mener, V.; Miranda, M.; et al. Evaluation of the safety, immunogenicity, and pharmacokinetic profile of a new, highly purified, heat-treated equine rabies immunoglobulin, administered either alone or in association with a purified, Vero-cell rabies vaccine. *Acta Trop.* **1998**, *70*, 317–333. [CrossRef]

16. Both, L.; Banyard, A.C.; van Dolleweerd, C.; Horton, D.L.; Ma, J.K.; Fooks, A.R. Passive immunity in the prevention of rabies. *Lancet Infect. Dis.* **2012**, *12*, 397–407. [CrossRef]

17. Gutierrez, J.M.; Lomonte, B.; Sanz, L.; Calvete, J.J.; Pla, D. Preclinical analysis of the efficacy of a polyspecific antivenom through antivenomics and neutralization assays. *J. Proteomics* **2014**, *105*, 340–350. [CrossRef] [PubMed]

18. Suthar, M.S.; Ma, D.Y.; Thomas, S.; Lund, J.M.; Zhang, N.; Daffis, S.; Rudensky, A.Y.; Bevan, M.J.; Clark, E.A.; Kaja, M.K.; et al. IPS-1 is essential for the control of West Nile virus infection and immunity. *PLoS Pathog.* **2010**. [CrossRef] [PubMed]

19. Kato, H.; Sato, S.; Yoneyama, M.; Yamamoto, M.; Uematsu, S.; Matsui, K.; Tsujimura, T.; Takeda, K.; Fujita, T.; Takeuchi, O.; et al. Cell type-specific involvement of RIG-I in antiviral response. *Immunity* **2005**, *23*, 19–28. [CrossRef] [PubMed]

20. Kumar, H.; Kawai, T.; Kato, H.; Sato, S.; Takahashi, K.; Coban, C.; Yamamoto, M.; Uematsu, S.; Ishii, K.J.; Takeuchi, O.; et al. Essential role of IPS-1 in innate immune responses against RNA viruses. *J. Exp. Med.* **2006**, *203*, 1795–1803. [CrossRef] [PubMed]

21. Brien, J.D.; Lazear, H.M.; Diamond, M.S. Propagation, quantification, detection, and storage of West Nile virus. *Curr. Protoc. Microbiol.* **2013**, *31*, 15D.3.1–15D.3.18. [PubMed]

22. Wang, C.; Zheng, X.; Gai, W.; Zhao, Y.; Wang, H.; Wang, H.; Feng, N.; Chi, H.; Qiu, B.; Li, N.; et al. MERS-Cov virus-like particles produced in insect cells induce specific humoural and cellular imminity in rhesus macaques. *Oncotarget* **2016**. [CrossRef] [PubMed]

23. Zhao, J.; Vijay, R.; Zhao, J.; Gale, M.; Diamond, M.S.; Perlman, S. MAVS expressed by hematopoietic cells is critical for control of West Nile virus infection and pathogenesis. *J. Virol.* **2016**, *90*, 7098–7108. [CrossRef] [PubMed]

24. Murgue, B.; Zeller, H.; Deubel, V. The ecology and epidemiology of West Nile virus in Africa, Europe and Asia. *Curr. Top. Microbiol. Immunol.* **2002**, *267*, 195–221. [PubMed]

25. Zhao, Z.; Yan, F.; Chen, Z.; Luo, D.; Duan, Y.; Yang, P.; Li, Z.; Peng, D.; Liu, X.; Wang, X. Cross clade prophylactic and therapeutic efficacy of polyvalent equine immunoglobulin F(ab')$_2$ against highly pathogenic avian influenza H5N1 in mice. *Int. Immunopharmacol.* **2011**, *11*, 2000–2006. [CrossRef] [PubMed]

26. Zhou, L.; Ni, B.; Luo, D.; Zhao, G.; Jia, Z.; Zhang, L.; Lin, Z.; Wang, L.; Zhang, S.; Xing, L.; et al. Inhibition of infection caused by severe acute respiratory syndrome-associated coronavirus by equine neutralizing antibody in aged mice. *Int. Immunopharmacol.* **2007**, *7*, 392–400. [CrossRef] [PubMed]

27. Lu, J.H.; Guo, Z.M.; Han, W.Y.; Wang, G.L.; Zhang, D.M.; Wang, Y.F.; Sun, S.Y.; Yang, Q.H.; Zheng, H.Y.; Wong, B.L.; et al. Preparation and development of equine hyperimmune globulin F(ab')$_2$. *Acta Pharmacol. Sin.* **2005**, *26*, 1479–1484. [CrossRef] [PubMed]

28. Luo, D.; Ni, B.; Zhao, G.; Jia, Z.; Zhou, L.; Pacal, M.; Zhang, L.; Zhang, S.; Xing, L.; Lin, Z.; et al. Protection frominfection with severe acute respiratory syndrome coronavirus in a Chinese hamster model by equine neutralizing F(ab')2. *Viral Immunol.* **2007**, *20*, 495–502. [CrossRef] [PubMed]

29. Wang, X.; Ni, B.; Du, X.; Zhao, G.; Gao, W.; Shi, X.; Zhang, S.; Zhang, L.; Wang, D.; Luo, D.; et al. Protection of mammalian cells from severe acute respiratory syndrome coronavirus infection by equine neutralizing antibody. *Antivir. Ther.* **2005**, *10*, 681–690. [PubMed]

30. Zhao, G.; Ni, B.; Jiang, H.; Luo, D.; Pacal, M.; Zhou, L.; Zhang, L.; Xing, L.; Zhang, L.; Jia, Z.; et al. Inhibition of severe acute respiratory syndrome-associated coronavirus infection by equine neutralizing antibody in golden Syrian hamsters. *Viral Immunol.* **2007**, *20*, 197–205. [CrossRef] [PubMed]

viruses

MDPI

Article

4EBP-Dependent Signaling Supports West Nile Virus Growth and Protein Expression

Katherine D. Shives [1], Aaron R. Massey [2], Nicholas A. May [1], Thomas E. Morrison [1] and J. David Beckham [1,2,*]

[1] Department of Immunology and Microbiology, University of Colorado Graduate School, Aurora, CO 80045, USA; katherineshives@gmail.com (K.D.S.); nicholas.may@ucdenver.edu (N.A.M.); thomas.morrison@ucdenver.edu (T.E.M.)
[2] Department of Medicine, Division of Infectious Diseases, University of Colorado School of Medicine, Aurora, CO 80045, USA; aaron.massey@ucdenver.edu
* Correspondence: david.beckham@ucdenver.edu; Tel.: +1-303-724-4927

Academic Editor: Michael R. Holbrook
Received: 8 July 2016; Accepted: 7 October 2016; Published: 18 October 2016

Abstract: West Nile virus (WNV) is a (+) sense, single-stranded RNA virus in the *Flavivirus* genus. WNV RNA possesses an $^{m7}GpppN_m$ 5' cap with 2'-O-methylation that mimics host mRNAs preventing innate immune detection and allowing the virus to translate its RNA genome through the utilization of cap-dependent translation initiation effectors in a wide variety of host species. Our prior work established the requirement of the host mammalian target of rapamycin complex 1 (mTORC1) for optimal WNV growth and protein expression; yet, the roles of the downstream effectors of mTORC1 in WNV translation are unknown. In this study, we utilize gene deletion mutants in the ribosomal protein kinase called S6 kinase (S6K) and eukaryotic translation initiation factor 4E-binding protein (4EBP) pathways downstream of mTORC1 to define the role of mTOR-dependent translation initiation signals in WNV gene expression and growth. We now show that WNV growth and protein expression are dependent on mTORC1 mediated-regulation of the eukaryotic translation initiation factor 4E-binding protein/eukaryotic translation initiation factor 4E-binding protein (4EBP/eIF4E) interaction and eukaryotic initiation factor 4F (eIF4F) complex formation to support viral growth and viral protein expression. We also show that the canonical signals of mTORC1 activation including ribosomal protein s6 (rpS6) and S6K phosphorylation are not required for WNV growth in these same conditions. Our data suggest that the mTORC1/4EBP/eIF4E signaling axis is activated to support the translation of the WNV genome.

Keywords: West Nile virus; RNA; translation; protein synthesis

---

1. Introduction

West Nile virus (WNV) is a prototypical (+) sense, single-stranded RNA virus from the family *Flaviviridae*. The viral nonstructural protein 3 contains methyltransferase [1] and 5' RNA triphosphatase activity [2,3] to generate a 5' capped genomic viral RNA. WNV genomic RNA possesses a canonical 7-methylguanosine ($^{m7}GpppN_m$)5' cap with a 2'-O-methylation that mimics host mRNAs preventing innate immune detection [4] and allowing the virus to translate its RNA genome through the utilization of cap-dependent translation initiation effectors in a wide variety of hosts [5,6]. However, it is unclear whether flaviviruses are obligated to utilize specific host cap-dependent translation initiation factors in a canonical manner [7].

Flaviviruses do not shut down the translation of host mRNA during infection, so work to understand the mechanisms that flaviviruses manipulate to compete with host mRNA for access to the translational apparatus will provide important insight into the function of the viral RNA.

We have previously shown that WNV activates mammalian target of rapamycin (mTOR) and that inhibition of mTOR results in a significant WNV growth defect associated with decreased viral protein production [8]. mTOR is a major regulator of cellular RNA translation initiation and gene expression and is highly conserved from yeast to mammals [9]. Initiation of translation is a major checkpoint of cellular control of host cell gene expression and is likely to be a major checkpoint for the control of viral gene expression. Thus, the study of viral interactions with the TOR pathway provides insight into the central features of the viral lifecycle and processes that promote the translation of viral RNA [10].

mTOR is a highly evolutionarily-conserved serine/threonine kinase that forms two distinct multi-protein complexes in higher eukaryotes: mTORC1 and mTORC2. mTORC1 includes the cofactor regulatory-associated protein of mTOR (Raptor) resulting in mTOR catalytic activity for specific targets (Figure 1). Raptor interacts with 70 kDa S6 kinase 1 (p70S6K) and the eukaryotic initiation factor 4E binding proteins (4EBP) [11–14].

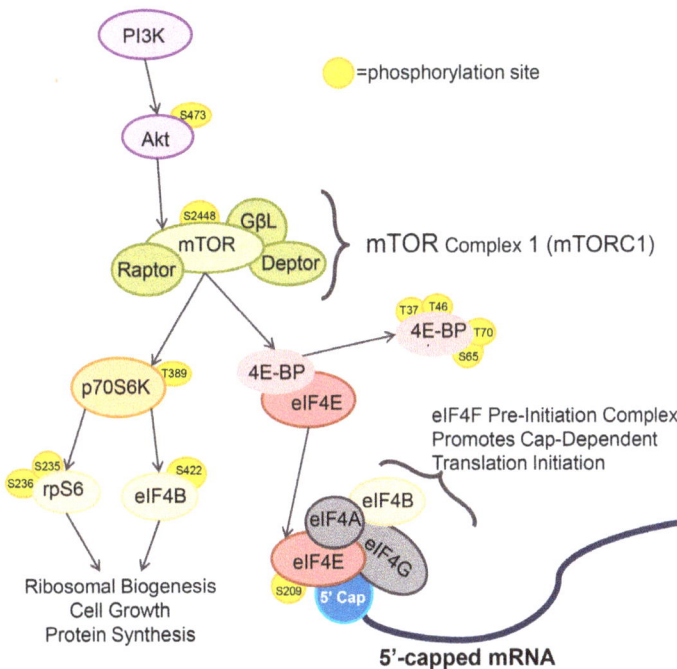

Figure 1. The mammalian target of rapamycin complex 1 (mTORC1) pathway and translation initiation. Cellular mTORC1 activity is regulated in part by phosphatidylinositol-3-kinase/protein kinase B (PI3K/Akt) signaling. Phosphorylation of mTOR on residue serine 2448 leads to activation of mTOR and phosphorylation of mTORC1 effector proteins, 70 kDa ribosomal protein S6 kinase 1 (p70S6K) and eukaryotic initiation factor 4E binding protein (4EBP). Phosphorylation-induced activity of p70S6K phosphorylates ribosomal protein S6 (rpS6) and eukaryotic initiation factor 4B (eIF4B) resulting in initiation of specific cap-dependent translation events. mTORC1-dependent phosphorylation of 4EBP leads to dissociation of 4EBP from eIF4E. Free eIF4E binds to the 5′ cap of host mRNAs and forms the eIF4F pre-initiation complex.

P70S6K is a member of the S6 kinase family, which includes two constitutively-expressed genes, *S6K1* and *S6K2* [15,16]. Due to alternative splicing events, a larger protein called p85S6K1 is produced, but has an unknown function, while the smaller p70S6K protein is thought to be the functional protein for translational control [17,18]. Activation of p70S6K by mTORC1 via phosphorylation at threonine 389 promotes phosphorylation and activation of ribosomal protein S6 (rpS6) and eukaryotic initiation factor 4B (eIF4B). The phosphorylation of rpS6 enhances translation initiation by promoting the recruitment of rpS6 to the 7-methylguanosine cap complex and facilitating the assembly of the pre-initiation complex [19]. When phosphorylated, eIF4B promotes translation initiation by enhancing the helicase activity of eIF4A during scanning in the mRNA 5′ untranslated region (UTR) [20,21].

mTOR also regulates cap-dependent translation initiation rates through phosphorylation of the eIF4E binding proteins: 4EBP1, 4EBP2 and 4EBP3. 4EBP1 and 4EBP2 are ubiquitously expressed in higher eukaryotes and function by binding the 5′ cap binding protein eIF4E, a core component of the eukaryotic pre-initiation eIF4F complex. When hypophosphorylated, as in conditions of nutrient starvation (a known repressor of mTORC1 activity), 4EBP1 binds to eIF4E, sequesters eIF4E in the cell and limits cap-dependent translation initiation events. With the addition of nutrients, mTORC1 phosphorylates 4EBP in an ordered event, beginning with threonine 37 and threonine 46, followed by threonine 70 and serine 65 and resulting in dissociation of 4EBP from eIF4E [22,23]. The 4EBP family of proteins also regulates the activity of eIF4E by altering the cellular localization of eIF4E [24,25]. The host factor eIF4E is a key component in the eIF4F pre-initiation complex and is required for cap-dependent translation initiation (Figure 1).

The eIF4F complex is commonly targeted by viruses that shutoff host protein synthesis to promote viral translation, but few data exist for the role of eIF4E during WNV infection [10,26,27].

Our previous work suggested that the mTOR-dependent activation of S6K1 and/or 4EBP play important roles in (+) strand RNA virus gene expression [8]. In this study, we used cellular gene deletion studies of the S6 kinase (S6K) and 4EBP pathways to define the role of mTOR-dependent signaling in WNV gene expression and growth. We now show that S6K/rpS6 activity does not impact WNV growth or protein expression, while loss of 4EBP function and disruption of eIF4F complex activity has a significant impact on WNV growth and protein expression. These data suggest that mTORC1/4EBP/eIF4E signaling plays an important role in modulating the intracellular environment to support WNV gene expression.

2. Materials and Methods

2.1. Virus Propagation and Titration

West Nile virus stocks were obtained from the clone-derived 385-99 (NY99) strain and propagated in C6/36 *Aedes albopictus* (American Type Culture Collection, ATCC® CRL-1660™, Manassas, VA, USA) cells as previously described [28]. Chikungunya virus (CHIKV) stocks were derived from the La Reunion 2006 OPY-1 (CHIKV-LR) strain and propagated on C6/36 cells as previously described [29]. Encephalomyocarditis virus (EMCV), strain EC9, was generated in HeLa cells (ATCC®, CCL-2™) transfected with pEC9 (gift from David Barton, University of Colorado, Aurora, CO, USA) to produce infectious virus stocks [30]. Syrian golden hamster kidney cells (BHK-21 [C-13]; ATCC®, CCL-10™) were used to measure viral titer by standard plaque assay [31]. For infection, cells were inoculated with WNV, CHIKV or EMCV and incubated at 37 °C and 5% CO_2 for 1 h. All time points were measured from the end of the 1-h adsorption period and reported as hours post-infection (hpi).

2.2. Cell Lines

All cell lines were maintained at 37 °C and 5% CO_2, except C6/36 cells, which were maintained at 28 °C. Vero (ATCC® CCL-81™), C6/36 and BHK-21 cells were maintained in Dulbecco's Modified Eagle Medium (Gibco, Grand Island, NY, USA) supplemented with 1% penicillin/streptomycin (Pen/Strep; Gibco), 10% heat-inactivated fetal bovine serum (Gibco), 1% non-essential amino acids (Gibco) and 1% sodium pyruvate (Gibco). Murine embryonic fibroblasts (MEFs, gift of Dr. Michael Hall, Freidreich Meischer Institute, Switzerland) were maintained in Dulbecco's Modified Eagle Medium supplemented with 1/% Pen/Strep and 10% heat-inactivated fetal bovine serum.

2.3. Inducible Raptor Murine Embryonic Fibroblasts

Inducible *Raptor* knockout (iRapKO) MEFs were gifts from Michael Hall of the Freidreich Meischer Institute and have been previously described in detail [32]. For induction, 2 mM 4-hydroxytamoxifen (Sigma-Aldrich, St. Louis, MO, USA) suspended in 95% ethanol was diluted in MEM to a final concentration of 1 μM. At 24 h post-induction, cells were treated with trypsin (Gibco), counted and plated at 80,000 cells/well in a 6-well plate format or 5000 cells/well in an 8-well chamber slide format. Cells were allowed 48 additional hours before experimental manipulation to allow for full induction of *Raptor* gene knockout. Cells were infected with 1×10^6 plaque-forming units (pfu) of WNV/CHIKV/EMCV per well (6-well plate format, multiplicity of infection (MOI) = 3) or 5×10^4 pfu WNV/well (8-well chamber slide format, MOI = 3) for 1 h at 37 $^\circ$C and 5% CO_2.

2.4. Western Blots

Cells were harvested by dissociating with trypsin at indicated times post-infection, suspended in ice-cold 1X phosphate buffered saline (PBS; Gibco) centrifuged at $3000 \times g$ for 5 min, and pellets suspended in lysis buffer (Cell Signaling Technology, Boston, MA, USA), plus Halt protease and phosphatase inhibitor cocktail (Thermo Scientific, Rockford, IL, USA) and disrupted using an ultrasonic processor VCX130 (Sonics & Materials, Newtown, CT, USA) for one cycle $\times$ 10 s. Whole cell extracts were run on standard 10% sodium dodecyl sulfate-polyacrylamide gel electrophoresis (SDS-PAGE) gels (Criterion system; Bio-Rad, Hercules, CA, USA). The separated proteins were electrically transferred to polyvinylidene fluoride 0.45-μm pore membranes (Millipore, Billerica, MA, USA) at 100V for 1 h. For all Western blot (WB) analyses, membranes were activated for 10–15 s in methanol, blocked for 1 h with 5% Blocking Grade Buffer (Bio-Rad) post-transfer, and probed with primary antibodies: rabbit monoclonal antibody to ribosomal protein S6, Raptor, 4EBP1, phospho-4EBP1 (T37/46), mTOR and β-actin; rabbit polyclonal to phospho-p70 S6 kinase (T389), phospho-ribosomal protein S6 (S235/236), p70 S6 kinase, phospho-eIF4B (S422) and eIF4B (Cell Signaling Technology). Additional polyclonal rabbit WNV NS3 antibody was a generous gift from Aaron Brault (Centers for Disease Control (CDC), Ft. Collins, CO, USA). For CHIKV antibody, we obtained mouse immunoglobulin from ascites to CHIKV antigen (V-548-701-562; ATCC® VR-1241AF™). After washing, membranes were probed with appropriate horseradish peroxidase (HRP)-conjugated secondary antibodies (Jackson ImmunoResearch, West Grove, PA, USA) and images obtained after incubation with Western Lightning ECL Pro (Perkin Elmer, Waltham, MA, USA) and visualized with the ChemiDoc XRS+ system (Bio-Rad). Band density was calculated using ImageJ analysis (version 1.47, Rasband, W.S., ImageJ, U. S. National Institutes of Health, Bethesda, Maryland, USA, http://imagej.nih.gov/ij/, 1997-2016) and corrected for β-actin band density to provide mean band density for each Western blot replicate.

2.5. RNA Isolation and qRT-PCR

Cells were harvested at the indicated time post-infection and RNA isolated using the RNeasy Mini Kit (Qiagen, Hilden, Germany) per the manufacturer's instructions and stored at -80 $^\circ$C until cDNA library generation. cDNA libraries were generated from total RNA isolates using the Super Script III First-Strand Synthesis System (Life Technologies, Carlsbad, CA, USA) as per the manufacturer's protocol using random hexamer primers. CHIKV quantitative reverse-transcription polymerase chain reaction (qRT-PCR) was accomplished using previously described methods [33].

2.6. MTT Assay

MTT (3-(4,5-dimethylthiazol-2-yl)-2,5-diphenyltetrazolium bromide) assays were conducted as per the manufacturers' protocols (Sigma-Aldrich, Cat. No. M5655). In brief, the MTT assay was conducted in 6-well plates, formazan crystal solubilized in acidified isopropanol and read in 96-well plates using the VictorX5 plate reader (Perkin Elmer). Uninfected, non-treated Vero cells were utilized as a viable-cell control, and a cell-free media control was included to correct for background absorbance.

2.7. Biochemical Inhibitor Studies

The eIF4F complex-formation inhibitor 4EGI-1 was obtained from Tocris Biosciences (Avonmouth, Bristol, UK) (CAS No. 315706-13-9/0). The compound was suspended in dimethyl sulfoxide (DMSO; Sigma) to make a 1000X stock at a concentration of 100 mM and diluted to final concentrations in DMEM (Gibco). Fresh solutions of the compound were prepared for individual experiments. A DMSO vehicle control was included in all studies.

2.8. Statistical Analysis

All data were analyzed using Prism software (GraphPad Prism6, La Jolla, CA, USA) using the indicated statistical analysis tests. Statistical significance was determined with an alpha less than 0.05.

3. Results

3.1. Raptor Deletion Reduces the Growth of 5′-Capped Viruses, but Not an IRES-Translated Virus

We previously demonstrated that WNV activates mTORC1 activity in several cell types and that mTORC1 activity supports viral growth and protein expression [8]. To investigate the role of cap-dependent signals in viral protein expression, we used an iRapKO MEF model that was previously described [32,34]. By knocking out the cofactor Raptor, we were able to specifically eliminate mTORC1 activity with no effect on mTOR or mTORC2 activity. Using the iRapKO system, we determined the effect of mTORC1 activity on multiple rounds of WNV replication using a low multiplicity of infection. Next, we determined whether the effect of Raptor deletion on viral growth and protein expression was specific for just WNV or played a role in pathogenesis for other capped RNA viruses. Thus, we used CHIKV (a 5′-capped RNA virus) and EMCV as additional controls. CHIKV utilizes a cap-dependent translation system for the translation of nonstructural proteins and uses a capped-subgenomic mRNA to express structural proteins. EMCV initiates translation using an internal ribosome entry site (IRES).

First, we completed a multi-step growth curve with WNV infection to evaluate the role of Raptor expression during multiple WNV replication cycles. iRapKO cells were treated with 4-hydroxytamoxifen (4-OHT) to induce knockout or mock-induced with ethanol as a control. Cells were then inoculated with WNV (MOI 0.001) at 72 h post-induction, and the virus-containing supernatant was collected and titered. WNV growth was significantly reduced in iRapKO cells at 12 h (5 pfu/mL ± 1.6 (mean ± standard error of the mean (SEM)), 24 h (52.3 pfu/mL ± 19.8) and 48 h (20 pfu/mL ± 5.2) post-infection, compared to the ethanol-treated control cells at 12 h (205.3 pfu/mL ± 42.8), 24 h (1.4×10^4 pfu/mL ± 3619) and 48 h (6.7×10^4 pfu/mL ± 1.5×10^4) post-infection ($p < 0.0001$, $n = 6$ per group, two-way ANOVA; Figure 2A).

Next, we evaluated the role of Raptor expression on CHIKV growth to define the role of mTORC1 on a different positive-strand RNA virus. iRapKO MEFs were treated with the inducing agent 4-OHT or ethanol vehicle control as above, were inoculated 72 h post-induction with CHIKV-LR (MOI = 3), and the supernatant was collected for the standard viral titer determination. In iRapKO cells, CHIKV-LR growth was significantly ($p = 0.005$) inhibited at 12 h (3×10^2 pfu/mL (mean) ± 1.23 (SEM)) and 24 h ($p < 0.0001$, 8×10 pfu/mL ± 1.6) post-infection compared to ethanol-treated, control cells at 12 h (1.6×10^3 pfu/mL ± 16) and 24 h post-infection (1.6×10^4 pfu/mL ± 1.2, $n = 3$, two-way ANOVA; Figure 2B). In CHIKV-LR-inoculated, iRapKO cells, the virus did not show evidence of growth following the t = 0 h titer of 4×10^2 pfu/mL.

Since CHIKV-LR did not exhibit evidence of viral growth in iRapKO cells, we next evaluated iRapKO cells for the ability to support viral growth of an IRES-driven EMCV isolate. Using the same treatment as described above, we inoculated control and iRapKO MEF cells with EMCV (MOI = 3) and harvested the supernatant at specific time points post-infection. We found no evidence of a growth defect in EMCV-inoculated iRapKO cells at 6 h (2.6×10^5 pfu/mL ± 8.1×10^4), 12 h (2.6×10^6 pfu/mL ± 5.5×10^8) or 24 h (3.5×10^6 pfu/mL ± 7.5×10^5) post-infection when compared to EMCV-inoculated ethanol-control treated MEF cells at 6 h (2×10^6 pfu/mL ± 6×10^5), 12 h

$(3.3 \times 10^6$ pfu/mL $\pm 2.4 \times 10^5)$ and 24 h $(8.9 \times 10^6$ pfu/mL $\pm 1.3 \times 10^6)$ post-infection $(p = 0.5988,$ $n = 3$, two-way ANOVA; Figure 2C). These data demonstrated that EMCV, a virus that depends on IRES-initiated translation, exhibits no significant growth defect when mTORC1 signaling is disrupted through Raptor knockout.

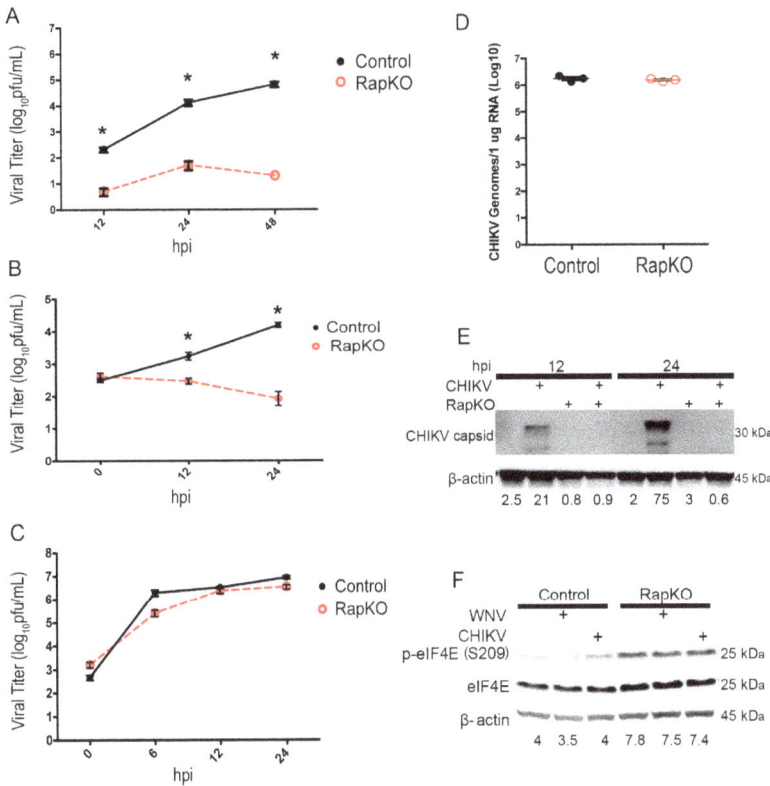

Figure 2. Raptor expression supports the growth and protein expression of 5′-capped RNA viruses. (**A**) Multi-step growth curve (MOI = 0.001) of West Nile virus (WNV) in control and Raptor knockout (RapKO) cells. $N = 6$ replicates per time point, * $p < 0.0001$; a single-step growth curve (MOI = 3) is shown for (**B**) Chikungunya virus La Reunion 2006 OPY-1 (CHIKV-LR) and (**C**) encephalomyocarditis virus (EMCV) in control and RapKO cells as determined by standard plaque assay of the supernatant. The viral titer is presented as log10 pfu/mL at indicated times (hours post-infection (hpi)). $N = 6$ replicates per time point, * $p < 0.005$; (**D**) Cell-associated CHIKV-LR genome copies in RapKO and control murine embryonic fibroblast (MEF) cells at 0 hpi determined by quantitative reverse-transcriptoin PCR (qRT-PCR). Data are presented as log10 CHIKV-LR genomes/μg of RNA; (**E**) Western blot (WB) analysis of CHIKV-LR protein synthesis in RapKO cells. Cellular lysates were harvested at the indicated time (hpi), normalized for total protein and subjected to WB analysis using antibodies to CHIKV capsid protein and a β-actin loading control. Images are representative of two independent experiments. Densitometry values are provided for CHIKV capsid corrected for β-actin expression; (**F**) Expression of phospho-eIF4E (serine 209), total eIF4E and β-actin in control and RapKO cells as determined by Western blot analysis. Total protein lysate was collected from cells either mock, WNV or CHIKV-LR infected (MOI = 3), separated by gel electrophoresis and probed for the indicated antibody targets. Images are representative of two independent experiments. Densitometry values are provided for p-eIF4E expression corrected for β-actin expression.

To further evaluate the CHIKV-LR growth defect in iRapKO cells, we determined if iRapKO cells were infected at the same level as control cells. iRapKO MEFs were treated with 4-OHT or ethanol vehicle control as above, and then inoculated 72 h post-induction with CHIKV (MOI = 3); cells were washed twice with PBS after one hour absorption as above, and harvested at t = 0 post-infection to evaluate CHIKV RNA using quantitative-reverse transcriptase PCR. We found a mean of 1.3×10^6 copies of CHIKV RNA/1 µg of RNA in both iRapKO cells and vehicle control cells (Figure 2D). This experiment was repeated in triplicate using a low pH wash instead of PBS to remove bound virus that had not entered the cell. We found no difference in the entry of CHIKV when comparing CHIKV copies/µL at t = 0 post-infection in iRapKO cells (3.5×10^4 (mean)) to vehicle control cells (3.42×10^4).

While CHIKV-LR-inoculated iRapKO cells were RNA-positive, the virus seemed unable to replicate in these cells. To determine if this was related to an effect on CHIKV protein production, we inoculated the same treatment groups described above with CHIKV-LR (MOI = 3) and harvested cells at 12 h and 24 h post-infection for total protein analysis. Western blot analysis of whole cell lysates revealed that CHIKV-inoculated, iRapKO cells were not expressing CHIKV capsid protein following infection (Figure 2). However, control cells expressed high levels of CHIKV capsid protein. These data suggest that targeted Raptor gene deletion to eliminate mTORC1 activity in MEF cells diminished CHIKV protein production.

Previous reports suggest differing roles of mTORC1 activity in CHIKV growth and eIF4E phosphorylation, so we examined the phosphorylation status of eIF4E in iRapKO MEFs [35,36]. Inhibition of mTORC1 may enhance eIF4E phosphorylation at serine 209 in certain cell types and may play a role in viral RNA translation [35,37]. To assess whether eIF4E phosphorylation occurred under our experimental conditions, we harvested cellular lysates from control and iRapKO MEFs infected with CHIKV-LR and WNV (MOI = 3) at 12 hpi. We found that eIF4E phosphorylation occurs in iRapKO conditions at 12 hpi as previously reported and is independent of infection status [35]; but phosphorylation of eIF4E in iRapKO cells did not compensate for the lack of mTORC1 signaling to promote CHIKV protein production in our studies (Figure 2F).

3.2. Impact of Raptor Knockout on mTORC1 Effectors

Our data show that mTORC1 has a significant impact on viral growth and viral protein production. We next defined the role of the primary mTORC1 downstream targets, S6K and 4EBP, in viral protein production. We inoculated ethanol-induced control cells and iRapKO cells as above with mock or WNV (MOI = 3) inoculums and harvested whole cell lysates at 0, 3, 12, 24 and 48 h post-infection. Western blot analysis for Raptor protein expression and eIF4E expression revealed that knockdown of Raptor expression in iRapKO cells had no significant impact on total eIF4E expression in WNV-infected cells compared to mock-inoculated cells (Figure 3A).

We previously reported that WNV infection induced p70S6K activation in serum-starved cells [8]. We extended these data by determining the role of downstream effectors of mTORC1 activity. Control and iRapKO cells were inoculated with mock or WNV (MOI = 3) as above, and cells were harvested for Western blot analysis at 12, 24 and 48 h post-infection (Figure 3B–D). On Western blot analysis, control cells exhibit phosphorylation of mTORC1 effector p70S6K at T389, while iRapKO cells exhibit loss of p-p70S6K expression, but unchanged total p70S6K expression in both the mock- and WNV-infected groups (Figure 3B). Thus, infection with WNV does not stimulate phosphorylation of p70S6K via an alternative pathway.

Following WNV infection of control MEF cells, total rpS6 expression was increased at 24 and 48 h post-infection, while phosphorylated rpS6 at S235/236 was increased at 48 h post-infection when compared to mock-infected cells (Figure 3C). In WNV-inoculated iRapKO cells, phosphorylated rpS6 expression was decreased at all time points, and total rpS6 expression increased across all time points and treatments (Figure 3C). While S6K and rpS6 are exclusively phosphorylated by mTORC1, eIF4B is phosphorylated by mTORC1 and other cellular kinases. Thus, we evaluated the phosphorylation status of eIF4B at serine 422 in the previously-described treatment groups and time points. We found that total and phosphorylated eIF4B expression was not significantly altered by WNV infection or Raptor expression (Figure 3D). These data show that iRapKO has targeted effects on the expression of downstream S6K-dependent signaling events with little off-target effects.

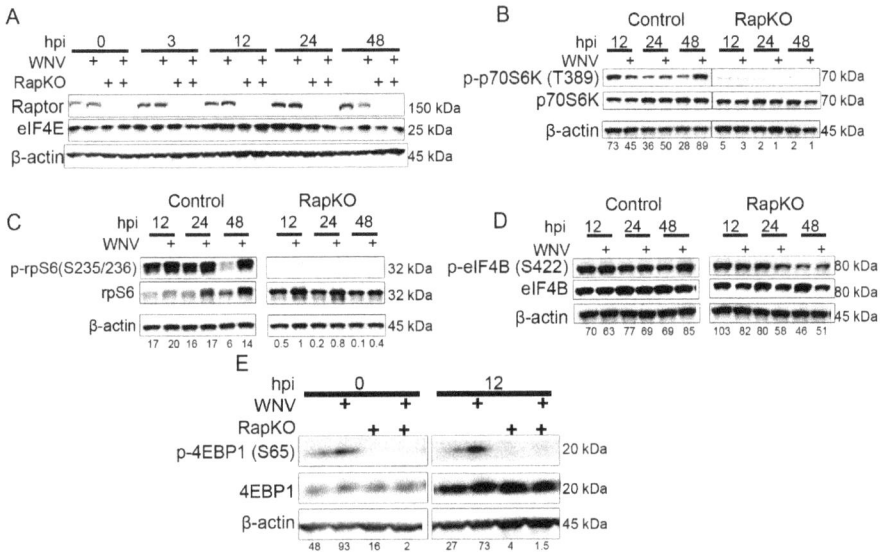

Figure 3. Raptor knockout results in a loss of p70S6K and 4EBP phosphorylation. Cells were either mock- or WNV-inoculated (MOI = 3) in control and iRapKO MEF cells, total cellular protein lysates were collected at the indicated hpi, and probed using antibodies for the indicated target proteins by western blot analysis. (**A**) Expression of Raptor and eIF4E; (**B**) Expression of phospho-p70S6K (tyrosine 389) and total p70S6K; (**C**) Expression of phospho-rpS6 (serine 235/236) and total rpS6. (**D**) Expression of phospho-eIF4B (serine 422) and total eIF4B; (**E**) Expression of phospho-4EBP1 (serine 65) and total 4EBP1. All images are representative of two independent experiments. β-actin was used as the loading control for all experiments. Densitometry values are provided for phospho-protein expression and corrected for β-actin expression.

We next defined the role of Raptor expression and WNV infection on 4EBP expression, the other primary target of mTORC1 activity. Control and iRapKO cells were inoculated with mock or WNV (MOI = 3) and whole cell lysates harvested at 0 h and 12 h post-infection (Figure 3E). Western blot analysis revealed loss of 4EBP phosphorylation signal at 0 h and 12 h post-infection in WNV-inoculated iRapKO cells, while total 4EBP levels remained stable between the two conditions (Figure 3E). In WNV-inoculated control cells, we also found evidence of increased phosphorylation of 4EBP at serine 65. Since this is the terminal phosphorylation event in sequential 4EBP phosphorylation, these data suggest that WNV-induced mTORC1 activity results in increased 4EBP phosphorylation with a subsequent increase in unbound eIF4E, cap-binding protein. Overall, these data show that loss of mTORC1 activity in this system results in decreased signaling events through the S6K/rpS6 and 4EBP pathways in iRapKO cells.

3.3. Loss of S6K/rpS6 Activity Does Not Impact WNV Growth

Next, we defined the role of the mTORC1/S6K/rpS6 signaling pathway in WNV protein expression. We first determined the role of rpS6 phosphorylation at serine 235 and serine 236 by using a MEF cell line that expresses an alanine substitution at positions 235 and 236 in the rpS6 (rpS6$^{P-/-}$) protein [38].

RpS6$^{p-/-}$ and matched, wild-type MEFs (rpS6$^{p+/+}$) were infected with mock or WNV (MOI = 3); supernatants and whole cell lysates were collected for viral titer analysis and Western blot at 6, 24 and 48 h post-infection. Viral titer analysis revealed no significant difference in WNV growth over a 48-h period (p = 0.30, two-way ANOVA, n = 11; Figure 4A). Western blot analysis of whole cell lysates at the same time points revealed no significant change in WNV envelope (ENV) protein expression despite loss of phosphorylation of rpS6 at both residues 235 and 236 (Figure 4B).

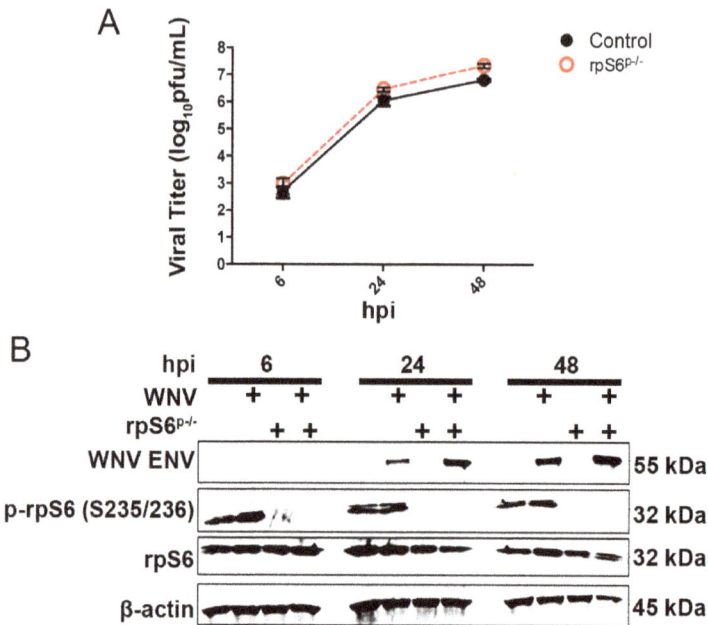

Figure 4. Loss of rpS6 phosphorylation does not impact WNV growth or protein expression. (**A**) Single-step growth curve of WNV (MOI = 3) in rpS6$^{-/-}$ and wild-type control MEFs as determined by standard plaque assay of the cell supernatant; (**B**) expression of the WNV envelope (WNV ENV), phospho-rpS6 (serine 235/236) and total rpS6 as determined by Western blot analysis of mock- and WNV-infected (MOI = 3) rpS6$^{-/-}$ and wild-type cells. Total cellular protein lysates were collected at the indicated hpi and probed using antibodies against the indicated targets. β-actin was used as the loading control. Images are representative of two independent experiments.

Although rpS6 is involved in the initiation complex, these results indicated that rpS6 phosphorylation is dispensable for the support of WNV protein expression [19].

Next, we obtained S6K1/2$^{-/-}$ MEFs as previously described to determine the contribution of S6K1 and S6K2 signaling to WNV protein expression [39]. We used an S6K1 and S6K2 double knockout MEF cell line (S6K1/2$^{-/-}$) because S6K1 and S6K2 are capable of compensatory upregulation [16]. Syngeneic control MEF cells and S6K1/2$^{-/-}$ MEFs were inoculated with WNV (MOI = 3) and supernatants obtained at 12, 24 and 48 h post-infection for viral titer analysis. We found no significant difference in WNV growth comparing S6K1/2$^{-/-}$ and S6K1/2$^{+/+}$ control MEF cells (p = 0.17, two-way ANOVA, n = 6, Figure 5A).

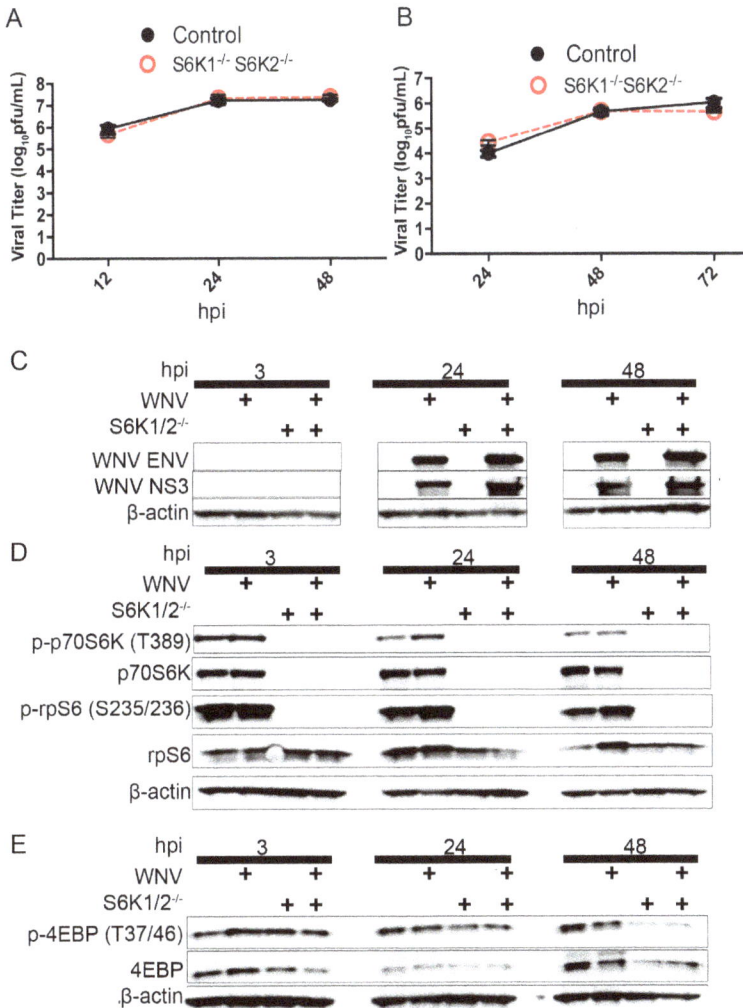

Figure 5. P70S6K activity does not support WNV growth or protein expression. (**A**) Single-step growth curve of WNV (MOI = 3) in S6K1/2$^{-/-}$ and wild-type control MEFs; (**B**) Multi-step growth curve (MOI = 0.001) of WNV in S6K1/2$^{-/-}$ and wild-type control MEFs. Supernatants were harvested at the indicated hpi for standard plaque assay. $N = 6$ replicates per time point; (**C**) WB analysis of mock- and WNV-infected (MOI = 3) S6K1/2$^{-/-}$ and wild-type cells for WNV protein expression. Cellular lysates were collected at the indicated hpi and probed using antibodies for WNV ENV, WNV NS3, and β-actin loading control. Representative of 2 independent experiments; (**D**) WB analysis of mock- and WNV-infected (MOI = 3) S6K1/2$^{-/-}$ and wild-type cells for rpS6 and p70S6K expression and phosphorylation. Cellular lysates were collected at the indicated hpi and probed using antibodies against p-p70S6K [389], total p70S6K, p-rpS6 [S235/236], total rpS6, and β-actin loading control. Representative of two independent experiments; (**E**) WB analysis of mock- and WNV-infected (MOI = 3) S6K1/2$^{-/-}$ and wild-type cells for 4EBP expression and phosphorylation. Cellular lysates were collected at the indicated hpi and probed using antibodies for p-4EBP [T37/46], total 4EBP, and β-actin loading control. Representative of two independent experiments.

145

Next, we evaluated the role of S6K expression in WNV growth following multiple steps of virus entry, replication, and egress using a multi-step growth curve. Syngeneic control MEF cells and S6K1/2$^{-/-}$ MEFs were inoculated with WNV (MOI = 0.001) and supernatants obtained at 24, 48 and 72 h post-infection for viral titer analysis. We found no significant difference in WNV growth comparing S6K1/2$^{-/-}$ and S6K1/2$^{+/+}$ control MEF cells ($p = 0.2$, two-way ANOVA, $n = 6$, Figure 5B). Next, we inoculated S6K1/2$^{-/-}$ and S6K1/2$^{+/+}$ MEF cells with mock or WNV (MOI = 3) and harvested whole cell lysates for Western blot analysis at 3, 24 and 48 h post-infection. We analyzed the expression of both nonstructural protein 3 (NS3) and ENV proteins from WNV. Both proteins are translated from the same polyprotein, but the individual fate of these proteins is different, with NS3 remaining intracellular; and ENV structural protein is packaged during viral egress. This approach allows for an additional control by comparing ENV, a protein exported out of the cell, to NS3, a protein that accumulates within the cell. We found that control and S6K1/2$^{-/-}$ MEFs exhibited similar levels of WNV NS3 and ENV protein expression at 24 and 48 h post-infection (Figure 5C). We also completed Western blot analysis for S6K and 4EBP signaling events in mock- and WNV-infected (MOI = 3) control and S6K1/2$^{-/-}$ MEFs at 3, 24 and 48 h post-infection. We found that knockout of S6K1 and S6K2 eliminated the expression of phosphorylated p70S6K, total p70S6K and phosphorylated rpS6 at all time points and in all treatment groups (Figure 5D). Total rpS6 expression was unchanged by S6K1/2 knockout. In the absence of S6K protein expression, we also determined if 4EBP phosphorylation was increased due to decreased competition for mTORC1-dependent phosphorylation of 4EBP from S6K. In the same treatment groups, we found no evidence of increased 4EBP phosphorylation in the S6K1/2$^{-/-}$ MEFs compared to controls (Figure 5E).

3.4. Knockout of 4EBP Expression Leads to Decreased WNV Growth and Decreased Viral Protein Production

We have shown that mTORC1 activity supports the expression of WNV proteins; however, loss of the mTORC1 effectors S6K and phosphorylated rpS6, both of which regulate translation initiation rates, had no significant impact on WNV growth or protein expression. Next, we inoculated 4EBP1/2$^{-/-}$ MEF cells and matched 4EBP1/2$^{+/+}$ control cells with WNV (MOI = 3) and collected the supernatant for viral titer analysis. Input virus was assayed using reverse transcription PCR (RT-PCR) and found to be equivalent at t = 0 (data not shown). WNV growth was significantly ($p < 0.0001$) decreased in WNV-infected 4EBP1/2$^{-/-}$ MEF cells at 12 h (4.3×10^3 pfu/mL $\pm 3.8 \times 10^2$), 24 h (4.4×10^5 pfu/mL $\pm 7 \times 10^4$), 36 h (8.6×10^5 pfu/mL $\pm 2 \times 10^5$) and 48 h (2.2×10^6 pfu/mL $\pm 1.7 \times 10^5$) compared to control cells at 12 h (5.2×10^4 pfu/mL $\pm 9.6 \times 10^3$), 24 h (5.4×10^6 pfu/mL $\pm 1.2 \times 10^6$), 36 h (1.4×10^7 pfu/mL $\pm 2.5 \times 10^6$) and 48 h (1.8×10^7 pfu/mL $\pm 2.5 \times 10^6$, $n = 6$, two-way ANOVA; Figure 6A).

Thus, WNV-infected 4EBP1/2$^{-/-}$ MEF cells exhibit an 8–16-fold reduction in viral growth at all time points compared to wild-type control MEF cells. To evaluate the effect of 4EBP expression on WNV growth over multiple cycles of viral entry, replication and egress, we used a multi-step growth curve analysis. We inoculated 4EBP1/2$^{-/-}$ MEF cells and matched 4EBP1/2$^{+/+}$ control cells with WNV (MOI = 0.001) and collected supernatant for viral titer analysis. We found that WNV growth was significantly ($p < 0.0001$) decreased at the 48-h and 72-h time points of infection with a mean difference of 2.8×10^6 pfu/mL $\pm 5 \times 10^5$ and 2.9×10^6 pfu/mL $\pm 5 \times 10^6$, respectively (two-way ANOVA, $n = 6$; Figure 6B).

Next, we inoculated 4EBP1/2$^{-/-}$ and control MEFs with WNV (MOI = 3) or mock treatments as above and harvested cellular lysates at 24 h post-infection for Western blot analysis. We found that 4EBP1/2$^{-/-}$ MEF cells exhibit loss of terminal phosphorylation (serine 65) of 4EBP and loss of total 4EBP expression with no change in the expression of eIF4E (Figure 6C). Interestingly, these data imply that 4EBP binding has no effect on the turnover of total eIF4E. Despite ongoing expression of eIF4E, we found significantly decreased expression of the WNV envelope protein and to a lesser extent NS3 in 4EBP knockout cells compared to control cells (Figure 6D). Despite significant changes to 4EBP signaling, 4EBP1/2$^{-/-}$ MEFs exhibit no significant alterations in the expression of phosphorylated p70S6K at threonine 389, total p70S6K, phosphorylated rpS6 at serine 235 and 236 and total rpS6

(Figure 6E). These data suggest that mTORC1-dependent interactions with 4EBP/eIF4E support WNV growth in host cells.

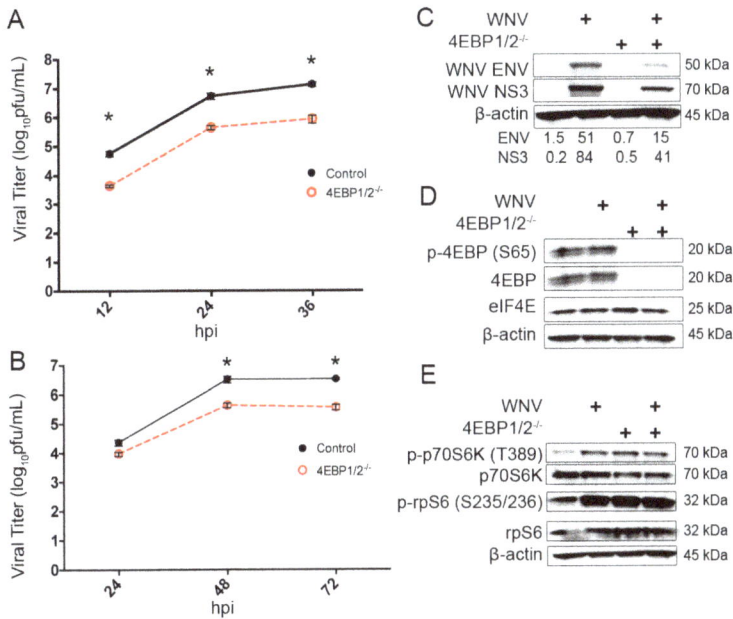

Figure 6. Expression of 4EBP1 and 4EBP2 support WNV growth and protein expression. (**A**) Single-step growth curve of WNV (MOI = 3) in 4EBP1/2$^{-/-}$ and wild-type control MEFs; (**B**) Multi-step growth curve of WNV (MOI = 0.001) in 4EBP1/2$^{-/-}$ and wild-type control MEFs. Supernatants were harvested at indicated hpi for standard plaque assay analysis. $n = 6$, * $p < 0.0001$; (**C**) Confirmation of 4EBP1/2$^{-/-}$ (knockout) phenotype. Cellular lysates were harvested at 24 hpi and WB analysis completed using antibodies for total 4EBP, phospho-4EBP (serine 65) and eIF4E. (**D**) WNV protein synthesis in 4EBP1/2$^{-/-}$ and matched control MEF cells. Cellular lysates were harvested at 24 hpi and subjected to WB analysis using antibodies against WNV NS3 and WNV ENV. (**E**) S6K pathway activity in 4EBP1/2$^{-/-}$ and matched control MEF cells. Cellular lysates were harvested at 24 hpi for WB analysis using antibodies against total p70S6K, p-p70S6K (threonine 389), total rpS6, and p-rpS6 (serine 235/236). All images representative of two independent experiments. β-actin is shown as a loading control for all western blots. Densitometry values provided for indicated viral proteins corrected for β-actin expression.

3.5. eIF4F Complex Formation Supports WNV Virus Growth and Protein Expression

Since 4EBP is a major regulator of initiation complex (eIF4F) formation, we determined the role of eIF4F complex formation on WNV growth and protein expression by using the inhibitor 4EGI-1. 4EGI-1 blocks the interaction between eIF4E and eIF4G, thereby preventing the formation of the pre-initiation complex on mRNA [40]. This compound is a valuable tool for defining how viruses interact with eIF4E and the eIF4F complex to facilitate translation [41].

Vero cells were inoculated with WNV, CHIKV-LR or EMCV at an MOI of three, and viruses were allowed to adsorb on the monolayer for 1 h at 37 °C. After incubation, cells were washed with warmed 1× PBS and media containing 4EGI-1 at concentrations of 100, 75, 50 and 10 μM, a DMSO vehicle control (0 μM) or an untreated control infection. Supernatants were harvested at 0, 6 and 12 hpi and analyzed to determine viral titer. We found that treatment with 4EGI-1, in a dose-dependent manner, significantly ($p < 0.0001$) decreased WNV growth at 12 h with 4EGI-1 concentrations of 75 μm (2.1×10^3 pfu/mL ± 482) compared to untreated controls at 12 h treated with PBS (5.1×10^5 pfu/mL ± 6.7×10^4) and DMSO (6.7×10^5 pfu/mL ± 1.3×10^5, $n = 12$, two-way ANOVA with multiple comparisons; Figure 7A).

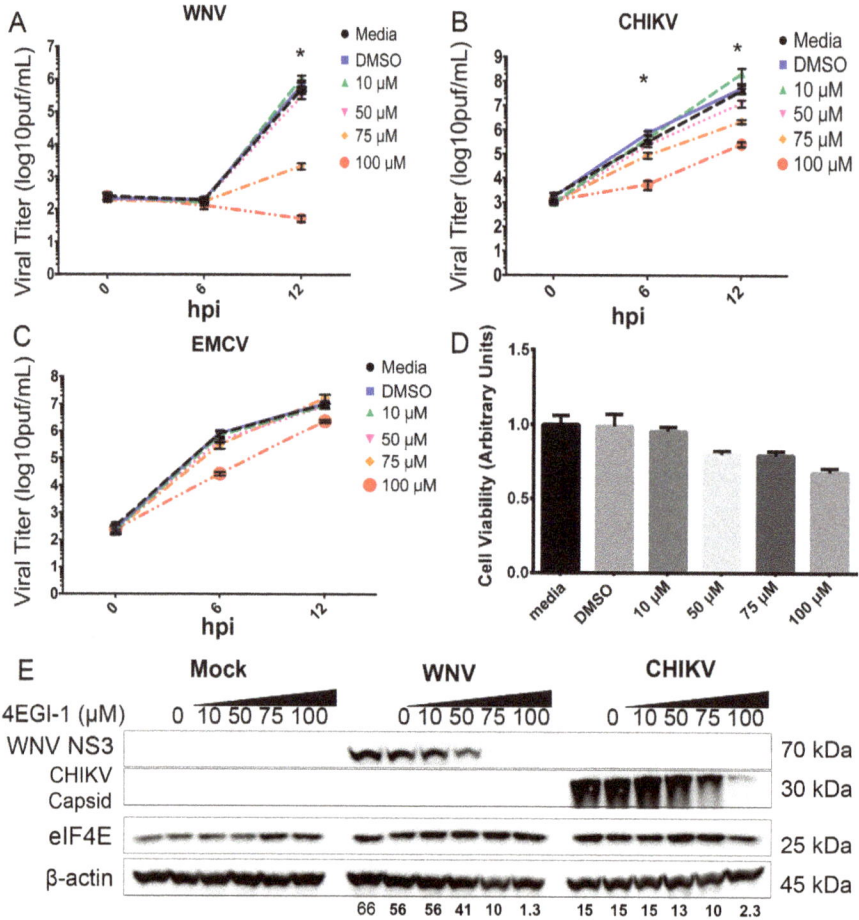

Figure 7. eIF4F complex formation supports 5′-capped positive-strand RNA viral growth and protein expression. Single-step growth curves (MOI = 3) for (**A**) WNV, (**B**) CHIKV-LR and (**C**) EMCV in 4EGI-treated Vero cells as determined by standard plaque assay of supernatants. 4EGI-1 was added at the indicated micromolar (μM) concentrations at t = 0 hpi, and supernatants were analyzed by plaque assay at the indicated times. Viral titer data are presented as log10 plaque forming units per mL of supernatant (log10 pfu/mL). $N = 9$–12 replicates per time point, * $p < 0.05$; (**D**) MTT assay for cellular viability of 4EGI-treated Vero cells. MTT cleavage products were solubilized and measured by spectrophotometry at 570 nm. Dimethyl sulfoxide (DMSO) was used as a solvent control, and all groups were normalized to a media control of 1.0 arbitrary units; (**E**) Western blot analysis of viral protein synthesis in mock, WNV-infected (MOI = 3) and CHIKV-infected (MOI = 3) Vero cells treated with the 4EGI-1 inhibitor. Vehicle control denoted by 4EGI-1 concentration of 0 μM. Images are representative of two independent experiments. Densitometry values provided for indicated viral proteins corrected for β-actin expression.

Thus, there was a 242-fold decrease in WNV growth in the 75 μm 4EGI-1 treatment group. As described above, we used CHIKV infection as an additional control for capped-RNA virus growth to ensure that the broad effect of 4EGI-1 was not specific to WNV. We found that treatment with 4EGI-1, in a dose-dependent manner, significantly ($p < 0.05$) decreased CHIKV-LR growth at 6 h and 12 h with a 4EGI-1 concentration of 75 μm (2.3×10^6 pfu/mL $\pm 4.7 \times 10^5$) compared to controls at 12 hpi that were treated with PBS (4.3×10^7 pfu/mL $\pm 1.1 \times 10^7$) and DMSO (5.4×10^7 pfu/mL $\pm 1 \times 10^7$, $n = 9$, two-way ANOVA with multiple comparisons; Figure 7B). Next, we utilized EMCV infection as a control for toxicity and cap-dependent translation since EMCV protein production is regulated by IRES-dependent initiation and 4EGI-1 should have no effect on EMCV growth. We found that EMCV growth was independent of 4EGI-1 treatment and exhibited no significant ($p > 0.05$) decrease in viral growth between treatment groups ($n = 12$, two-way ANOVA; Figure 7C). Previous work has suggested that EMCV growth is independent of eIF4F complex formation [42]. Since the 6-h time point appeared to exhibit decreased EMCV growth at the 100-μm dose of 4EGI-1, we completed a multiple comparisons test of our two-way ANOVA and found no significant difference between the vehicle-treated control and the 100-μm dosed 4EGI-1 inhibitor despite 12 replicate experiments. However, the decrease in EMCV growth at this time point may be related to toxicity at the higher dose of 4EGI-1. Thus, we completed toxicity assays for the 4EGI-1 inhibitor.

To evaluate cellular viability following 4EGI-1 treatments, we completed MTT assays at 12 h post-treatment with indicated doses of inhibitor. As shown in Figure 7D, 4EGI-1 had no statistically-significant impact on cellular viability at concentrations below 100 μM in our Vero cell system. Cells treated with the highest levels of 4EGI-1 (100 μM) exhibited 65% viability compared to untreated controls at 12 h post-treatment. Based on this toxicity data, we analyzed only 4EGI-1 doses below the 100 μM dose for viral growth studies. These data show that following 4EGI-1 treatments below the 100-μm dose, cells were viable and capable of biosynthetic processes. This conclusion is supported by evidence of EMCV growth following 4EGI-1 treatment at doses of inhibitor below 100 μM.

Next, we determined the effect of 4EGI-1 treatment on WNV and CHIKV protein production. Cells were inoculated with mock inoculum, WNV (MOI = 3) or CHIVK (MOI = 3) and incubated for 1 h as above. After washing cells, medium was added with the indicated concentrations of 4EGI-1 inhibitor (Figure 7E). Cells were harvested at 12 h post-infection and lysates processed for Western blot analysis using antibodies to WNV NS3 protein, CHIKV capsid and total eIF4E. We found that treatment with 4EGI-1 results in a dose-responsive loss of viral protein production for both WNV- and CHIKV-inoculated cells (Figure 7E). This effect was independent of the effects on total eIF4E production indicating that the 4EGI-1 inhibitor disrupted initiation complex formation without altering the expression of cap-binding proteins like eIF4E. These data are in agreement with viral growth data that show 4EGI-1 dose-responsive inhibition of CHIKV and WNV growth.

4. Discussion

Flaviviruses have limited genomic coding capacity and, thus, have evolved mechanisms designed to utilize the host RNA translation machinery to successfully express viral proteins. The mechanisms by which (+) sense, capped RNA viruses recruit translation initiation factors to the viral RNA to promote translation has remained incompletely defined. Moreover, arboviruses, such as CHIKV and WNV, must translate their genomes in evolutionally-distant vertebrate and invertebrate hosts and, therefore, must be able to manipulate the translational apparatus in very disparate systems [6]. We now show that WNV growth and protein expression is dependent on specific host-cell initiation events that involve interactions with the 4EBP and the eIF4F complex. We think these interactions are specific for the virus because other cellular translation events downstream of S6 kinase are not required for WNV gene expression. These data show that WNV utilizes specific translation initiation signals to support viral protein production. From an evolutionary standpoint, some RNA viruses, like the arboviruses, may modulate the TORC1/4EBP/eIF4E pathway because this system regulates cellular cap-binding events and is highly conserved in eukaryotic cells from yeast to mammals [9,43].

Previously published results have demonstrated that WNV infection induces the activation of the mTORC1/S6K signaling pathway as early as 3 h post-infection, but the cap-dependent translation initiation effectors downstream of TORC1 responsible for viral growth were not known [8]. In this study, we demonstrate that WNV depends on mTORC1/Raptor activity for viral growth over multiple life cycles of the virus. This effect may apply broadly to 5′-capped-RNA viruses, such as CHIKV, since we found that CHIKV is unable to sustain growth in the absence of mTORC1 activity. However, we show that viruses like EMCV that initiate viral RNA translation using an IRES replicate in the absence of TORC1 activity. Recent work studying the interactions between CHIKV and mTOR signaling have revealed differing conclusions for the role of mTORC1 activity in CHIKV growth [35,36]. Our studies build upon this existing body of work and demonstrate that capped RNA viruses, such as WNV and CHIKV are dependent on mTORC1 signaling to support viral growth independent of eIF4E phosphorylation status. We found evidence of phosphorylation of eIF4E in iRapKO MEF models; however, phosphorylation of eIF4E occurred in all Raptor knockout cells whether infected with virus or mock infected. Thus, we conclude that phosphorylation of eIF4E is not sufficient to overcome loss of mTORC1 activity in the Raptor knockout MEF system. Differences in the experimental results between groups may be due to approaches used to inhibit mTORC1. We use an inducible Raptor gene deletion system that results in a total loss of mTORC1 activity while other studies used small interfering RNA (siRNA) approaches and biochemical inhibitors, which may have different effects on cellular translation initiation for the duration of an experiment. For example, P70S6K and 4EBP compete for Raptor binding prior to phosphorylation by mTORC1 [13], and phosphorylation of 4EBP1 recovers from mTORC1 inhibition much more rapidly and effectively than p70S6K phosphorylation. Thus, treatment with an mTOR catalytic site inhibitor blocks phosphorylation for both species, but 4EBP is more resistant to mTOR inhibition because it recovers quickly [14]. This implies that studies that use transient pharmacologic inhibition of mTOR signaling may disproportionately inhibit p70S6K activity more than 4EBP activity.

Our data also show that canonical markers of mTORC1 activation, phosphorylation of p70S6K and rpS6 have no impact on WNV growth and no effect on viral protein production for positive strand viruses, such as WNV. We used an MEF cell line with gene knockout of *S6K1* and *S6K2* with resulting loss of p70S6K activity. Based on these studies, S6K signaling downstream of mTORC1 is dispensable for WNV growth despite the established role of S6K in promoting protein translation [20]. This is important since p70S6K is a major, central signaling kinase, and these findings support two important conclusions. First, different mRNAs require different support from host translational signaling, and RNA viruses like WNV utilize specific host signals to support translation, while other host signals, such as S6K and rpS6, are dispensable for the virus. Second, the reason for decreased WNV growth in the Raptor knockout cells cannot be attributed to loss of central, signaling kinases since knockout of S6K had no effect on WNV growth or protein expression.

Conversely, we found that WNV growth and protein production were dependent on 4EBP expression and eIF4F complex formation. Without 4EBP expression to regulate the availability of eIF4E in host cells, we found a significant WNV growth defect and decreased viral protein production. This was an unexpected result as 4EBP is considered a negative regulator of eIF4E function, and loss of the negative regulator should result in a larger free pool of eIF4E. Instead, we found no change in total eIF4E expression and found that knockout of 4EBP decreased viral growth and viral protein synthesis in both a single-step and a multi-step viral growth curve. We propose several possible explanations for these findings. Interactions between mTORC1, 4EBP and eIF4E may result in different localization of the eIF4E pool to support translation events at different cellular sites such as cytoplasmic or membrane-bound ribosomes. Alternatively, the results with 4EBP may be complicated by involvement with changes in host translation events of innate immune transcripts. In fact, translation of specific subsets of mRNA defined by the 5′ UTR structure may be partially determined by specific interactions with eIF4E and 4EBP [38,44–46].

Using multi-step WNV growth curves, we also found that Raptor knockout MEF cells exhibited a more profound effect on WNV growth (Figure 2A) when compared to the WNV multi-step growth curve in 4EBP knockout cells (Figure 6B). These data imply that 4EBP may be an important output of control for WNV-induced mTORC1 activity, but may not be the only outcome of mTORC1 activity. Our prior data show that WNV activates mTORC1 activity independent of autophagy and cell cycle [8,31]. Thus, the loss of mTORC1-dependent phosphorylation and inhibition of 4EBP has a more profound effect on WNV growth than loss of 4EBP. This is most likely due to decreased availability of eIF4E in the Raptor knockout system, especially as infection progresses. This potential mechanism is supported by the WNV multi-step growth curve in Raptor knockout MEF cells. While limiting eIF4E is available, WNV growth occurs early in Raptor knockout MEFs. However, as infection continues, sequestration of eIF4E by unphosphorylated 4EBP results in a quick plateau and then fall in WNV growth. This effect is even more profound following CHIKV infection, which is likely due to virus reliance on eIF4E binding for both genomic and subgenomic RNA translation.

Since deletion of 4EBP in our studies is not equivalent to preventing mTORC1-dependent phosphorylation of 4EBP, we evaluated the role of direct interactions between eIF4E and eIF4G at the cap using the 4EGI-1 inhibitor. We found that eIF4F complex formation supports (+) strand RNA virus growth and protein production in a dose-dependent fashion. Thus, we conclude that the mTORC1, 4EBP and eIF4E pathway supports viral protein production and viral growth. Based on our work to date, we propose a model for WNV-induced changes in host cell translation initiation signaling (Figure 8).

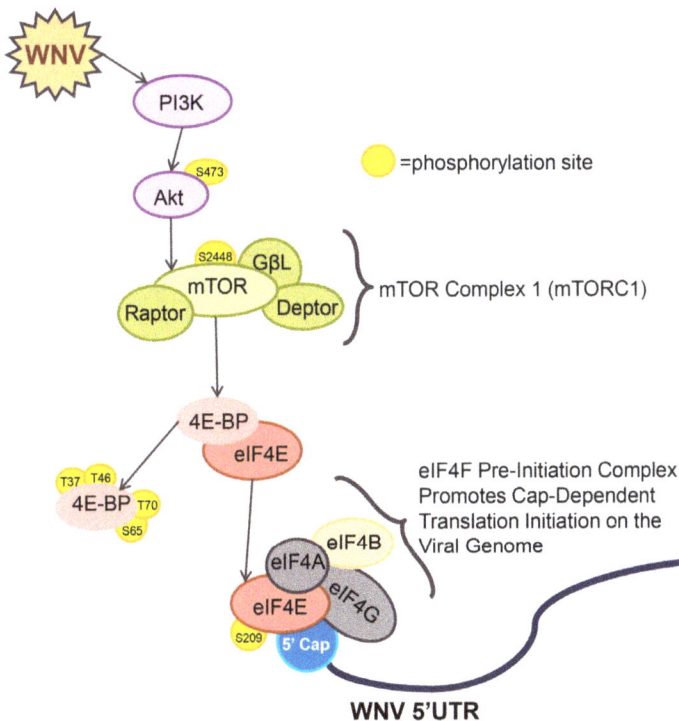

Figure 8. Working model for WNV activation of mTOR and the effect on viral RNA translation initiation events. WNV infection of host cells leads to activation of mTORC1 resulting in increased phosphorylation of the translation-initiation repressor 4EBP and dissociation from eIF4E. Free eIF4E binds to the 5′ cap on WNV genomic RNA and assembles the eIF4F pre-initiation complex required for cap-dependent translation of viral RNA, resulting in viral protein expression and viral growth.

Our observations suggest that the loss of mTORC1 impacts WNV replication at the level of translation, but not genomic replication and that these are two distinct, but linked processes in the flavivirus replication cycle. As the (+) sense RNA is the template for both genomic packaging and viral RNA translation, viral (+) sense RNA must be actively recruited for each function independently, as these disparate functions cannot occur concurrently on the same template. Based on our experimental observations, the deletion of Raptor and inactivation of mTORC1 may be impacting WNV growth in two ways. Deletion of mTORC1 activity may inhibit recruitment of the 40S pre-initiation complex and initiation of (+) sense RNA scanning or loss of mTORC1 cap-dependent initiation signals may result in a reduced pool of viral RNA in heavy polysomes. The polysome recruitment model would provide a definitive competitive advantage for viral RNA translation in a cellular environment in which viral RNA actually represents the minority of total RNAs species. At early stages of WNV infection, a single viral RNA in a heavy polysome could produce more functional protein than the equivalent genome load when limited to monosomes under conditions of reduced translation initiation. Further studies will be necessary to determine if mTORC1 activity is driving the formation of polysomes on 5′-capped (+) sense RNA viruses, such was WNV, and if loss of polysome formation on the viral genome is the mechanistic basis for our observed growth defects in this model.

Maintenance of proteostasis via mTORC1/4EBP/eIF4F is a key point of disruption in diverse disease mechanisms; whether it be virally induced, oncogenic or neurodegenerative [47–53]. Many of these pathogenic states are marked by the hyperactivation of Akt and mTORC1 and the downstream effectors S6K and 4EBPs [26,47,49,51,54–56]. The initiation of canonical cap-dependent translation depends on eIF4E and eIF4F complex formation, making the mTORC1/4EBP/eIF4E signaling cascade a potential target for future therapeutics for diverse viral infections [26,48,57]. This approach is aided by the development of new technologies, such as transcriptome-wide analysis, which has allowed researchers to probe the requirement for different parts of the eIF4F complex to translate RNAs with specific sequences or structures in the 5′ UTR [58].

We used EMCV as a control in our studies for cap-dependent virus translation since this virus uses an IRES. Prior work has shown that 4EGI may still inhibit EMCV and other IRES-controlled RNA viruses despite the fact that the mechanism of 4EGI-inhibition is thought to be specific to cap-binding events [41]. However, our data show that EMCV viral growth is largely resistant to the effects of the 4EGI inhibitor, but EMCV may exhibit decreased growth at higher doses of 4EGI-1 due to cell toxicity. We believe that our studies add to the existing literature on this topic since we used live virus infections and assayed viral titer, while prior studies have used reporter systems. These data may imply that other elements in the RNA genome of EMCV may be required to support IRES-independent activity to escape the effects of the 4EGI inhibitor. Additional studies with viruses that utilize different initiation strategies are required to understand the complex interactions between *cis*-acting RNA elements with translation initiation events for RNA viruses.

There are important weaknesses of the current study that must be taken into account. First, our studies as reported do not directly assay for RNA translation events. The logical extension of this work is to utilize new technologies such as ribosome profiling and other technical approaches to evaluate the effect of specific host signaling events during viral infection on viral and host RNA translation directly. Second, the effect of the 4EGI-1 inhibitor on the eIF4F complex may occur distal from the cap by competing with 4EBP. Thus, the effect of the 4EGI-1 inhibitor may be related to the viral requirement for 4EBP activity. This is an important consideration in these studies, and future studies defining the effect of 4EGI-1 on cap formation in vivo are required to determine the effect of this inhibitor on eIF4F formation.

We used CHIKV in several experiments as an additional 5′ cap-dependent RNA virus control to show that the effects of cellular manipulation on viral growth are not specific to just WNV. It is important to note that many of the features of CHIKV growth following knockout of mTORC1 signaling are very different from WNV. For example, CHIKV viral growth and protein production are virtually nonexistent in Raptor knockout cells, while WNV seems to grow at a reduced rate with reduced protein production. It is important to note that CHIKV has different requirements for cap-dependent translation since the structural proteins are expressed from a subgenomic transcript. Thus, it is plausible that CHIKV has different interactions with specific host-translational events, since CHIKV growth in our system exhibits increased sensitivity to mTORC1-dependent signaling. Additional studies examining the interactions between 5′-capped RNA viruses and host translational control systems will lead to novel insight into translational control mechanisms and provide novel insights into potential broad-spectrum, host-targeted therapies that specifically inhibit viral translation while protecting host immune regulation.

Acknowledgments: J.D.B. is supported by the University of Colorado School of Medicine Neuroscience Center Pilot Award, University of Colorado School of Medicine and Department of Medicine Funds. T.E.M. and N.A.M. were supported by National Institute of Allergy and Infectious Diseases (NIAID) R01 AI108725. K.D.S. is supported by University of Colorado T32 Microbiology Training grant. The authors would like to thank the many researchers who generously provided the cell lines utilized in the manuscript. Many thanks are owed to Oded Meyuhas (The Hebrew University of Jerusalem, Israel) for the rpS6$^{P-/-}$ and rpS6$^{P+/+}$ MEFs; Sara Kozma (Freidreich Meischer Institute, Switzerland) for S6K1$^{-/-}$ S6K2$^{-/-}$ and S6K1$^{+/+}$ S6K2$^{+/+}$ MEFs; Nahum Sonenberg (McGill University, Canada) for 4EBP1$^{-/-}$ 4EBP2$^{-/-}$ MEFs; and Michael N. Hall (Freidreich Meischer Institute, Switzerland) for the inducible Raptor knockout (iRapKO) MEF system.

Author Contributions: J.D.B., T.E.M. and K.D.S. conceived of and designed the experiments. K.D.S., A.R.M. and N.A.M. performed experiments. J.D.B. and T.E.M. analyzed data and contributed reagents and materials. K.D.S. and J.D.B. wrote the paper.

Conflicts of Interest: The authors declare no conflict of interest.

References

1. Egloff, M.-P.; Benarroch, D.; Selisko, B.; Romette, J.-L.; Canard, B. An RNA cap (nucleoside-2′-O-)-methyltransferase in the flavivirus RNA polymerase NS5: Crystal structure and functional characterization. *EMBO J.* **2002**, *21*, 2757–2768. [CrossRef] [PubMed]
2. Wengler, G. The NS 3 nonstructural protein of flaviviruses contains an RNA triphosphatase activity. *Virology* **1993**, *197*, 265–273. [CrossRef] [PubMed]
3. Bartelma, G.; Padmanabhan, R. Expression, purification, and characterization of the RNA 5′-triphosphatase activity of dengue virus type 2 nonstructural protein 3. *Virology* **2002**, *299*, 122–132. [CrossRef] [PubMed]
4. Szretter, K.J.; Daniels, B.P.; Cho, H.; Gainey, M.D.; Yokoyama, W.M.; Gale, M., Jr.; Virgin, H.W.; Klein, R.S.; Sen, G.C.; Diamond, M.S. 2′-O methylation of the viral mRNA cap by West Nile virus evades IFIT1-dependent and -independent mechanisms of host restriction in vivo. *PLoS Pathog.* **2012**, *8*, e1002698. [CrossRef] [PubMed]
5. Bowen, R.A.; Nemeth, N.M. Experimental infections with West Nile virus. *Curr. Opin. Infect. Dis.* **2007**, *20*, 293–297. [CrossRef] [PubMed]
6. Weaver, S.C.; Barrett, A.D.T. Transmission cycles, host range, evolution and emergence of arboviral disease. *Nat. Rev. Microbiol.* **2004**, *2*, 789–801. [CrossRef] [PubMed]
7. Edgil, D.; Polacek, C.; Harris, E. Dengue virus utilizes a novel strategy for translation initiation when cap-dependent translation is inhibited. *J. Virol.* **2006**, *80*, 2976–2986. [CrossRef] [PubMed]
8. Shives, K.D.; Beatman, E.L.; Chamanian, M.; O'Brien, C.; Hobson-Peters, J.; Beckham, J.D. West Nile virus-induced activation of mammalian Target of Rapamycin (mTOR) Complex 1 supports viral growth and viral protein expression. *J. Virol.* **2014**, *88*, 9458–9471. [CrossRef] [PubMed]
9. Dann, S.G.; Thomas, G. The amino acid sensitive TOR pathway from yeast to mammals. *FEBS Lett.* **2006**, *580*, 2821–2829. [CrossRef] [PubMed]
10. Montero, H.; García-Román, R.; Mora, S.I. eIF4E as a Control Target for Viruses. *Viruses* **2015**, *7*, 739–750. [CrossRef] [PubMed]
11. Schalm, S.S.; Fingar, D.C.; Sabatini, D.M.; Blenis, J. TOS motif-mediated raptor binding regulates 4E-BP1 multisite phosphorylation and function. *Curr. Biol.* **2003**, *13*, 797–806. [CrossRef]
12. Schalm, S.S.; Blenis, J. Identification of a conserved motif required for mTOR signaling. *Curr. Biol.* **2002**, *12*, 632–639. [CrossRef]

13. Dennis, M.D.; Kimball, S.R.; Jefferson, L.S. Mechanistic target of rapamycin complex 1 (mTORC1)-mediated phosphorylation is governed by competition between substrates for interaction with raptor. *J. Biol. Chem.* **2013**, *288*, 10–19. [CrossRef] [PubMed]

14. Choo, A.Y.; Yoon, S.-O.; Kim, S.G.; Roux, P.P.; Blenis, J. Rapamycin differentially inhibits S6Ks and 4E-BP1 to mediate cell-type-specific repression of mRNA translation. *Proc. Natl. Acad. Sci. USA* **2008**, *105*, 17414–17419. [CrossRef] [PubMed]

15. Dufner, A.; Thomas, G. Ribosomal S6 Kinase Signaling and the Control of Translation. *Exp. Cell Res.* **1999**, *253*, 100–109. [CrossRef] [PubMed]

16. Shima, H.; Pende, M.; Chen, Y.; Fumagalli, S.; Thomas, G.; Kozma, S.C. Disruption of the p70(s6k)/p85(s6k) gene reveals a small mouse phenotype and a new functional S6 kinase. *EMBO J.* **1998**, *17*, 6649–6659. [CrossRef] [PubMed]

17. Reinhard, C.; Fernandez, A.; Lamb, N.J.; Thomas, G. Nuclear localization of p85s6k: Functional requirement for entry into S phase. *EMBO J.* **1994**, *13*, 1557–1565. [PubMed]

18. Reinhard, C.; Thomas, G.; Kozma, S.C. A single gene encodes two isoforms of the p70 S6 kinase: Activation upon mitogenic stimulation. *Proc. Natl. Acad. Sci. USA* **1992**, *89*, 4052–4056. [CrossRef] [PubMed]

19. Roux, P.P.; Shahbazian, D.; Vu, H.; Holz, M.K.; Cohen, M.S.; Taunton, J.; Sonenberg, N.; Blenis, J. RAS/ERK signaling promotes site-specific ribosomal protein S6 phosphorylation via RSK and stimulates cap-dependent translation. *J. Biol. Chem.* **2007**, *282*, 14056–14064. [CrossRef] [PubMed]

20. Dennis, M.D.; Jefferson, L.S.; Kimball, S.R. Role of p70S6K1-mediated phosphorylation of eIF4B and PDCD4 proteins in the regulation of protein synthesis. *J. Biol. Chem.* **2012**, *287*, 42890–42899. [CrossRef] [PubMed]

21. Parsyan, A.; Svitkin, Y.; Shahbazian, D.; Gkogkas, C.; Lasko, P.; Merrick, W.C.; Sonenberg, N. mRNA helicases: The tacticians of translational control. *Nat. Rev. Mol. Cell Biol.* **2011**, *12*, 235–245. [CrossRef] [PubMed]

22. Gingras, A.C.; Raught, B.; Gygi, S.P.; Niedzwiecka, A.; Miron, M.; Burley, S.K.; Polakiewicz, R.D.; Wysłouch-Cieszyńska, A.; Aebersold, R.; Sonenberg, N. Hierarchical phosphorylation of the translation inhibitor 4E-BP1. *Genes Dev.* **2001**, *15*, 2852–2864. [PubMed]

23. Livingstone, M.; Bidinosti, M. Rapamycin-insensitive mTORC1 activity controls eIF4E:4E-BP1 binding. *F1000Research* **2012**, *1*, 4. [CrossRef] [PubMed]

24. Sukarieh, R.; Sonenberg, N.; Pelletier, J. The eIF4E-binding proteins are modifiers of cytoplasmic eIF4E relocalization during the heat shock response. *Am. J. Physiol. Cell Physiol.* **2009**, *296*, C1207–C1217. [CrossRef] [PubMed]

25. Rong, L.; Livingstone, M.; Sukarieh, R.; Petroulakis, E.; Gingras, A.C.; Crosby, K.; Smith, B.; Polakiewicz, R.D.; Pelletier, J.; Ferraiuolo, M.A.; et al. Control of eIF4E cellular localization by eIF4E-binding proteins, 4E-BPs. *RNA* **2008**, *14*, 1318–1327. [CrossRef] [PubMed]

26. Walsh, D. Manipulation of the host translation initiation complex eIF4F by DNA viruses. *Biochem. Soc. Trans.* **2010**, *38*, 1511–1516. [CrossRef] [PubMed]

27. Aumayr, M.; Fedosyuk, S.; Ruzicska, K.; Sousa-Blin, C.; Kontaxis, G.; Skern, T. NMR analysis of the interaction of picornaviral proteinases Lb and 2A with their substrate eukaryotic initiation factor 4GII. *Protein Sci.* **2015**, *24*, 1979–1996. [CrossRef] [PubMed]

28. Brault, A.C.; Huang, C.Y.; Langevin, S.A.; Kinney, R.M.; Bowen, R.A.; Ramey, W.N.; Panella, N.A.; Holmes, E.C.; Powers, A.M.; Miller, B.R. A single positively selected West Nile viral mutation confers increased virogenesis in American crows. *Nat. Genet.* **2007**, *39*, 1162–1166. [CrossRef] [PubMed]

29. Burrack, K.A.S.; Hawman, D.W.; Jupille, H.J.; Oko, L.; Minor, M.; Shives, K.D.; Gunn, B.M.; Long, K.M.; Morrison, T.E. Attenuating mutations in nsP1 reveal tissue specific mechanisms for control of Ross River virus infection. *J. Virol.* **2014**, *88*, 3719–3732. [CrossRef] [PubMed]

30. Hahn, H.; Palmenberg, A.C. Encephalomyocarditis viruses with short poly(C) tracts are more virulent than their mengovirus counterparts. *J. Virol.* **1995**, *69*, 2697–2699. [PubMed]

31. Beatman, E.; Oyer, R.; Shives, K.D.; Hedman, K.; Brault, A.C.; Tyler, K.L.; Beckham, J.D. West Nile virus growth is independent of autophagy activation. *Virology* **2012**, *433*, 262–272. [CrossRef] [PubMed]

32. Cybulski, N.; Zinzalla, V.; Hall, M.N. Inducible raptor and rictor knockout mouse embryonic fibroblasts. *Methods Mol. Biol.* **2012**, *821*, 267–278. [PubMed]

33. Hawman, D.W.; Stoermer, K.A.; Montgomery, S.A.; Pal, P.; Oko, L.; Diamond, M.S.; Morrison, T.E. Chronic joint disease caused by persistent Chikungunya virus infection is controlled by the adaptive immune response. *J. Virol.* **2013**, *87*, 13878–13888. [CrossRef] [PubMed]

34. Robitaille, A.M.; Christen, S.; Shimobayashi, M.; Cornu, M.; Fava, L.L.; Moes, S.; Prescianotto-Baschong, C.; Sauer, U.; Jenoe, P.; Hall, M.N. Quantitative Phosphoproteomics Reveal mTORC1 Activates de Novo Pyrimidine Synthesis. *Science* **2013**, *339*, 1320–1323. [CrossRef] [PubMed]

35. Joubert, P.-E.; Stapleford, K.; Guivel-Benhassine, F.; Vignuzzi, M.; Schwartz, O.; Albert, M.L. Inhibition of mTORC1 Enhances the Translation of Chikungunya Proteins via the Activation of the MnK/eIF4E Pathway. *PLoS Pathog.* **2015**, *11*, e1005091. [CrossRef] [PubMed]

36. Thaa, B.; Biasiotto, R.; Eng, K.; Neuvonen, M.; Götte, B.; Rheinemann, L.; Mutso, M.; Utt, A.; Varghese, F.; Balistreri, G. Differential Phosphatidylinositol-3-Kinase-Akt-mTOR Activation by Semliki Forest and Chikungunya Viruses Is Dependent on nsP3 and Connected to Replication Complex Internalization. *J. Virol.* **2015**, *89*, 11420–11437. [CrossRef] [PubMed]

37. Wang, X.; Yue, P.; Chan, C.-B.; Ye, K.; Ueda, T.; Watanabe-Fukunaga, R.; Fukunaga, R.; Fu, H.; Khuri, F.R.; Sun, S.Y. Inhibition of mammalian target of rapamycin induces phosphatidylinositol 3-kinase-dependent and Mnk-mediated eukaryotic translation initiation factor 4E phosphorylation. *Mol. Cell. Biol.* **2007**, *27*, 7405–7413. [CrossRef] [PubMed]

38. Colina, R.; Costa-Mattioli, M.; Dowling, R.J.O.; Jaramillo, M.; Tai, L.H.; Breitbach, C.J.; Martineau, Y.; Larsson, O.; Rong, L.; Svitkin, Y.V.; et al. Translational control of the innate immune response through IRF-7. *Nature* **2008**, *452*, 323–328. [CrossRef] [PubMed]

39. Pende, M.; Um, S.H.; Mieulet, V.; Sticker, M.; Goss, V.L.; Mestan, J.; Mueller, M.; Fumagalli, S.; Kozma, S.C.; Thomas, G. S6K1(−/−)/S6K2(−/−) mice exhibit perinatal lethality and rapamycin-sensitive 5'-terminal oligopyrimidine mRNA translation and reveal a mitogen-activated protein kinase-dependent S6 kinase pathway. *Mol. Cell. Biol.* **2004**, *24*, 3112–3124. [CrossRef] [PubMed]

40. Moerke, N.J.; Aktas, H.; Chen, H.; Cantel, S.; Reibarkh, M.Y.; Fahmy, A.; Gross, J.D.; Degterev, A.; Yuan, J.; Chorev, M.; et al. Small-molecule inhibition of the interaction between the translation initiation factors eIF4E and eIF4G. *Cell* **2007**, *128*, 257–267. [CrossRef] [PubMed]

41. Redondo, N.; García-Moreno, M.; Sanz, M.A.; Carrasco, L. Translation of viral mRNAs that do not require eIF4E is blocked by the inhibitor 4EGI-1. *Virology* **2013**, *444*, 171–180. [CrossRef] [PubMed]

42. Gingras, A.C.; Svitkin, Y.; Belsham, G.J.; Pause, A.; Sonenberg, N. Activation of the translational suppressor 4E-BP1 following infection with encephalomyocarditis virus and poliovirus. *Proc. Natl. Acad. Sci. USA* **1996**, *93*, 5578–5583. [CrossRef] [PubMed]

43. Foster, K.G.; Fingar, D.C. Mammalian target of rapamycin (mTOR): Conducting the cellular signaling symphony. *J. Biol. Chem.* **2010**, *285*, 14071–14077. [CrossRef] [PubMed]

44. Stumpf, C.R.; Ruggero, D. The cancerous translation apparatus. *Curr. Opin. Genet. Dev.* **2011**, *21*, 474–483. [CrossRef] [PubMed]

45. Pourdehnad, M.; Truitt, M.L.; Siddiqi, I.N.; Ducker, G.S.; Shokat, K.M.; Ruggero, D. Myc and mTOR converge on a common node in protein synthesis control that confers synthetic lethality in Myc-driven cancers. *Proc. Natl. Acad. Sci. USA* **2013**, *110*, 11988–11993. [CrossRef] [PubMed]

46. Grosso, S.; Pesce, E.; Brina, D.; Beugnet, A.; Loreni, F.; Biffo, S. Sensitivity of global translation to mTOR inhibition in REN cells depends on the equilibrium between eIF4E and 4E-BP1. *PLoS ONE* **2011**, *6*, e29136. [CrossRef] [PubMed]

47. Chong, Z.Z.; Shang, Y.C.; Zhang, L.; Wang, S.; Maiese, K. Mammalian target of rapamycin: Hitting the bull's-eye for neurological disorders. *Oxid. Med. Cell. Longev.* **2010**, *3*, 374–391. [CrossRef] [PubMed]

48. Hsieh, A.C.; Ruggero, D. Targeting eukaryotic translation initiation factor 4E (eIF4E) in cancer. *Clin. Cancer Res.* **2010**, *16*, 4914–4920. [CrossRef] [PubMed]

49. Hsieh, A.C.; Liu, Y.; Edlind, M.P.; Ingolia, N.T.; Janes, M.R.; Sher, A.; Shi, E.Y.; Stumpf, C.R.; Christensen, C.; Bonham, M.J.; et al. The translational landscape of mTOR signalling steers cancer initiation and metastasis. *Nature* **2012**, *485*, 55–61. [CrossRef] [PubMed]

50. Hopkins, K.C.; Tartell, M.A.; Herrmann, C.; Hackett, B.A.; Taschuk, F.; Panda, D.; Menghani, S.V.; Sabin, L.R.; Cherry, S. Virus-induced translational arrest through 4EBP1/2-dependent decay of 5'-TOP mRNAs restricts viral infection. *Proc. Natl. Acad. Sci. USA* **2015**, *112*, E2920–E2929. [CrossRef] [PubMed]

51. Maiese, K.; Chong, Z.Z.; Shang, Y.C.; Wang, S. mTOR: On target for novel therapeutic strategies in the nervous system. *Trends Mol. Med.* **2013**, *19*, 51–60. [CrossRef] [PubMed]

52. Chen, L.; Aktas, B.H.; Wang, Y.; He, X.; Sahoo, R.; Zhang, N.; Denoyelle, S.; Kabha, E.; Yang, H.; Freedman, R.; et al. Tumor suppression by small molecule inhibitors of translation initiation. *Oncotarget* **2012**, *3*, 869–881. [CrossRef] [PubMed]

53. De la Parra, C.; Borrero-Garcia, L.D.; Cruz-Collazo, A.; Schneider, R.J.; Dharmawardhane, S. Equol, an isoflavone metabolite, regulates cancer cell viability and protein synthesis initiation via c-Myc and eIF4G. *J. Biol. Chem.* **2015**, *290*, 6047–6057. [CrossRef] [PubMed]

54. Hsieh, A.C.; Costa, M.; Zollo, O.; Davis, C.; Feldman, M.E.; Testa, J.R.; Meyuhas, O.; Shokat, K.M.; Ruggero, D. Genetic dissection of the oncogenic mTOR pathway reveals druggable addiction to translational control via 4EBP-eIF4E. *Cancer Cell* **2010**, *17*, 249–261. [CrossRef] [PubMed]

55. Li, X.; Alafuzoff, I.; Soininen, H.; Winblad, B.; Pei, J.-J. Levels of mTOR and its downstream targets 4E-BP1, eEF2, and eEF2 kinase in relationships with tau in Alzheimer's disease brain. *FEBS J.* **2005**, *272*, 4211–4220. [CrossRef] [PubMed]

56. Martineau, Y.; Azar, R.; Müller, D.; Lasfargues, C.; El Khawand, S.; Anesia, R.; Pelletier, J.; Bousquet, C.; Pyronnet, S. Pancreatic tumours escape from translational control through 4E-BP1 loss. *Oncogene* **2014**, *33*, 1367–1374. [CrossRef] [PubMed]

57. Martineau, Y.; Azar, R.; Bousquet, C.; Pyronnet, S. Anti-oncogenic potential of the eIF4E-binding proteins. *Oncogene* **2013**, *32*, 671–677. [CrossRef] [PubMed]

58. Shatsky, I.N.; Dmitriev, S.E.; Andreev, D.E.; Terenin, I.M. Transcriptome-wide studies uncover the diversity of modes of mRNA recruitment to eukaryotic ribosomes. *Crit. Rev. Biochem. Mol. Biol.* **2014**, *49*, 164–177. [CrossRef] [PubMed]

viruses

MDPI

Communication

Analysis of the Langat Virus Genome in Persistent Infection of an *Ixodes scapularis* Cell Line

Luwanika Mlera [1], Wessam Melik [1,†], Danielle K. Offerdahl [1], Eric Dahlstrom [2], Stephen F. Porcella [2] and Marshall E. Bloom [1,*

[1]   Biology of Vector-Borne Viruses Section, Laboratory of Virology, National Institutes of Health, Hamilton, MT 59840, USA; Luwanika.Mlera@nih.gov (L.M.); Wessam.Melik@oru.se (W.M.); offerdahld@niaid.nih.gov (D.K.O.)

[2]   Genomics Unit, Research Technologies Branch, Hamilton, MT 59840, USA; eric.dahlstrom@nih.gov (E.D.); SPORCELLA@niaid.nih.gov (S.F.P.)

*   Correspondence: mbloom@niaid.nih.gov; Tel.: +1-406-375-9707

†   Present address: Faculty of Medicine, Örebro University, Örebro 70281, Sweden.

Academic Editor: Michael Holbrook
Received: 22 July 2016; Accepted: 7 September 2016; Published: 10 September 2016

Abstract: Tick-borne flaviviruses (TBFVs) cause a broad spectrum of disease manifestations ranging from asymptomatic to mild febrile illness and life threatening encephalitis. These single-stranded positive-sense (ss(+)) RNA viruses are naturally maintained in a persistent infection of ixodid ticks and small-medium sized mammals. The development of cell lines from the ixodid ticks has provided a valuable surrogate system for studying the biology of TBFVs in vitro. When we infected ISE6 cells, an *Ixodes scapularis* embryonic cell line, with Langat virus (LGTV) we observed that the infection proceeded directly into persistence without any cytopathic effect. Analysis of the viral genome at selected time points showed that no defective genomes were generated during LGTV persistence by 10 weeks of cell passage. This was in contrast to LGTV persistence in 293T cells in which defective viral genomes are detectable by five weeks of serial cell passage. We identified two synonymous nucleotide changes i.e., 1893A→C (29% of 5978 reads at 12 h post infection (hpi)) and 2284T→A (34% of 4191 reads at 12 hpi) in the region encoding for the viral protein E. These results suggested that the mechanisms supporting LGTV persistence are different between tick and mammalian cells.

Keywords: Langat virus genome; tick-borne flavivirus; persistent infection; *Ixodes scapularis*; ISE6 cells; deep-sequencing

---

The tick-borne flaviviruses (TBFVs) are associated with a variety of clinical diseases in humans, ranging from asymptomatic to mild febrile illness or severe, sometimes fatal meningoencephalitis or hemorrhagic fever [1–5]. In spite of the fact that there is an effective vaccine [6,7], there are 10,000–15,000 TBFV infections each year with mortalities as high as 20%, depending on the particular virus [2,8]. Although much of the TBFV morbidity and mortality results from acute infections, there is increasing evidence that chronic or persistent infection may lead to long-term illness and sequelae [9].

TBFV infection typically results from the bite of an infected ixodid or hard-bodied tick, and because the viruses have a global distribution, the principal vector species varies from region to region. For example, the principal vectors for Powassan virus in North America are *Ixodes scapularis* and *Ixodes cookei*, whereas *Haemaphysalis longicornis* transmits the virus in East Asia [10]. However, the Alkhurma virus utilizes a soft-bodied or argasid tick, *Ornithodoros savignyi*, as a vector, so perhaps TBFV vector competence is broader than generally appreciated [11]. Nevertheless, once the ticks are infected, the virus persists across the various life stages, and can be passed transovarially to progeny [12]. Co-feeding of infected and uninfected ticks on the same host animal demonstrates that horizontal transmission of the virus among ticks also occurs [2,13–15].

Clearly, the interactions of TBFV with invertebrate hosts present a complex vector-pathogen relationship, and the biology of virus persistence is an important facet. The study of these relationships has been greatly aided by the development of cell lines derived from ixodid ticks [16,17]. We have recently initiated studies to characterize viral persistence and determine its role in the biology of TBFV infections [18,19]. Infection of mammalian cells in vitro with TBFVs and some encephalitic mosquito-borne flaviviruses leads to an acute lytic crisis mediated by apoptosis [18,20,21]. However, a persistent infection is initiated in the few surviving cells and persistence is maintained indefinitely [19,22–25]. Using extensive unbiased next-generation sequencing, we demonstrated that defective genomes, which would be packaged to become defective interfering particles (DIPs) are not present at the initiation of viral persistence in mammalian cell lines, but are a prominent feature during the maintenance of viral persistence [19]. In marked contrast to mammalian cells, infection of cell lines derived from *Ixodes*, *Boophilus*, *Hyaloma*, *Ornithodoros* or *Rhipicephalus* tick species does not lead to apparent cell death or obvious cytopathological changes [23,26]. Furthermore, we showed that infection of ISE6 cells derived from *Ixodes scapularis* embryos [27] with a TBFV leads directly to viral persistence [23]. However, in that previous work we did not examine TBFV genomes in detail. Given the results of our studies on TBFV persistence in mammalian cells, we wanted to evaluate the viral genome stability of persistent TBFV infection in ISE6 cells with the same methodology. Therefore, in this publication, we have established a model system for TBFV persistence in ISE6 cells and have used unbiased deep-sequencing to investigate potential genomic evolution and alterations. During persistent TBFV infection of these cells, the TBFV genome was remarkably stable and no evidence of truncated genomes or DIPs was observed.

In order to do these studies, we infected 1.5×10^6 ISE6 cells in 25 cm^2 CellStar® flasks (Greiner Bio-One, Kremsmünster, Austria) with Langat TP21 virus [28] derived from a full length molecular clone [19] at a multiplicity of infection (MOI) of 5 for 1 h at 37 °C with rocking. The infecting medium was removed and cells were washed three times with phosphate-buffered saline (PBS) and maintained in a L-15C300 medium supplemented with 5% tryptose phosphate broth, 5% fetal bovine serum (FBS) (Invitrogen; Life Technologies, Carlsbad, CA, USA), and 0.1% bovine lipoprotein concentrate (MP Biomedicals, Santa Ana, CA, USA) at 34 °C. Cultures were studied at selected time points after infection.

Following infection, the infected ISE6 cells were observed closely for evidence of cytopathology or a lytic crisis, as was noted in our previous studies on mammalian cells [19]. At no time was evidence of cytopathology or crisis observed, a result consistent with our earlier studies [23].

Immunofluorescence was used to evaluate the extent of Langat virus (LGTV) infection in the ISE6 cultures. 10^5 ISE6 cells in 4-well Labtek chamber slides (Nunc®, Sigma-Aldrich, Atlanta, GA, USA) were infected at a MOI of 5, and prepared for immunofluorescent microscopy at 12, 96, and 1680 h post infection (hpi). At each time point, cells were washed twice with PBS, fixed with 4% paraformaldehyde, probed with a mouse monoclonal anti-E (11H12) antibody (a kind gift from Dr. Connie Schmaljohn, USAMRID, Fort Detrick, Frederick, MD, USA) and counterstained with 4',6-diamidino-2-phenylindole (DAPI). Examination of these preparations revealed that few cells were infected at 12 hpi. However, a higher number of cells were infected at 96 hpi as shown by positive staining for the LGTV E protein (Figure 1A). Furthermore, the fraction of ISE6 cells positive for E appeared to remain stable out to 1680 h (Figure 1A). These results indicated that most cells in the cultures were expressing LGTV proteins by 96 hpi and maintained expression for an extended period.

In order to confirm that E protein expression corresponded to a persistent infection, we determined the course of LGTV titer and genome copies. Supernatants were harvested at 12, 48, 96, 336 and 1680 hpi for virus titration using an immunofocus assay as described before [19,23]. Virus titer peaked to 2.0×10^5 ffu/mL at 96 hpi. At two weeks post infection (336 hpi), virus titer declined to 5.1×10^3 ffu/mL, but showed a modest increase to 3.7×10^4 ffu/mL at 1680 hpi (Figure 1B). Thus, a persistent infection had been initiated and was maintained.

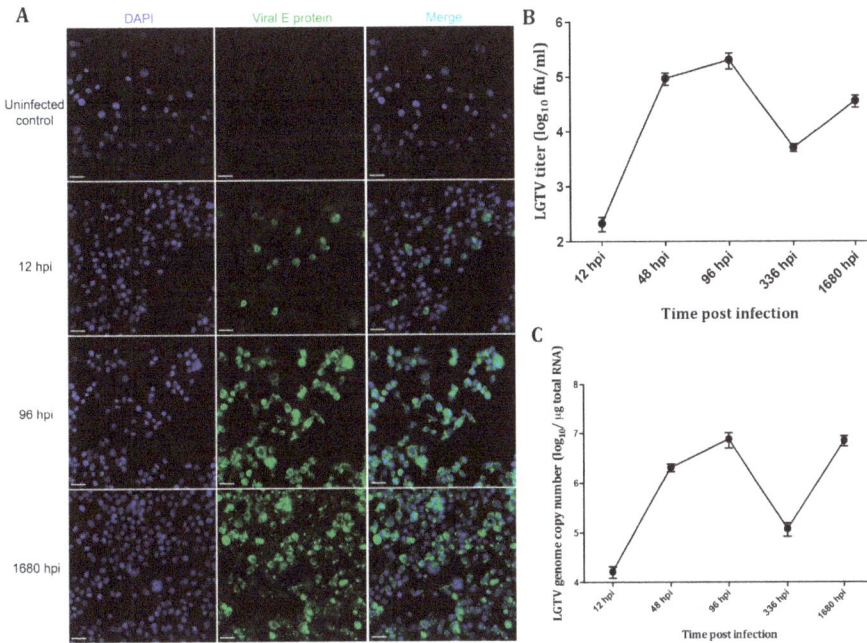

Figure 1. Langat virus (LGTV) replication kinetics in ISE6 cells. (**A**) Detection of the expression of LGTV E protein by confocal microscopy. Few cells were infected at 12 h post infection (hpi) as indicated by viral E protein staining in a low number of cells, but almost all of the cells were infected at 96 and 1680 hpi. The scale bar represents 10 μm; (**B**) LGTV TP21 titer obtained by an immunofocus assay using an anti-E antibody; (**C**) LGTV RNA genome copy numbers measured by quantitative PCR (qPCR). DAPI: 4′,6-diamidino-2-phenylindole.

To determine LGTV genome copy numbers, total cellular RNA was extracted at 12, 48, 96, 336 and 1680 hpi using an RNeasy kit (Qiagen, Los Angeles, CA, USA) as per the manufacturer's instructions. Complementary DNA (cDNA) was synthesized from 1 μg of total RNA using a VILO cDNA kit (Invitrogen) according to the manufacturer's instructions. 2 μL of the cDNA synthesis reaction was added to a quantitative PCR (qPCR) reaction mix containing LGTV-specific primers/probe (forward primer: GGATTGTTGCCCAGGATTCTC; probe: FAM-CATTGGCACCGGCCTACGCT-NFQ; and reverse primer: TTCCAGGTGGGTGCATCTC), IX Platinum qPCR Supermix UDG with Rox (Invitrogen, Life Technologies, Carlsbad, CA, USA). The qPCR assay was performed using a 7900HT fast real time PCR system (Applied Biosystems, Foster City, CA, USA).

The number of genome copies showed a similar pattern, but was several logs higher, again confirming persistent infection. Interestingly, we noted a decline in viral titer at 336 hpi that was associated with a corresponding dip in genome copy numbers (Figure 1C), suggesting that the replication rate was low at this time point. The 2-week time point corresponds to the time at which we normally split the cells, and this could have contributed to the lower rate of virus replication. Overall, these titers were comparable to those in our previous observations in ISE6 cells [23], as well as studies with tick-borne encephalitis virus (TBEV) in IDE2 cells, also derived from *Ixodes scapularis* [26], suggesting that the replication kinetics of these viruses was similar in cells from this tick species.

As mentioned, our recent studies in HEK 293T cells show that DIPs were not present at the initiation of persistent infection, but were a feature once persistence was established [19]. Consequently, we were curious to see if DIPs were present in persistently infected ISE6 cells. Therefore, we deep-sequenced the LGTV genome extracted from total cellular RNA on a HiSeq 2500 sequencer

(Illumina, San Diego, CA, USA) as described before [19]. The sequence reads were aligned and visualized with Integrated Genomics Viewer software (version 2.2.10, Broad Institute, Cambridge, MA, USA) [29,30], and 253,809 sequence read pairs were aligned at 12 hpi resulting in a depth of coverage of 2300-fold. 2,884,383 sequence pairs were obtained at 96 hpi to achieve a sequencing depth of 26,000-fold. At 1680 hpi, we obtained 881,289 sequence pairs to achieve an average depth of coverage of 8000-fold. Interestingly, analysis of the LGTV genome alignments failed to identify truncated genomes at any of these time points (Figure 2). This contrasted with our observations of 293T cells, in which truncations could be detected as early as 5 weeks of passaging persistently infected cells [19]. These results suggested that DIPs are not generated during persistent TBFV infection of ISE6 cells.

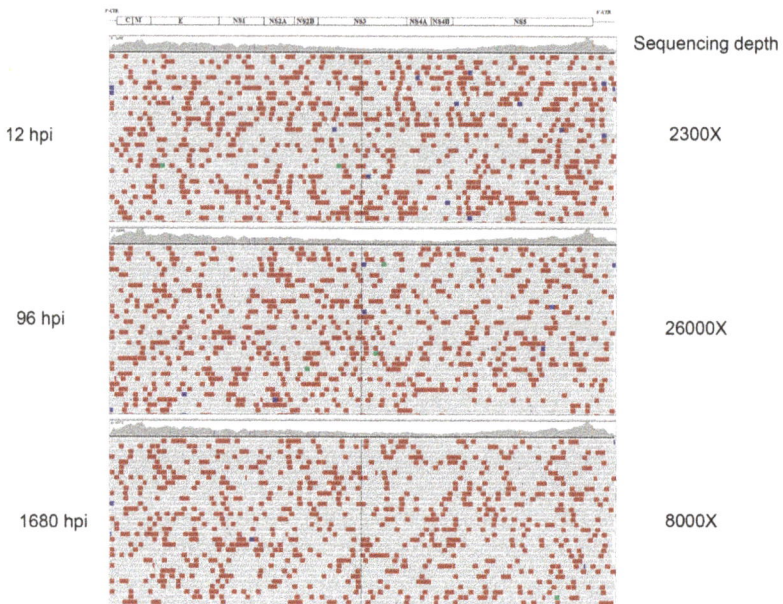

Figure 2. Integrative Genomic Viewer alignment of LGTV TP21 sequence reads obtained at 12, 96 and 1680 hpi. The horizontal gray bars represent sequence read alignments and the colored bars were read pairs of unexpected size or orientation. Genome truncations would have appeared as clear regions interspaced between horizontal gray bars [9]. The LGTV genome map was added as a schematic above the panels.

We also compared the sequences at 12, 96 and 1680 hpi for any nucleotide sequence changes to an LGTV reference sequence (GenBank accession No. EU790644). The same nucleotide sequence changes, that we attributed to the rescue of LGTV in Vero cells [19], were also observed in the viral genome at all time-points studied. However, we identified two synonymous nucleotide changes i.e., 1893A→C (29% of 5978 reads at 12 hpi) and 2284T→A (34% of 4191 reads at 12 hpi) in the region encoding for the viral protein E. Interestingly, the synonymous nucleotide change 4299C→T in the region encoding for NS2B, which we detected in late LGTV persistence in 293T cells [19], was also detected only at 1680 hpi. The significance of this nucleotide change during viral persistence is unclear.

In summary, LGTV readily initiated a persistent infection in ISE6 cells with no evidence of an acute lytic phase. In contrast to mammalian cells, viral persistence was not associated with DIPs. These results suggested that the mechanisms supporting viral persistence may differ between tick and mammalian cells.

Acknowledgments: This study was supported by the Division of Intramural Research of the NIAID, NIH. We thank Kimmo Virtaneva and Stacey Ricklefs for technical assistance.

Author Contributions: L.M. and W.M. contributed equally; L.M., W.M., S.F.P. and M.E.B. conceived and designed the experiments; L.M., W.M. and D.K.O. performed the experiments; L.M., W.M. and E.D. analyzed the data; L.M. wrote the paper.

Conflicts of Interest: The authors declare no conflict of interest.

References

1. Gritsun, T.S.; Frolova, T.V.; Zhankov, A.I.; Armesto, M.; Turner, S.L.; Frolova, M.P.; Pogodina, V.V.; Lashkevich, V.A.; Gould, E.A. Characterization of a siberian virus isolated from a patient with progressive chronic tick-borne encephalitis. *J. Virol.* **2003**, *77*, 25–36. [CrossRef] [PubMed]

2. Gritsun, T.S.; Lashkevich, V.A.; Gould, E.A. Tick-borne encephalitis. *Antivir. Res.* **2003**, *57*, 129–146. [CrossRef]

3. Chambers, T.J.; Diamond, M.S. Pathogenesis of flavivirus encephalitis. *Adv. Virus Res.* **2003**, *60*, 273–342. [PubMed]

4. Mansfield, K.L.; Johnson, N.; Phipps, L.P.; Stephenson, J.R.; Fooks, A.R.; Solomon, T. Tick-borne encephalitis virus - a review of an emerging zoonosis. *J. Gen. Virol.* **2009**, *90*, 1781–1794. [CrossRef] [PubMed]

5. Gritsun, T.S.; Nuttall, P.A.; Gould, E.A. Tick-borne Flaviviruses. *Adv. Virus Res.* **2003**, *61*, 317–371. [PubMed]

6. Orlinger, K.K.; Hofmeister, Y.; Fritz, R.; Holzer, G.W.; Falkner, F.G.; Unger, B.; Loew-Baselli, A.; Poellabauer, E.M.; Ehrlich, H.J.; Barrett, P.N.; et al. A tick-borne encephalitis virus vaccine based on the European prototype strain induces broadly reactive cross-neutralizing antibodies in humans. *J. Infect. Dis.* **2011**, *203*, 1556–1564. [CrossRef] [PubMed]

7. Lehrer, A.T.; Holbrook, M.R. Tick-borne Encephalitis Vaccines. *J. Bioterror. Biodef.* **2011**, *2011* (Suppl. 1), 3. [CrossRef] [PubMed]

8. Dobler, G. Zoonotic tick-borne flaviviruses. *Vet. Microb.* **2010**, *140*, 221–228. [CrossRef] [PubMed]

9. Mlera, L.; Melik, W.; Bloom, M.E. The role of viral persistence in flavivirus biology. *Pathog. Dis.* **2014**, *71*, 137–163. [CrossRef] [PubMed]

10. LaSala, P.R.; Holbrook, M. Tick-Borne Flaviviruses. *Clin. Lab. Med.* **2010**, *30*, 221–235. [CrossRef] [PubMed]

11. Charrel, R.N.; Fagbo, S.; Moureau, G.; Alqahtani, M.H.; Temmam, S.; de Lamballerie, X. Alkhurma Hemorrhagic Fever Virus in Ornithodoros savignyi Ticks. *Emerg. Infect. Dis.* **2007**, *13*, 153–155. [CrossRef] [PubMed]

12. Havlikova, S.; Lickova, M.; Klempa, B. Non-viraemic transmission of tick-borne viruses. *Acta Virol.* **2013**, *57*, 123–129. [CrossRef] [PubMed]

13. Charrel, R.N.; Attoui, H.; Butenko, A.M.; Clegg, J.C.; Deubel, V.; Frolova, T.V.; Gould, E.A.; Gritsun, T.S.; Heinz, F.X.; Labuda, M.; et al. Tick-borne virus diseases of human interest in Europe. *Clin. Microbiol. Infect.* **2004**, *10*, 1040–1055. [CrossRef] [PubMed]

14. Labuda, M.; Kozuch, O.; Zuffova, E.; Eleckova, E.; Hails, R.S.; Nuttall, P.A. Tick-borne encephalitis virus transmission between ticks cofeeding on specific immune natural rodent hosts. *Virology* **1997**, *235*, 138–143. [CrossRef] [PubMed]

15. Labuda, M.; Nuttall, P.A.; Kozuch, O.; Eleckova, E.; Williams, T.; Zuffova, E.; Sabo, A. Non-viraemic transmission of tick-borne encephalitis virus: A mechanism for arbovirus survival in nature. *Experientia* **1993**, *49*, 802–805. [CrossRef] [PubMed]

16. Bell-Sakyi, L.; Zweygarth, E.; Blouin, E.F.; Gould, E.A.; Jongejan, F. Tick cell lines: tools for tick and tick-borne disease research. *Trends Parasitol.* **2007**, *23*, 450–457. [CrossRef] [PubMed]

17. Oliver, J.D.; Chávez, A.S.O.; Felsheim, R.F.; Kurtti, T.J.; Munderloh, U.G. An Ixodes scapularis cell line with a predominantly neuron-like phenotype. *Exp. Appl. Acarol.* **2015**, *66*, 427–442. [CrossRef] [PubMed]

18. Mlera, L.; Lam, J.; Offerdahl, D.K.; Martens, C.; Sturdevant, D.; Turner, C.V.; Porcella, S.F.; Bloom, M.E. Transcriptome Analysis Reveals a Signature Profile for Tick-Borne Flavivirus Persistence in HEK 293T Cells. *mBio* **2016**, *7*, e00314–e00316. [CrossRef] [PubMed]

19. Mlera, L.; Offerdahl, D.K.; Martens, C.; Porcella, S.F.; Melik, W.; Bloom, M.E. Development of a Model System for Tick-Borne Flavivirus Persistence in HEK 293T Cells. *mBio* **2015**, *6*, e00614–e00615. [CrossRef] [PubMed]

20. Ruzek, D.; Vancova, M.; Tesarova, M.; Ahantarig, A.; Kopecky, J.; Grubhoffer, L. Morphological changes in human neural cells following tick-borne encephalitis virus infection. *J. Gen. Virol.* **2009**, *90*, 1649–1658. [CrossRef] [PubMed]
21. Kleinschmidt, M.C.; Michaelis, M.; Ogbomo, H.; Doerr, H.-W.; Cinatl, J. Inhibition of apoptosis prevents West Nile virus induced cell death. *BMC Microbiol.* **2007**, *7*, 1–8. [CrossRef] [PubMed]
22. Ghosh Roy, S.; Sadigh, B.; Datan, E.; Lockshin, R.A.; Zakeri, Z. Regulation of cell survival and death during Flavivirus infections. *World J. Biol. Chem.* **2014**, *5*, 93–105. [PubMed]
23. Offerdahl, D.K.; Dorward, D.W.; Hansen, B.T.; Bloom, M.E. A Three-Dimensional Comparison of Tick-Borne Flavivirus Infection in Mammalian and Tick Cell Lines. *PLoS ONE* **2012**, *7*, e47912. [CrossRef] [PubMed]
24. Lancaster, M.U.; Hodgetts, S.I.; Mackenzie, J.S.; Urosevic, N. Characterization of Defective Viral RNA Produced during Persistent Infection of Vero Cells with Murray Valley Encephalitis Virus. *J. Virol.* **1998**, *72*, 2474–2482. [PubMed]
25. Schmaljohn, C.; Blair, C.D. Persistent infection of cultured mammalian cells by Japanese encephalitis virus. *J. Virol.* **1977**, *24*, 580–589. [PubMed]
26. Ruzek, D.; Bell-Sakyi, L.; Kopecky, J.; Grubhoffer, L. Growth of tick-borne encephalitis virus (European subtype) in cell lines from vector and non-vector ticks. *Virus Res.* **2008**, *137*, 142–146. [CrossRef] [PubMed]
27. Munderloh, U.G.; Liu, Y.; Wang, M.; Chen, C.; Kurtti, T.J. Establishment, Maintenance and Description of Cell Lines from the Tick Ixodes scapularis. *J. Parasitol.* **1994**, *80*, 533–543. [CrossRef] [PubMed]
28. Gordon Smith, C.E. A Virus Resembling Russian Spring-Summer Encephalitis Virus from an Ixodid Tick in Malaya. *Nature* **1956**, *178*, 581–582. [CrossRef]
29. Robinson, J.T.; Thorvaldsdottir, H.; Winckler, W.; Guttman, M.; Lander, E.S.; Getz, G.; Mesirov, J.P. Integrative genomics viewer. *Nat. Biotechnol.* **2011**, *29*, 24–26. [CrossRef] [PubMed]
30. Thorvaldsdóttir, H.; Robinson, J.T.; Mesirov, J.P. Integrative Genomics Viewer (IGV): High-performance genomics data visualization and exploration. *BriefBioinform* **2013**, *14*, 178–192. [CrossRef] [PubMed]

Article

Epitope Identification and Application for Diagnosis of Duck Tembusu Virus Infections in Ducks

Chenxi Li [1,†], Junyan Liu [2,†], Wulin Shaozhou [1], Xiaofei Bai [1], Qingshan Zhang [1], Ronghong Hua [1], Jyung-Hurng Liu [3,*], Ming Liu [1] and Yun Zhang [1,*]

[1]   State Key Laboratory of Veterinary Biotechnology, Harbin Veterinary Research Institute of Chinese Academy of Agricultural Sciences, Harbin 150001, China; lichenxihsy@126.com (C.L.); luke0871@aliyun.com (W.S.); baixiaofei_1@163.com (X.B.); zhangqingshan91@126.com (Q.Z.); huaronghong@163.com (R.H.); liuming04@126.com (M.L.)

[2]   School of Electrical Engineering & Automation, Harbin Institute of Technology, Harbin 150006, China; junyanliu@126.com

[3]   Institute of Genomics and Bioinformatics, National Chung Hsing University, Taichung 402, Taiwan

*   Correspondence: jhliu@nchu.edu.tw (J.-H.L.); yunzhang03@yahoo.com (Y.Z.)

†   These authors contributed equally to this work.

Academic Editor: Michael R. Holbrook
Received: 3 August 2016; Accepted: 4 November 2016; Published: 10 November 2016

Abstract: Duck Tembusu virus (DTMUV) causes substantial egg drop disease. DTMUV was first identified in China and rapidly spread to Malaysia and Thailand. The antigenicity of the DTMUV E protein has not yet been characterized. Here, we investigated antigenic sites on the E protein using the non-neutralizing monoclonal antibodies (mAbs) 1F3 and 1A5. Two minimal epitopes were mapped to [221]LD/NLPW[225] and [87]YAEYI[91] by using phage display and mutagenesis. DTMUV-positive duck sera reacted with the epitopes, thus indicating the importance of the minimal amino acids of the epitopes for antibody-epitope binding. The performance of the dot blotting assay with the corresponding positive sera indicated that YAEYI was DTMUV type-specific, whereas [221]LD/NLPW[225] was a cross-reactive epitope for West Nile virus (WNV), dengue virus (DENV), and Japanese encephalitis virus (JEV) and corresponded to conserved and variable amino acid sequences among these strains. The structure model of the E protein revealed that YAEYI and LD/NLPW were located on domain (D) II, which confirmed that DII might contain a type-specific non-neutralizing epitope. The YAEYI epitope-based antigen demonstrated its diagnostic potential by reacting with high specificity to serum samples obtained from DTMUV-infected ducks. Based on these observations, a YAEYI-based serological test could be used for DTMUV surveillance and could differentiate DTMUV infections from JEV or WNV infections. These findings provide new insights into the organization of epitopes on flavivirus E proteins that might be valuable for the development of epitope-based serological diagnostic tests for DTMUV.

Keywords: duck Tembusu virus; E protein epitopes; type specific and cross-reactive epitopes; E protein 3D structure; diagnosis

1. Introduction

Flaviviruses are positive-sense RNA viruses that are classified in the genus *Flavivirus*, family *Flaviviridae* [1]. Duck Tembusu virus (DTMUV) is a newly identified flavivirus that was first isolated in southeastern China in 2010 [2] and then subsequently spread to Malaysia and Thailand [3,4]. Genomic sequencing revealed that the virus was a mosquito-borne Ntaya group flavivirus [2,5–7]. DTMUV-infected ducks develop devastating egg production drop disease, and multiple bird species have been suggested as DTMUV hosts [5,8,9]. Postmortem examination demonstrated that infected ducks exhibited severe ovarian hemorrhage, ovaritis, and regression. The unknown transmission routes, quick spread and zoonotic nature have raised the concern of the public concerning the potential of DTMUV as a human pathogen.

In a manner similar to that of other flaviviruses, the DTMUV genome encodes three structural proteins (C, prM/M, and E) and seven nonstructural proteins (NS1, NS2A, NS2B, NS3, NS4A, NS4B, and NS5) [1,5,10]. Flavivirus structural proteins are reportedly involved in cellular attachment, membrane fusion and virion assembly, whereas the nonstructural proteins are responsible for genome replication [11]. The glycosylated E protein is located on the virion surface in most flaviviruses and plays an important role in virulence, antigenicity, host range, and tissue tropism [12–14]. The flavivirus E protein consists of three structurally distinct domains (D): DI, DII, and DIII. DI contains predominantly type-specific non-neutralizing epitopes [15]. DII is involved in virus-mediated membrane fusion and contains many cross-reactive epitopes that elicit neutralizing and non-neutralizing antibodies [16]. DIII contains multiple type- and subtype-specific epitopes that elicit only virus neutralizing antibodies [15–17].

Birds are the natural reservoirs or amplifying hosts for some flaviviruses, such as West Nile virus (WNV) and Japanese encephalitis virus (JEV). Laboratory diagnosis of WNV and JEV infection is predominantly serological [18,19], but caution is advised due to the high degree of cross-reactivity among flaviviruses [20,21]. An epitope-blocking enzyme immunoassay has been successfully used for the detection of virus-specific antibodies in bird serum samples [22]. Therefore, serotype-specific B cell epitopes should be identified and used to diagnose DTMUV infections in birds or to differentiate DTMUV from other flaviviruses.

In this study, we identified two E protein epitopes and assessed their cross-reactivity to other flaviviruses and their localization on the E protein 3D structure. These findings will extend our understanding of the structure-function relationships and the cross-reaction functions in the immune response. Moreover, our results provide insights into the improvement of the flavivirus serodiagnosis and the understanding of the viral pathogenesis.

2. Materials and Methods

2.1. Virus, E-Specific Monoclonal Antibodies, and JEV-, DENV-, and WNV-Positive Sera

DTMUV TA strain was grown on duck embryo fibroblasts (DEF) or embryonated eggs as previously described [6]. The E-specific monoclonal antibodies (mAbs) 1F3 and 1A5 were developed in our lab and characterized previously [23]. JEV and WNV-positive rabbit sera were donated by Dr. Ronghong Hua, (Harbin Veterinary Research Institute of Chinese Academy of Agricultural Sciences (CAAS), Harbin, China) and DENV-positive sera was donated by Dr. Xian Qi (Nanjing Municipal Centers for Disease Control and Prevention, Nanjing, China).

2.2. Affinity Purification of Monoclonal Antibodies

The mAbs were purified from mouse ascites fluid using Protein G Agarose (Invitrogen, Carlsbad, CA, USA) according to the manufacturer's instructions. The purified immunoglobulin (Ig) G antibody concentrations were determined by measuring the absorbance at 278 nm.

2.3. Epitope Mapping

The epitopes were mapped with purified 1A5 and 1F3 using the Ph.D-12™ Phage Display Peptide Library Kit (New England BioLabs Inc., Ipswich, MA, USA) as previously described [24,25]. Briefly, each well of a 96-well plate was coated with 10 µg/mL of purified mAb and then blocked with blocking buffer. The phage library was added to the plate and incubated for 1 h. After five washes with Tris-buffered saline (TBS) (50 mM Tris-HCl, 150 mM NaCl; pH 7.5), 1 M Tris-HCl (pH 9.1) was added to the plate [24,25]. The eluted phages were amplified and titered on lysogeny broth (LB)/isopropyl β-D-1-thiogalactopyranoside (IPTG)/5-bromo-4-chloro-3-indolyl-D-galactoside (X-Gal) plates for selection. Three rounds of biopanning were performed. The ratio of output to input was calculated as the titer of the amplified output phages/the titer of the input phages.

2.4. Phage Enzyme-Linked Immunosorbent Assay and Phage Clone Sequencing

After the three rounds of biopanning described above and elsewhere [24,25], individual phage clones were selected for target binding in the enzyme-linked immunosorbent assay (ELISA). Briefly, 96-well plates were coated with 100 ng of the mAbs (1F3 and 1A5) or an anti-porcine interferon (IFN)-c mAb (Sigma, St. Louis, MO, USA) as a negative control. After the coated wells were blocked, the phages (10^{10} pfu/100 µL/well) were added. The coated plates were washed ten times with phosphate buffered saline (PBS) + 0.5% (*v*/*v*) Tween-20 (PBST), and the bound phages were reacted with an horseradish peroxidase (HRP)-conjugated sheep anti-M13 antibody (Pharmacia, Piscataway, NY, USA) as previously described [24,25]. Colored precipitation was achieved by adding substrate solution containing *o*-phenylenediamine (OPD). The positive phage clones were sequenced using a previously reported sequencing primer [24,25].

2.5. Identification of the Essential Amino Acids in the Epitopes by Dot Blotting and Western Blot Analysis

To precisely define the epitopes, we designed and synthesized two groups of fragments corresponding to the roughly mapped epitopes. Complementary oligonucleotide primers specific for each peptide fragment were designed as previously described [26]. Nucleotide segments with Eco RI/Xho I site sticky ends were produced after direct annealing and then cloned into the pGEX6p-1 vector (GE Healthcare, Beijing, China) as previously described [26]. The expressed truncated fragments were purified using the Glutathione S-transferase (GST) Purification Kit (TaKaRa, Dalian, China). Dot blotting was performed by spotting purified peptide solution onto a nitrocellulose (NC) membrane (Millipore, Bedford, MA, USA) Approximately 1 µg of purified synthesized peptide or the unrelated control peptide YIRTPACWD (from the duck reovirus σB protein) [26] diluted with Tris sodium chloride EDTA buffer (TNE) (10 mM NaCl, 10 mM Tris-Hcl,1 mM EDTA; pH 7.4) was spotted onto the NC membrane. Then, the NC membrane was incubated with the mAbs (diluted 1:2000 in PBS) at 37 °C for 1 h. After three washes with PBST, the NC was probed with a 1:500 dilution of an HRP-conjugated goat anti-mouse IgG (KPL, Gaithersburg, MD, USA) at 37 °C for 1 h. Western blot was performed by subjecting the purified GST peptides to electrophoresis in 10% acrylamide gels, followed by electro-transfer to a NC. The membrane was probed with a duck anti-DTMUV antibody diluted 1:150 in PBST, followed by a reaction with a horseradish peroxide-conjugated goat anti-duck antibody (1:500) (KPL) for 90 min at room temperature.

2.6. Sequence Analysis

To assess the level of conservation of the epitopes among the DTMUVs and other representative flaviviruses, we constructed a sequence alignment of the epitopes and determined the corresponding locations in the E proteins of the DTMUV strains and other flaviviruses using the DNASTAR Lasergene program (DNASTAR Inc., Madison, WI, USA) [27].

2.7. Cross-Reactivity of the Epitopes to WNV-, JEV-, and DENV-Positive Sera

Mapped epitope cross-reactions with other flaviviruses were determined by the dot blotting assay as described above. Briefly, approximately 1 µg of each synthesized epitope peptide or the control peptide YIRTPACWD diluted with TNE buffer was spotted onto the NC. Then, the NC membrane was incubated with WNV-, JEV-, and DENV-positive sera at 37 °C for 1 h. After three washes with PBST, the NC membrane was probed with a HRP-conjugated antibody targeting the corresponding IgG (KPL) at 37 °C for 1 h.

2.8. Protein E Modeling and Prediction

To analyze the epitope locations, we built a structure model of the DTMUV E protein. Because the DTMUV E protein shared 62% sequence identity with the JEV E protein, we chose the JEV E protein crystal structure (PDB ID: 3P54) [28] as the modeling template using MODELLE [29]. ProSA [30] and PROCHECK [31] were used to validate the stereochemical qualities of the final model. GlycoEP [32] and NGlycPred [33] were used to predict the *N*-glycosylation sites on the DTMUV E protein. The final structure was visualized and analyzed with PyMOL [34].

2.9. Competitive Inhibition Binding Assay of Monoclonal Antibody 1A5 to a Synthetic Peptide

To test for synthetic peptide inhibition of mAb 1A5 binding to the E protein, 100 μL of E antigen (10 μg/mL) was used to coat 96-well plates at 4 °C overnight. Then, the plates were blocked with 1% bovine serum albumin (BSA) as previously described. The peptides were synthesized in the multiple-antigen peptide (MAP) form [35]. YAEYI (in final peptide concentrations of 0, 10, 20, 40, 80, 160, and 640 μg/mL) or the unrelated control peptide YIRTPACWD was mixed with the blocking mAb 1A5 (0.2 μg/mL diluted in PBST) and incubated at room temperature for 45 min; these peptide/antibody mixtures were added to the E antigen-coated 96-well plates and incubated at room temperature for 1 h. After the plate was washed with PBST, HRP-conjugated goat anti-mouse IgG was added and the binding was assessed. The mean optical density at 405 nm (OD405) plus three times the standard deviation was used to determine the cutoff value.

2.10. Detection of DTMUV Infection in Duck Serum Samples

ELISA plates were coated with 50 μL/well of 640 μg/mL of synthetic peptide antigen and incubated at 4 °C overnight. After washing with PBST, the plates were blocked with PBST containing 5% (*w*/*v*) skimmed milk for 1 h at 37 °C. The diluted duck DTMUV-positive/negative sera and the WNV- and JEV-positive sera diluted in blocking solution were added and incubated for 1 h at 37 °C. After washing, 100 μL/well of the diluted goat anti-duck IgG conjugate (1:400 dilution) in blocking solution was added and incubated for 1 h at 37 °C. The plates were washed three times and then incubated with 50 μL of *p*-nitrophenyl phosphate (PNPP) substrate (Shanghai Biomedicine, Shanghai, China) for 20 min. The reactions were stopped by adding 3 M NaOH, and the plate was read on a microplate reader (Bio-Rad, Beijing, China) at 405 nm. An aliquot (100 μL/well) of the 1:400 diluted conjugate was added after washing.

3. Results

3.1. Epitope Prediction

To map the locations of the E epitopes, we screened a phage-displayed 12-mer random peptide library using mAbs. After three rounds of biopanning, phage clones were selected and their reactivity with the mAbs and the negative control anti-porcine IFN-c mAb was evaluated. Sixteen of twenty-one clones (A1–A16) reacted with mAb 1F3, and 12 of 15 clones (B1–B12) reacted with 1A5 (OD450 nm $\geq$ 1.20); the other clones were less reactive with 1F3 and 1A5 (OD450 nm < 0.36), respectively. None of the selected clones reacted with the anti-porcine IFN-c mAb (OD450 nm < 0.27) (Figure 1). Sequencing of the phage clones with high OD values revealed the consensus sequences DLD/NLPWT (mapped with 1F3) and YAEYI (mapped with 1A5) (Table 1). These amino acid sequences are identical to the DLNLPWT (aa 220 to 226) and YAEYI (aa 87 to 91) sequences of the DTMUV TA strain E protein.

Table 1. Peptide sequences of the selected phage clones.

Phage Clone	Sequence	Phage Clone	Sequence
A1	S A E N E L N L P W Q R N A L V	B1	S R N L S Y A E Y I Q I
A2	M A N A E I D L T L P W T T	B2	G N Y S E Y I V G K L V
A3	D L P W T K	B3	S S Y A N Y I Q F R N T
A4	H P H D L N D L T S P F	B4	S S Y T A Y I M A R G Q
A5	E F W T A L S D P W Y F	B5	N S M S E Y I N Y I L T
A6	A H L H D P F T T L S P	B6	V D Y S T Y I S R L T S
A7	L D F H D L N R P F N N	B7	N F M N Y A E Y Y V Q K K
A8	T H D P L D S P W N F S	B8	V D Y S T Y I S R L T S
A9	F N D L D L P F G K R A	B9	T V H S Y E E Y T A R R
A10	S Y D L D L P W I A R K	B10	V S P Y A E Y W L S Q M
A11	S F L E L D P P W T T N	B11	W D Y N L Y I K Y V A R
A12	Q H S F L D L P W H L T	B12	V D Y A T Y I S R L T S
A13	H P H D L N L P T S P F		
A14	H P H D L N L P T S P F		
A15	M A N A D L N L P W T K		
A16	T S H S W D L N L P S G		
Consensus	D L D/N L P W T	Consensus	Y A E Y I
Virus TA	219 H D L N L P W T 226	Virus TA	84 K A T Y A E Y I C K K D 97

Consensus amino acids are shown in bold.

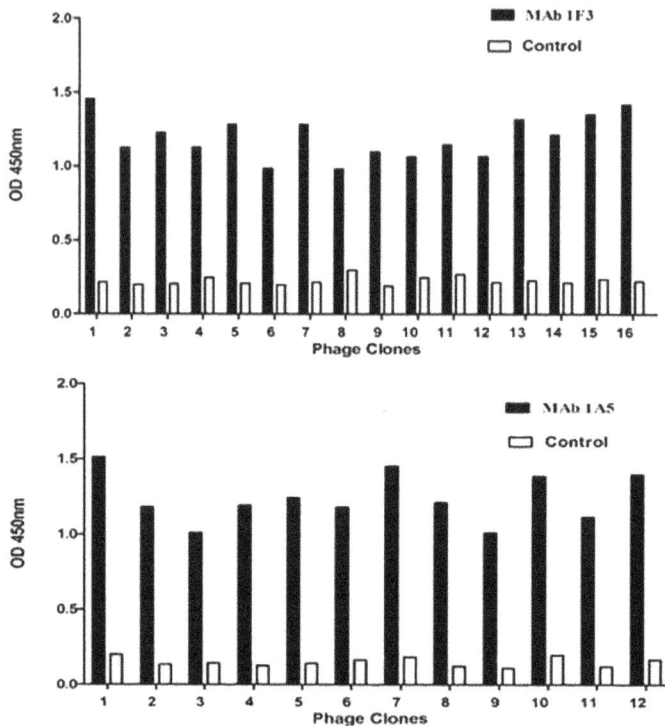

Figure 1. Detection of selected phages for monoclonal antibody (mAb) binding in the phage enzyme-linked immunosorbent assay (ELISA). The selected phage clones were detected by 1F3 and 1A5 or the anti-porcine interferon (IFN)-c mAb (negative control) after three rounds of biopanning. OD, optical density.

3.2. Mapping of the Minimal Epitopes by Dot Blotting and Western Blot

To confirm that the identified epitopes were recognized by the corresponding 1F3 and 1A5 mAbs, we expressed and purified fragments representing DLD/NLPWT and YAEYI (Table 2). Dot blotting showed that the DLD/NLPWT and YAEYI E protein fragments were recognized by the 1F3 and 1A5 mAbs, respectively (Figure 2A,B), whereas the control peptide YIRTPACWD did not react with the mAbs. This result suggests that the DLD/NLPWT and YAEYI fragments may be B cell epitopes of the DTMUV E protein. To define the epitopes more precisely, we synthesized substitutions or C- or N-terminal deletion mutants of the DLD/NLPWT and YAEYI peptides (Table 2). The amino acid substitution of ^{222}N with ^{222}D in ^{220}DLD/NLPWT^{226} did not abolish the 1F3 antibody binding activity (Figure 2A), suggesting that ^{222}D or ^{222}N in ^{220}DLD/NLPWT^{226} were mutual replicable amino acids in this position. Deletion of the amino acid ^{220}D or ^{226}T at the N- or C-terminus of ^{220}DLNLPWT^{226} did not affect 1F3 antibody binding activity but deletion of the amino acid ^{225}W abolished 1F3 binding activity, suggesting that LD/NLPW was the minimal epitope mapped by 1F3. Deletion of the amino acid ^{87}Y or ^{91}I at the N- or C-terminus of ^{87}YAEYI^{91} abolished 1A5 binding activity, which suggested that YAEYI was the minimal epitope recognized by 1A5 (Figure 2B). The identified minimal epitopes LD/NLPW (Figure 3A) and YAEYI (Figure 3B) were confirmed by Western blot analysis with duck DTMUV-positive serum.

Table 2. Primers for the truncated epitope fragments.

Primers	Sequence	Truncated Peptide
1F3-1-F	5′-aattcgatctcaacttaccatggacac-3′	GST-DLNLPWT
1F3-1-R	5′-tcgagtgtccatggtaagttgagatcg-3′	
1F3-2-F	5′-aattcgatctcgacttaccatggacac-3′	GST-DLDLPWT
1F3-2-R	5′-tcgagtgtccatggtaagtcgagatcg-3′	
1F3-3-F	5′-aattcgatctcaacttaccatggc-3′	GST-DLNLPW
1F3-3-R	5′-tcgagccatggtaagttgagatcg-3′	
1F3-4-F	5′-aattcctcaacttaccatggc-3′	GST-LNLPW
1F3-4-R	5′-tcgagccatggtaagttgagg-3′	
1F3-5-F	5′-aattcctcaacttaccac-3′	GST-LNLP
1F3-5-R	5′-tcgagtggtaagttgagg-3′	
1F3-6-F	5′-aattcaacttaccatggc-3′	GST-NLPW
1F3-6-R	5′-tcgagccatggtaagttg-3′	
1A5-1-F	5′-aattctacgctgaatacatac-3′	GST-YAEYI
1A5-1-R	5′-tcgagtatgtattcagcgtag-3′	
1A5-2-F	5′-aattcgctgaatacatac-3′	GST-AEYI
1A5-2-R	5′-tcgagtatgtattcagcg-3′	
1A5-3-F	5′-aattctacgctgaatacc-3′	GST-YAEY
1A5-3-R	5′-tcgaggtattcagcgtag-3′	

GST, Glutathione S-transferase.

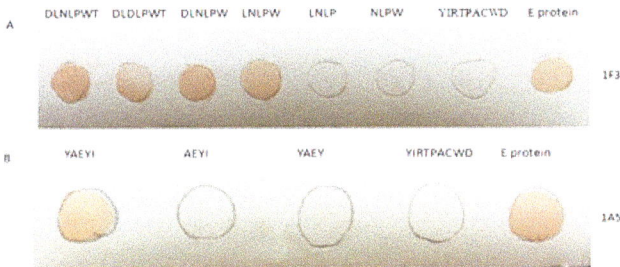

Figure 2. Identification of the E protein epitopes based on mAbs IF3 (**A**) and 1A5 (**B**) reactivity with the synthesized peptides in the dot blotting assay. YIRTPACWD and the E protein were used as the negative and positive control, respectively.

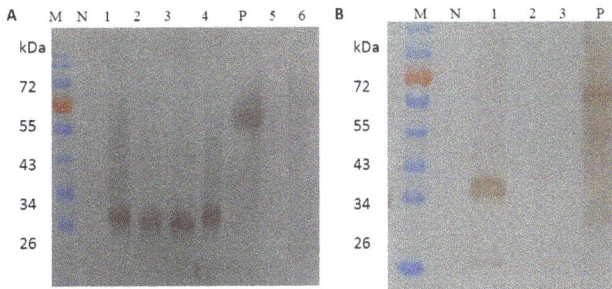

Figure 3. The reactivity of the synthesized mutations of the LD/NLPW (**A**) and YAEYI (**B**) peptides to duck anti-duck Tembusu virus (DTMUV) serum by Western blot. YIRTPACWD and the E protein were used as the negative (N) and positive (P) controls, respectively. M: Protein marker; N: GST negative control; (**A**) lane 1, GST-DLNLPWT; lane 2, GST-DLDLPWT; lane 3, GST-DLNLPW; lane 4, GST-LNLPW; lane 5, GST-LNLP; lane 6, NLPW; (**B**) lane 1, GST-YAEYI; lane 2, GST-AEYI; lane 3, GST-YAEY.

3.3. Sequence Analysis of the Identified Epitopes among the DTMUV Strains and Other Flaviviruses

To determine the conservation of LNLPW and YAEYI in the E proteins, we aligned the epitope region sequences of DTMUV with sequences from other flaviviruses. Information for the DTMUV, WNV, DENV, and JEV E protein sequences obtained from GenBank is provided in Table 3. This sequence alignment revealed that the amino acids in the [221]LDLPW[225] (Figure 4A) and [87]YAEYI[91] (Figure 4B) epitope regions were identical among the DTMUV strains, indicating that these motifs represented conserved epitopes in the DTMUV E protein. LNLPW was conserved in both DTMUV and WNV (Figure 5A) but was divergent compared to DENV and JEV. The YAEYI epitope was completely conserved in the DTMUV species but was highly divergent compared to the WNV, DENV, and JEV E protein sequences (Figure 5B), suggesting that YAEYI is a DTMUV type-specific epitope.

Table 3. Flavivirus strains used in the sequence analysis in this study.

Species	Strain	GenBank No.	Location/Year of Isolation	
DTMUV	TA	JQ289550.1	China	2010
DTMUV	DK/TH/CU-1	KR061333.1	Thailand	2013
DTMUV	GDHD2014-3	KT159713.1	China	2014
DTMUV	WFZ-2012	KC990545.1	China	2012
DTMUV	SH001	KP742476.1	China	2015
DTMUV	AH2011	KJ958533.1	China	2012
DTMUV	GDLH01	KT824876.1	China	2015
DTMUV	WZDu	AB917089.1	China	2012
DTMUV	ZJ GH-2	JQ314465.1	China	2010
DTMUV	SDLC	KJ740747.1	China	2013
DENV-1	HNRG24827	KC692511.1	Argentina	2010
DENV-2	DENV-2/Pk/Swat-02	KJ701507.1	Pakistan	2013
DENV-3	D83-144	KJ737430.1	Thailand	1983
DENV-4	EHI310A129SY10	JX024758.1	Singapore	2010
WNV	USA	AY646354.1	USA	2002
WNV	FtC-3699	KR868734.1	USA	2012
JEV	HEN0701	FJ495189.1	China	2007
JEV	CC27-S6	AY303797.1	Taiwan	2003

DTMUV, duck Tembusu virus; DENV, dengue virus; WNV, West Nile virus; JEV, Japanese encephalitis virus.

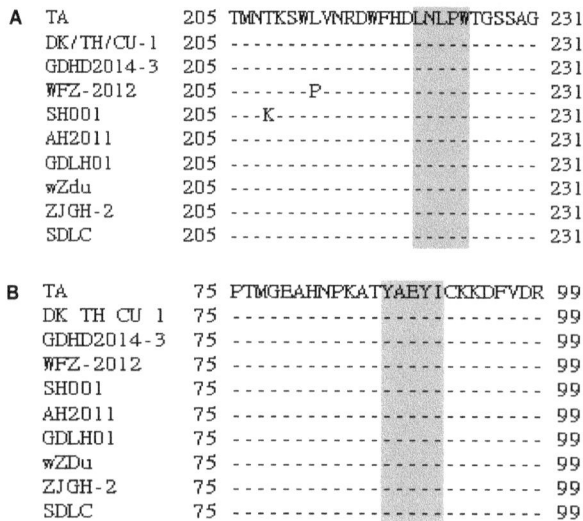

Figure 4. Sequence alignment of the epitope-coding regions LNLPW (**A**) and YAEYI (**B**) in the DTMUV strain E proteins. The amino acid positions for each sequence are numbered on both sides. The DTMUV TA strain sequence is shown at the top; the dashes indicate identical amino acids. The identified epitope region is boxed in grey.

```
A  DTMUV TA              205 TMNTKSWLVNRDWFHDLNLPWTGSSAG 231
   DENV1 HNRG24827       200 --KE-----HKQ--L--P----SGAST 226
   DENV2 Pk/Swat-02      200 Q-EN-A---H-Q--L--P---LPGADI 226
   DENV3 D83-144         198 --KN-A-M-H-Q--F--P----SGATT 224
   DENV4 EHI310A129SY10  200 K-KK-T---HKQ--L--P---SAGADT 226
   WNV   USA             205 -VG--TF--H-E--M------SSAGST 231
   WNV   FtC-3699        205 -VG--TF--H-E--M------SSAGST 231
   JEV   HEN0701         206 -VGS-----H-E-----S----SP-ST 232
   JEV   CC27-S6         206 -VGS-----H-E-----A----SP-ST 232


B  DTMUV TA              75 PTMGEAHNPKATYAEYICKKDFVDR 99
   DENV1 HNRG24827       75 --Q---TLVEEQD-NFV-RRT---- 99
   DENV2 Pk/Swat-02      75 --Q--PSLNEEQDKRFV--HSM--- 99
   DENV3 D83-144         75 --Q---IL-EEQDQN-V--HTY--- 99
   DENV4 EHI310A129SY10  75 --Q--PYLKEEQDQQ---RR-V--- 99
   WNV   USA             75 --------D---DPAFV-RQGV--- 99
   WNV   FtC-3699        75 --------D-RADPAFV-RQGV--- 99
   JEV   HEN0701         75 --T-----E-RADSS-V--QG-T-- 99
   JEV   CC27-S6         75 --T-----E-RADSS-V--QG-T-- 99
```

Figure 5. Sequence alignment of the epitope-coding regions LNLPW (**A**) and YAEYI (**B**) in the flavivirus strain E proteins. The amino acid positions for each sequence are numbered on both sides. The DTMUV TA strain sequence is shown at the top; the dashes indicate identical amino acids. The identified epitope region is boxed in grey.

3.4. YAEYI and LNLPW Peptide Fragment Reactivity to WNV-, JEV-, and DENV-Positive Sera

To demonstrate the epitope cross-reactivity with other flaviviruses, purified YAEYI and LNLPW fragments were used to test their cross-reactivity with WNV-, JEV-, and DENV-positive sera in the dot blotting assay. The WNV-, JEV-, and DENV-positive sera reacted with the LNLPW peptide and E protein but did not react with YAEYI and the negative control peptide (Figure 6).

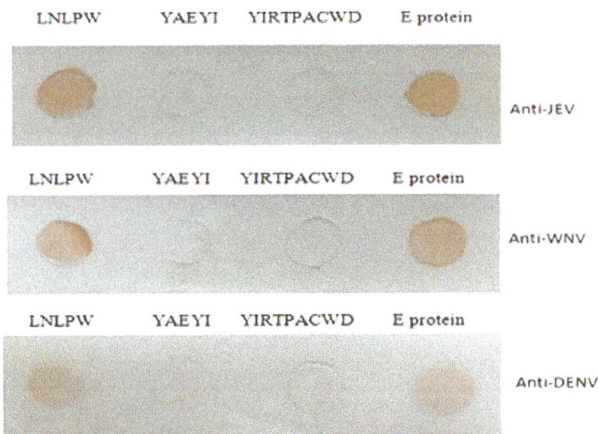

Figure 6. The cross-reactivity of the epitopes to the Japanese encephalitis virus (JEV)-, West Nile virus (WNV)-, and dengue virus (DENV)-positive sera in the dot blotting assay. YIRTPACWD and the E protein were used as the negative and positive controls, respectively.

3.5. Location of Two Epitopes on the E Protein 3D Structure

The resulting structure was evaluated by ProSA [30] and PROCHECK [31] to determine the stereochemical quality. These validation results revealed that the model structure of the DTMUV E protein shown in Figure S1 was reliable for further study. The overall structure of the DTMUV E protein resembled the structure of the previously reported flavivirus E protein [28] and had three distinct domains: a central β-barrel (domain 1), an elongated finger-like structure (domain II), and a C-terminal immunoglobulin-like module (domain III) (Figure 7). Based on the protein sequence, GlycoEP [32] suggested that the DTMUV E protein might have two potential N-glycosylation sites (^{154}N and ^{314}N), with prediction scores of 0.838 and 0.438, respectively. However, according to the structural and residue pattern information, NGlycPred [33] predicted ^{154}N as the only glycosylation site with a score of 0.427. Therefore, the possibility of ^{314}N serving as a glycosylation is questionable. The 3D structure of the E protein showed that two epitopes possessed loop conformations (Figure 7). Epitope YAEYI was located near the lateral ridge of domain II, and LNLPW was located close to the domain II central interface.

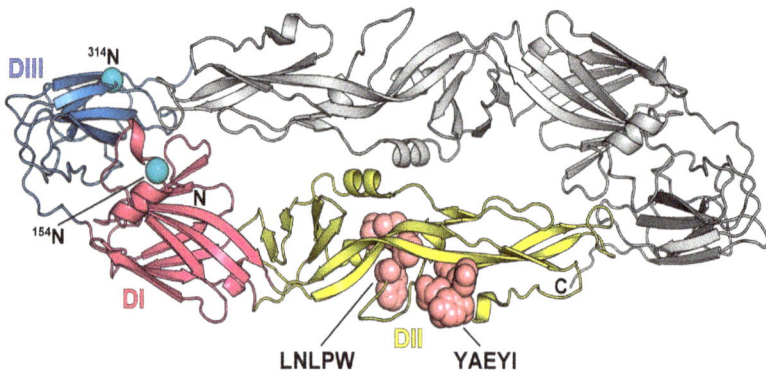

Figure 7. The locations of epitopes on the DTMUV E protein dimer. The DTMUV E protein structure is modeled based on the JEV E crystal structure protein using MODELLER [34]. Domains I, II and III in are colored in magenta, yellow and blue, respectively, in one monomer. The other monomer is colored grey. The locations of the two epitopes are depicted as spheres and labeled. The locations of two epitopes are depicted as spheres and labeled. Two predicted N-glycosylation sites by GlycoEP [32] and NGlycPred [33] are colored as cyan.

3.6. Competitive Inhibition of Synthetic Peptide YAEYI Binding to Monoclonal Antibody 1A5

The peptides were synthesized in the MAP form because the binding efficiency of an eight-chain MAP is greater than the binding efficiency of a single-chain peptide [35]. Competitive binding assays were performed to confirm that the E protein peptide YAEYI was an E protein epitope. These assays showed that the reactivity of mAb 1A5 with the E protein was markedly inhibited by the synthetic antigen peptide YAEYI in a dose-dependent manner ($p < 0.05$; Figure 8) after Student's *t*-test statistical method analysis.

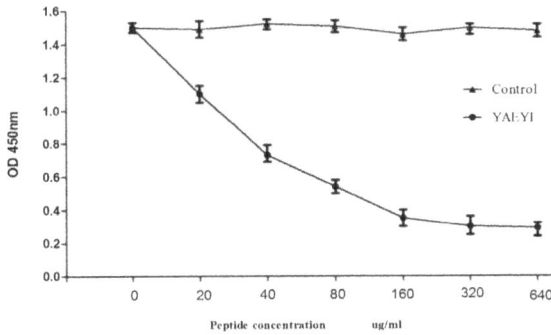

Figure 8. Competitive inhibition of synthetic peptide YAEYI binding to mAb 1F3. A competitive ELISA was performed using the antigen peptide YAEYI as the competitor for the E protein. Values represent three independent experiments with triplicate determinations included in each experiment ($p < 0.05$, Student's *t*-test).

3.7. Sensitivity and Reactivity of an Epitope-Based Peptide Applied for the Diagnosis of DTMUV in Serum Samples

Mean optical density at 405 nm (0.235) plus three times the standard deviation (0.0465) was used to determine the cutoff value (0.375). The synthetic antigen peptide YAEYI was able to detect DTMUV infections in 24 serum samples collected from 25 DTMUV infections confirmed by ELISA (Figure 9). One serum (D15) was detected negative by peptide YAEYI (0.374). In contrast, 25 SPF duck sera, anti-WNV sera, and anti-JEV sera were seronegative using the same epitope-based peptide serologic test. The specificity of this test was 100% for the SPF duck sera without DTMUV infection (Figure 9). The sensitivity of this epitope-based peptide serologic test for DTMUV infection was 96%.

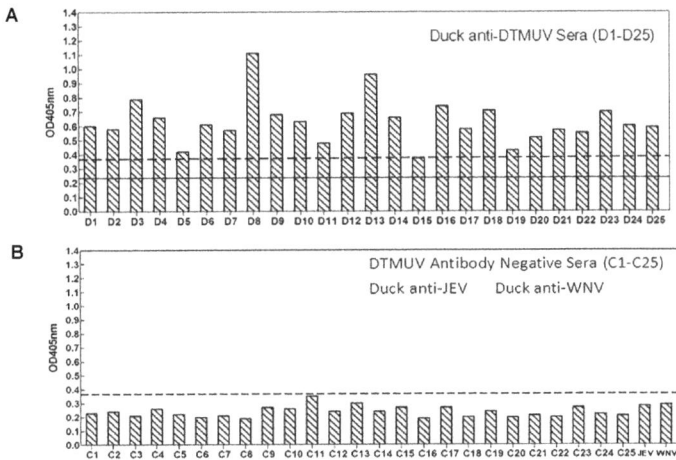

Figure 9. ELISA reactivity of the DTMUV synthetic peptide against serum samples from 25 DTMUV-infected duck (**A**) versus sera from 25 healthy ducks (**B**). The cutoff value (dashed lines) was calculated as 0.375. Solid line, mean OD405.

4. Discussion

Monoclonal antibodies against flaviviruses are powerful tools for mapping flavivirus epitopes and investigating antigenic structures. The flavivirus E protein was confirmed to be a strong immunogen

Viruses **2016**, *8*, 306

for antibody production. The aims of this study were to investigate the antigenic sites on the DTMUV structural envelop E protein using mAbs against E protein fragments and to define the B cell epitopes recognized by these antibodies. Epitopes for two mAbs were mapped precisely to clusters in two locations (i.e., [220]DLD/NLPWT[226] (1F3) and [87]YAEYI[91] (1A5)) on the E protein using a 12-mer random peptide phage display system. The dot blotting assay with mAbs and deleted peptide fragments from each group showed the minimal epitopes for IF3 and 1A5 on LNLPW and YAEYI. Anti-DTMUV duck sera recognized the two epitopes in the Western blot assay, indicating the importance of the minimal antigenic domains of the epitopes for antibody-epitope binding reactivity.

The sequence alignment revealed that epitope YAEYI was completely conserved among the various DTMUV strains but was highly divergent from the WNV, DENV, and JEV strains, which suggested the YAEYI was a DTMUV type-specific epitope mapped by mAb 1A5. This domain did not exhibit cross-reactivity with WNV, DENV, and JEV, suggesting that it might have an application for the specific serological detection of DTMUV infection. Moreover, the structure model of the DTMUV E protein showed that the YAEYI epitope was near the lateral ridge of DII and protruded from the surface of the E protein, which supported the hypothesis that the YAEYI peptide could be easily exposed and used as an antigen to detect DTMUV type-specific infection. The mAb 1A5 is expected to react broadly with various DTMUV strains; however, the YAEYI-based ELISA reactivity for the detection of DTMUV-infected duck serum samples was not as high as the detection obtained using whole virus or viral proteins, which had multiple epitopes. Therefore, identifying more DTMUV-specific epitopes and combining more epitope-based peptide antigens for DTMUV-positive sera detection will increase the sensitivity of this serologic diagnosis method. Using this epitope-based peptide antigen to detect DTMUV-positive sera is relatively simple and specific and does not require the paired serum samples needed for conventional tests. Furthermore, the WNV-/JEV-positive serum samples showed no ELISA reactivity with the YAEYI-based antigen, which suggested that the assay was capable of differentiating between DTMUV and WNV/JEV in duck serum samples. Further study on the application of this YAEYI-based ELISA with a large number of samples is in progress.

A previous report showed that the JEV epitope reacted with both JEV-positive and WNV-positive sera [36], suggesting that JEV and WNV were two members of the JEV serocomplex; this association caused the cross-reactivity among these flaviviruses. The sequence alignment showed that the LNLPW epitope was highly conserved in both DTMUV and WNV, indicating that DTMUV and WNV contained the same immunodominant epitope. This hypothesis was confirmed by the dot blotting assay, which showed that anti-WVN sera could bind LNLPW, suggesting that DTMUV and WNV were also part of a serocomplex.

The 3D structure of the E protein suggested that the LNLPW epitope did not protrude from the E protein surface. Previous studies in the DENV virion particle demonstrated that the binding of some E reactive antibodies relied on the dynamic motion of protein molecules ("breathing"), leading to transient exposure of buried epitopes [37–39]. Accordingly, whether the "breathing" of the DTMUV E protein will expose the two epitopes and allow mAb binding or inhibit the activity remains to be experimentally resolved.

5. Conclusions

In summary, we identified for the first time two novel epitopes of the DTMUV E protein: one DTMUV type-specific epitope and one widely cross-reactive epitope. We also reported for the first time the 3D structure of the E protein and epitope locations. This information has provided new insights into the structure and organization of epitopes on the DTMUV E protein and valuable epitope information for the development of diagnostic assays for the specific detection of DTMUV infection.

Supplementary Materials: The following are available online at www.mdpi.com/1999-4915/8/11/306/s1, Figure S1: The stereochemical quality of the structure evaluated by ProSA and PROCHECK.

Acknowledgments: This work was supported by National Key R and D Program (2016YFD0500100) and the Natural Science Foundation of China (31670153). Y.Z. supervised and provided the funding for the study.

Author Contributions: M.L. and Y.Z. designed the study. C.L., R.Z., J.L. and X.B. performed the experiments. Q.Z., R.H., J.L., S.W. and J.L. analyzed the data. M.L., C.L. and Y.Z. wrote the paper.

Conflicts of Interest: The authors declare no competing financial interests.

References

1. Lindenbach, B.D.; Thiel, H.J.; Rice, C.M. *Flaviviridae*: The viruses and their replication. In *Fields Virology*, 5th ed.; Knipe, D.M., Howley, P.M., Griffin, D.E., Lamb, R.A., Martin, M.A., Roizman, B., Straus, S.E., Eds.; Lippincott Williams & Wilkins: Philadelphia, PA, USA, 2007; pp. 1101–1152.
2. Su, J.J.; Li, S.; Hu, X.D.; Yu, X.L.; Wang, Y.Y.; Liu, P.P.; Lu, X.S.; Zhang, G.Z.; Hu, X.Y.; Liu, D.; et al. Duck egg drop syndrome caused by BYD virus, a new Tembusu related virus. *PLoS ONE* **2011**, *6*, 18106. [CrossRef] [PubMed]
3. Homonnay, Z.G.; Kovacs, E.W.; Banyai, K.; Albert, M.; Feher, E.; Mato, T.; Tatar-Kis, T.; Palya, V. Tembusu-like flavivirus (Perak virus) as the cause of neurological disease outbreaks in young Pekin ducks. *Avian Pathol.* **2014**, *43*, 552–560. [CrossRef] [PubMed]
4. Thontiravong, A.; Ninvilai, P.; Tunterak, W.; Nonthabenjawan, N.; Chaiyavong, S.; Angkabkingkaew, K.; Mungkundar, C.; Phuengpho, W.; Oraveerakul, K.; Amonsin, A. Tembusu-Related Flavivirus in Ducks, Thailand. *Emerg. Infect. Dis.* **2015**, *21*, 2164–2167. [CrossRef] [PubMed]
5. Liu, M.; Liu, C.G.; Li, G.; Li, X.J.; Yin, X.C.; Chen, Y.H.; Zhang, Y. Complete genomic sequence of duck flavivirus from China. *J. Virol.* **2012**, *86*, 3398. [CrossRef] [PubMed]
6. Liu, M.; Chen, S.Y.; Chen, Y.H.; Liu, C.G.; Chen, S.L.; Yin, X.C.; Li, G.; Zhang, Y. Adapted Tembusu-Like Virus in Chickens and Geese in China. *J. Clin. Microbiol.* **2012**, *50*, 2807–2809. [CrossRef] [PubMed]
7. Cao, Z.Z.; Zhang, C.; Liu, Y.H.; Ye, W.C.; Han, J.W.; Ma, G.M.; Zhang, D.D.; Xu, F.; Gao, Y.; Tang, X.H.; et al. Tembusu virus in ducks, China. *Emerg. Infect. Dis.* **2011**, *17*, 1873–1875. [CrossRef] [PubMed]
8. Chen, P.; Liu, J.; Jiang, Y.; Zhao, Y.; Li, Q.; Wu, L.; He, X.; Chen, H. The vaccine efficacy of recombinant duck enteritis virus expressing secreted E with or without PrM proteins of duck tembusu virus. *Vaccine* **2014**, *32*, 5271–5277. [CrossRef] [PubMed]
9. Huang, X.; Han, K.; Zhao, D.; Liu, Y.; Zhang, J.; Niu, H.; Zhang, K.; Zhu, J.; Wu, D. Identification and molecular characterization of a novel flavivirus isolated from geese 265 in China. *Res. Vet. Sci.* **2013**, *94*, 774–780. [CrossRef] [PubMed]
10. Bai, X.F.; Lv, R.; Liu, C.G.; Qiu, N.; He, Y.L.; Yin, X.C.; Li, X.J.; Liu, M.; Zhang, Y. Molecular characterization of a duck Tembusu virus from China. *Virus Genes* **2013**, *47*, 478–482. [CrossRef] [PubMed]
11. Avirutnan, P.; Fuchs, A.; Hauhart, R.E.; Somnuke, P.; Youn, S.; Diamond, M.S.; Atkinson, J.P. Antagonism of the complement component C4 by flavivirus nonstructural protein NS1. *J. Exp. Med.* **2010**, *207*, 793–806. [CrossRef] [PubMed]
12. Crill, W.D.; Chang, G.J. Localization and characterization of flavivirus envelope glycoprotein cross-reactive epitopes. *J. Virol.* **2004**, *78*, 13975–13986. [CrossRef] [PubMed]
13. Crill, W.D.; Roehrig, J.T. Monoclonal antibodies that bind to domain III of dengue virus E glycoprotein are the most efficient blockers of virus adsorption to Vero cells. *J. Virol.* **2001**, *75*, 7769–7773. [CrossRef] [PubMed]
14. Roehrig, J.T.; Diamond, M.S.; Kuhn, R.J.; Rossmann, M.G. Binding of a neutralizing antibody to dengue virus alters the arrangement of surface glycoproteins. *Nat. Struct. Mol. Biol.* **2008**, *15*, 312–317.
15. Roehrig, J.T.; Bolin, R.A.; Kelly, R.G. Monoclonal antibody mapping of the envelope glycoprotein of the dengue 2 virus, Jamaica. *Virology* **1998**, *246*, 317–328. [CrossRef] [PubMed]
16. Rey, F.A.; Heinz, F.X.C.; Mandl, C.; Harrison, S.G. The envelope glycoprotein from tick-borne encephalitis virus at 2 Å resolution. *Nature* **1995**, *375*, 291–298. [CrossRef] [PubMed]
17. Roehrig, J.T. Immunochemistry of the dengue viruses. In *Dengue and Dengue Hemorrhagic Fever*; Gubler, D.J., Kuno, G., Eds.; CAB International: New York, NY, USA, 1997; pp. 199–219.
18. Shi, P.Y.; Wong, S.J. Serologic diagnosis of West Nile virus infection. *Expert Rev. Mol. Diagn.* **2003**, *3*, 733–741. [CrossRef] [PubMed]
19. Hirota, J.; Shimizu, S.; Shibahara, T.; Kobayashi, S. Cross-reactivity of chicken anti-Japanese encephalitis virus serum and anti-West Nile virus serum in serological diagnosis. *J. Vet. Med. Sci.* **2012**, *74*, 1497–1499. [CrossRef] [PubMed]

20. Martin, D.A.; Muth, D.A.; Brown, T.; Johnson, A.J.; Karabatsos, N.; Roehrig, J.T. Standardization of immunoglobulin M capture enzyme-linked immunosorbent assays for routine diagnosis of arboviral infections. *J. Clin. Microbiol.* **2000**, *38*, 1823–1826. [PubMed]

21. Martin, D.A.; Biggerstaff, B.J.; Allen, B.; Johnson, A.J.; Lanciotti, R.S.; Roehrig, J.T. Use of immunoglobulin M cross-reactions in differential diagnosis of human fl aviviral encephalitis infections in the United States. *Clin. Diagn. Lab. Immunol.* **2002**, *9*, 544–549. [PubMed]

22. Hall, R.A.; Broom, A.K.; Hartnett, A.C.; Howard, M.J.; Mackenzie, J.S. Immunodominant epitopes on the NS1 protein of MVE and KUN viruses serve as targets for a blocking ELISA to detect virus-specific antibodies in sentinel animal serum. *J. Virol. Meth.* **1995**, *51*, 201–210. [CrossRef]

23. Bai, X.F.; Shaozhou, W.L.; Zhang, Q.S.; Li, C.X.; Qiu, N.; Meng, R.Z.; Liu, M.; Zhang, Y. Characterization of monoclonal antibodies against duck Tembusu virus E protein: An antigen-capture ELISA for the detection of Tembusu virus infection. *Arch. Virol.* **2015**, *160*, 757–764. [CrossRef] [PubMed]

24. Xue, M.; Shi, X.M.; Zhang, J.; Zhao, Y.; Cui, H.Y. Identification of a Conserved B-cell Epitope on Reticuloendotheliosis Virus Envelope Protein by Screening a Phage-displayed Random Peptide Library. *PLoS ONE* **2012**, *7*, e49842. [CrossRef] [PubMed]

25. Wu, X.Y.; Li, X.J.; Zhang, Q.S.; Wulin, S.Z.; Bai, X.F.; Zhang, Z.Z. Identification of a Conserved B-Cell Epitope on Duck Hepatitis A Type 1 Virus VP1 Protein. *PLoS ONE* **2015**, *10*, e0118041. [CrossRef] [PubMed]

26. Li, Y.F.; Yin, X.C.; Chen, X.D.; Li, X.J.; Li, J.Z.; Liu, C.G.; Liu, M.; Zhang, Y. Antigenic analysis monoclonal antibodies against different epitopes of σB protein of Muscovy duck reovirus. *Virus Res.* **2012**, *163*, 546–551. [CrossRef] [PubMed]

27. Burland, T.G. DNASTAR's Lasergene sequence analysis software. *Meth. Mol. Biol.* **2000**, *132*, 71–91.

28. Luca, V.C.; AbiMansour, J.; Nelson, C.A.; Fremont, D.H. Crystal structure of the Japanese encephalitis virus envelope protein. *J. Virol.* **2012**, *86*, 2337–2346. [CrossRef] [PubMed]

29. Eswar, N.; Webb, B.; Marti-Renom, M.A.; Madhusudhan, M.S.; Eramian, D.; Shen, M.Y.; Pieper, U.; Sali, A. Comparative protein structure modeling using MODELLER. *Curr. Protoc. Protein Sci.* **2007**. [CrossRef]

30. Wiederstein, M.; Sippl, M.J. ProSA-web: Interactive web service for the recognition of errors in three-dimensional structures of proteins. *Nucleic Acids Res.* **2007**, *35*, 407–410. [CrossRef] [PubMed]

31. Laskowski, R.A.; Macarthur, M.W.; Moss, D.S.; Thornton, J.M. PROCHECK: A program to check the stereochemical quality of protein structures. *J. Appl. Cryst.* **1993**, *26*, 283–291. [CrossRef]

32. Chauhan, J.S.; Rao, A.; Raghava, G.P. In silico platform for prediction of N-, O- and C-glycosites in eukaryotic protein sequences. *PLoS ONE* **2013**, *8*, e67008. [CrossRef] [PubMed]

33. Chuang, G.Y.; Boyington, J.C.; Joyce, M.G.; Zhu, J.; Nabel, G.J.; Kwong, P.D.; Georgiev, I. Computational prediction of N-linked glycosylation incorporating structural properties and patterns. *Bioinformatics* **2012**, *28*, 2249–2255. [CrossRef] [PubMed]

34. DeLano, W.L. *The PyMOL Molecular Graphics System*; v1.5.0.4; Schrödinger, LLC: New York, NY, USA, 2004.

35. Tam, J.P.; Zavala, F. Multiple antigen peptide: A novel approach to increase detection sensitivity of synthetic peptides in solid-phase immunoassays. *J. Immunol. Methods* **1989**, *124*, 53–61. [CrossRef]

36. Sun, E.C.; Zhao, J.; Yang, T.; Liu, N.H.; Geng, H.W.; Qin, Y.L.; Wang, L.F.; Bu, Z.G.; Yang, Y.H.; Lunt, R.A.; et al. Identification of a conserved JEV serocomplex B-cell epitope by screening a phage-display peptide library with a mAb generated against West Nile virus capsid protein. *Virol. J.* **2011**, *8*, 100. [CrossRef] [PubMed]

37. Wahala, W.M.; Silva, A.M. The human antibody response to dengue virus infection. *Viruses* **2011**, *3*, 2374–2395. [CrossRef] [PubMed]

38. Fibriansah, G.; Ng, T.S.; Kostyuchenko, V.A.; Lee, J.; Lee, S.; Wang, J.; Lok, S.M. Structural changes in dengue virus when exposed to a temperature of 37 degrees. *J. Virol.* **2013**, *87*, 7585–7592. [CrossRef] [PubMed]

39. Austin, S.K.; Dowd, K.A.; Shrestha, B.; Nelson, C.A.; Edeling, M.A.; Johnson, S.; Pierson, T.C.; Diamond, M.S.; Fremont, D.H. Structural basis of differential neutralization of DENV-1 genotypes by an antibody that recognizes a cryptic epitope. *PLoS Pathog.* **2012**, *8*, e1002930. [CrossRef] [PubMed]

viruses

MDPI

Article

Prevalence and Clinical Impact of Human Pegivirus-1 Infection in HIV-1-Infected Individuals in Yunnan, China

Zhijiang Miao [1,†], Li Gao [2,†], Yindi Song [1], Ming Yang [1], Mi Zhang [2], Jincheng Lou [2], Yue Zhao [1], Xicheng Wang [2], Yue Feng [1,*], Xingqi Dong [2,*] and Xueshan Xia [1,*]

[1]   Faculty of Life Science and Technology, Kunming University of Science and Technology, Kunming 650500, China; miaozhijiang@yeah.net (Z.M.); yindisong@163.com (Y.S.); qjyangming@163.com (M.Y.); zy19860908@yeah.net (Y.Z.)

[2]   Department of Infectious Diseases, Yunnan Provincial Hospital of Infectious Diseases, Kunming 650301, China; gaoli296@aliyun.com (L.G.); zm050306@sohu.com (M.Z.); ljc666yn@163.com (J.L.); wxch62597@foxmail.com (X.W.)

*   Correspondence: fyky2005@163.com (Y.F.); dongxq8001@126.com (X.D.); oliverxia2000@aliyun.com (X.X.); Tel.: +86-871-6592-0756 (Y.F.); +86-871-6872-8091 (X.D.); +86-871-6592-0562 (X.X.)

†   These authors contributed equally to this work.

Academic Editor: Michael R. Holbrook

Received: 6 December 2016; Accepted: 25 January 2017; Published: 15 February 2017

Abstract: Human Pegivirus-1 (HPgV-1) may have a beneficial impact on disease progression in human immunodeficiency virus-1 (HIV-1) infection. However, analysis of the genotypic diversity of HPgV-1 and its relevance to the progression of HIV-1 disease remains limited. A total of 1062 HIV-1-infected individuals were recruited in all sixteen prefectures of Yunnan province, China. The reverse transcription nested polymerase chain reaction (RT-nPCR), phylogenetic analyses, and clinical data analyses were used to detect HPgV-1 infection, determine genotype, and analyze HPgV-1 genotype impact on HIV-1 disease progression. The overall positive rate of HPgV-1 RNA was 23.4% (248/1062), and the frequency of HPgV-1 infection in injecting drug users (IDUs) (28.5%, 131/460) was significantly higher than in heterosexuals (19.4%, 117/602). Multiple genotypes were identified in 212 subjects with successful sequencing for the *E2* gene, including genotype 7 (55.7%), genotype 3 (34.9%), genotype 4 (4.7%), genotype 2 (3.3%), and an unclassified group (1.4%). Moreover, genotype 7 predominated in IDUs, whereas genotype 3 was the most common in heterosexuals. Our results revealed that HPgV-1 genotype 7 groups exhibited significantly lower HIV-1 viral load and higher CD4+ cell counts. This finding suggests that HPgV-1 genotype 7 may be associated with a better progression of HIV-1 disease.

Keywords: HPgV-1; HIV-1; HCV; co-infection; injecting drug users; heterosexuals; genotypic diversity; clinical effect

1. Introduction

Human pegivirus-1 (HPgV-1), also known as GB virus C (GBV-C) or hepatitis G virus (HGV), is a positive-sense single-stranded RNA (ssRNA) virus that has recently been classified under the *Pegivirus* (pe, persistent; g, GB or G) genus of the *Flaviviridae* family [1–4]. HPgV-1 possesses a genome of approximately 9.4 kb nucleotides in size that encodes a polyprotein of 2900 amino acids with characteristic structural proteins (E1 and E2) and non-structural motifs (NS2, NS3, NS4A, NS4B, NS5A, and NS5B) which is organized similarly to the genome of the hepatitis C virus (HCV) [4,5]. To date, seven HPgV-1 genotypes have been identified by phylogenetic analysis of the full-length and partial regions of the genome 5′ untranslated region (5′-UTR) and envelope protein 2 (*E2*) [6]. These genotypes display a specific geographical distribution; for instance, the HPgV-1 genotype 1 is predominant in West African, genotype 2 in the United States and Europe, genotype 3 in East Asia, genotype 4 in Myanmar and Vietnam, genotype 5 in South Africa, genotype 6 in Indonesia. More recently, genotype 7 has been found in Yunnan, China [7].

HPgV-1 is a nonpathogenic human virus that is transmitted efficiently via parenteral, vertical, and sexual routes [4,8,9]. As a result of the shared modes of transmission with the human immunodeficiency virus (HIV) and HCV, HPgV-1 infection is more prevalent among HIV-1- and/or HCV-positive individuals. It has been reported that up to 40% of HCV- and/or HIV-1-infected patients are HPgV-1 positive [10,11]. Consistent with these findings, our previous study revealed that 35.8% of HIV-1-infected injecting drug users (IDUs) in Yunnan, China were positive for HPgV-1 RNA [7].

Notably, several studies have confirmed that HPgV-1 does not cause any liver related disease in humans [4,8,12]. However, co-infection with HIV-1 may result in favorable outcomes such as lower mortality rate, slower disease progression, and longer survival [7–9]. In addition, a recent study showed that in these HCV/HIV-co-infected patients, HPgV-1 RNA was associated with a significant reduction in the severity of HCV-related liver disease [13–15]. However, several studies have failed to confirm a positive impact of HPgv-1 co-infection on HIV disease [14,16,17]. An explanation for this finding is that the genotypic diversity of HPgV-1 may play a role in modulating disease progression in HIV-1-infected individuals. Furthermore, HPgV-1 genotypes 2 and 5 have been associated with delayed progression of AIDS [7,14,18,19]. Therefore, it is necessary to study HPgV-1 genotypic diversity and examine its clinical impact in larger cohorts with HIV-1 infection. This present study investigates the prevalence and genotypic diversity of HPgV-1 and its impact on disease progression in 1062 HIV-1-positive patients in Yunnan, China.

2. Materials and Methods

2.1. Ethical Statements

All subjects gave their informed consent for inclusion before they participated in the study. The study was conducted in accordance with the Declaration of Helsinki, and the protocol was approved by the Ethics Committee of Yunnan Provincial Hospital of Infectious Disease, AIDS Care Center (Approval No. YNACC [2015]-12).

2.2. Study Population and Sample Collection

In this study, plasma samples were collected from a total of 1062 HIV-positive individuals from Yunnan Provincial Hospital of Infectious Disease, AIDS Care Center (YNACC) between August 2011 and August 2015; these individuals were from each of the 16 prefectures of the Yunnan Province (Figure 1). Blood samples from the individuals were eligible if they were residents of the Yunnan province, self-reported ever having HIV high-risk drug use and sexual behaviors, consent to use patient information in studies of viral epidemics, and able to provide informed consent. Each participant completed a face-to-face interview with trained interviewers to obtain the following information: (1) demographic data, including age, ethnicity, education and marital status etc.; (2) high risk sexual behaviors, including sexual orientation, sexual debut, number of male and/or female sex partners, and experiences of buying or selling sex with other partners in the past 6 months; (3) illicit drug use behaviors, including lifetime history of drug use, age at first injection, and needle sharing. After the interview, an HIV counselor met with the participant for pretest counseling. The interviewer was conducted by trained public health personnel with extensive experience interviewing facial features regarding sexual and drug-use behaviors, and the route of HIV transmission was determined based on analysis of interview data. The samples collected from the participants having multiple risk behaviors and only 18 samples from homosexuals with small sample sizes were excluded from analyses in this study. HIV-1 infection status was determined using an enzyme-linked immunosorbent assay (ELISA) (Abnova, Taibei, Taiwan) and confirmed by western blot assay (MP Biomedicals, Santa Ana, CA, USA).

The presence of HCV RNA was detected by RT-nPCR based on the NS5b sequence [20], HCV RNA loads were assessed using the Roche Cobas Amplicor 2.0 assay (Roche, Amsterdam, The Netherlands), and HCV genotype was determined by LIPA 2.0 assay (Siemens Healthcare, Berkeley, CA, USA). Demographic data relating to age, gender, and route of transmission were recorded via self-report questionnaires. The clinical parameters of disease progression, including the alanine aminotransferase (ALT) and aspartate aminotransferase (AST) levels, HIV-1 RNA load, and $CD4^+$ T cell counts were determined at sampling time.

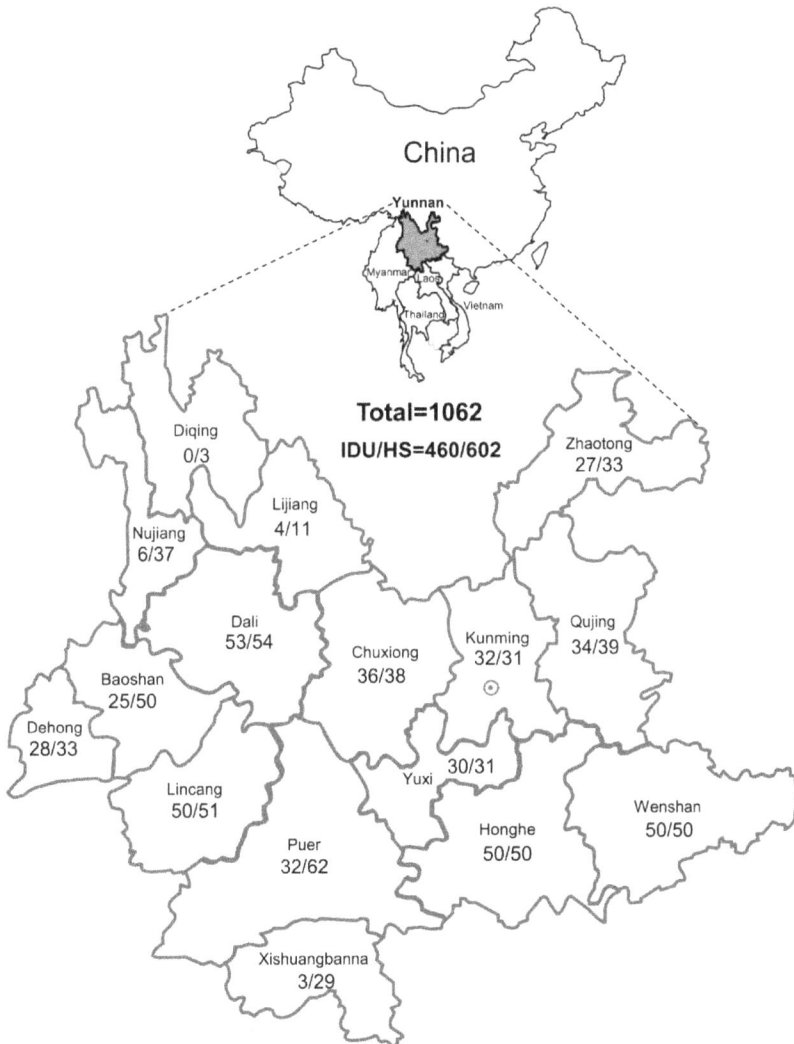

Figure 1. Maps of the study region and geographical distribution of subjects from all sixteen prefectures of the Yunnan province of southwestern China. The Yunnan province of southwestern China is marked in gray. The geographical location of the sixteen regions within this province and the number of samples (IDUs, injecting drug users; HS, heterosexual contact) from each region are shown. This map is modified by the authors according to the free map templates [21] using MapInfo Professional 11.0 (Pitney Bowes Inc., Troy, NY, USA).

2.3. RNA Extraction, HPgV-1 Gene Amplification and Sequencing

HPgV-1 RNA was isolated from 200-µL plasma samples using the High Pure Viral RNA Kit according to the procedure described in the manual (Roche, Basel, Switzerland)). Then, the 5'-UTR (U36380:119-497) and E2 (U36380:950-1844) sequences were amplified by nested PCR; the PCR primers (supplement Table S1) and conditions were as reported in previous study [7,9]. Owing to the high degree of conservation and amplification efficiency of the 5'-UTR, this region was used to evaluate the HPgV-1 infection rate.

The E2 region was used to determine the HPgV-1 genotype, as this sequence could analyze the different genotypes with the same consistency as the complete genome [7–9]. The first PCR reaction was performed using One Step reverse transcription PCR (Takara, Dalian, China) and the second using 2 × Taq PCR MasterMix (Tiangen, Beijing, China). The PCR products were firstly detected by agarose gel (1.0%) electrophoresis and visualized under ultraviolet (UV) illumination for the presence of an 895-nucleotide band and then purified by using a DNA purification kit (Tiangen, Beijing, China); subsequently, the purified products were sent to Shenzhen Invitrogen Biotechnology Co., Ltd. (Shenzhen, Guangdong, China) for sequencing by using an ABI 3730XL automated DNA sequencer (Applied Biosystems, Carlsbad, CA, USA).

2.4. Sequence Analyses

The sequencing data were initially checked via a NCBI BLAST search [22]. The resulting sequences were edited using BioEdit 7.2.5 software [23]. The reference sequences available in GenBank [24] were downloaded to conduct a comparative analysis of all the HPgV-1 E2 genomic sequences. Multiple alignments of the selected sequences were performed by Clustal Omega [25]. Subsequently, the data generated were processed using the BioEdit 7.1.5 software. Phylogenetic trees were constructed based on the obtained datasets using MEGA version 6.0.6 [26] with maximum-likelihood method using the general time reversible + gamma distribution + invariant sites (GTR + τ + I) model. Bootstrap values were calculated based on 1000 replications of the alignment. Principal coordinate analysis was performed using a principal coordinate analysis (PCOORD). All the HPgV-1 E2 genomic sequences obtained in this study have been deposited in GenBank under accession numbers KX430523-KX430734.

2.5. Statistical Analysis

Statistical analyses were conducted using the SPSS 21.0 statistical analysis software package (IBM, Armonk, NY, USA). For descriptive analyses, the means and standard deviations, frequency, and percentage values were reported. The tests of differences between the HPgV-1-infected group and HPgV-1-uninfected group were performed using a t test for the difference in means (for age, ATL, AST, CD4$^+$ counts, log HIV-1 RNA, and log HCV RNA) and Fisher exact test for gender, HIV-1 risk behavior, and HCV genotypes. All the p values below 0.05 were considered to indicate statistical significance.

3. Results

3.1. Epidemiologic and Demographic Characteristics

Blood samples were collected from a total of 1062 HIV-1-positive individuals from all 16 prefectures of the Yunnan province, from August 2011 to August 2015. Among these, 56.69% (602/1062) had become infected mainly via heterosexual contact and 43.31% (460/1062) by injecting drug user. The epidemiological characteristics of the 1062 subjects included in the present study are summarized in Figure 1. The mean age of the participants was 38.57 ± 10.22 years, and the ratio of males to females was 696:366. The following clinical characteristics were identified: the mean ALT (41.31 ± 47.22 IU/L), the mean AST (44.70 ± 64.59 IU/L), the mean CD4$^+$ cell count (301.54 ± 187.58 cells/uL), and HIV-1 RNA (4.05 ± 0.67 log copies/mL). In addition, the gender, ALT, and AST showed highly significant differences among different HIV-1 risk behaviors (IDU vs. heterosexual contact) (Supplement Table S2).

3.2. HPgV-1 Infection Status

Out of the 1062 patients with HIV-1 infection, the overall prevalence of HPgV-1 infection was 23.4% (248/1062) via 5′-UTR (378bp) amplification by nested RT-PCR. Among these, 117 (19.4%) of 602 heterosexuals were classified as HPgV-1 RNA-positive and 131 (28.5%) of 460 IDUs tested positive for HPgV-1 RNA. The HPgV-1 positive rate of HIV-1-infected IDUs is significant higher than that of heterosexuals ($p < 0.05$).

Notably, HPgV-1 infection rates among the HIV-1-infected individuals differed based on age and HIV-1 RNA load as follows: age (in years) 16–30 (29.5%), 31–50 (22.5%), > 50 (16.4%) ($p = 0.019$); HIV-1 RNA load 2.98–4.00 (27.2%), 4.00–5.00 (19.8%), >5 (15.8%) ($p = 0.005$). These results revealed that HPgV-1 infection is associated with increasing age and HIV-1 RNA.

In the sampled prefectures, the HPgV-1 RNA-positive rates among HIV-1-infected individuals were higher than 30%; the rate of HPgV-1-positive individuals was 37.3% in Baoshan, 32.6% in Nujiang, and 30.0% in Kunming. There were no significant differences between these prefectures. However, the HPgV-1 RNA-positive rates among individuals from these prefectures were significantly higher than in those from Xishuangbannan (12.5%), Yuxi (13.1%), and Lijiang (13.3%) ($p < 0.05$) (Supplemental Figure S1A). For the remaining prefectures, the HPgV-1 RNA-positive rate was intermediate (16.2%–27.0%) (Figure S1A). Only three HIV-1-positive subjects were from Diqing; among these, HPgV-1 RNA was not detected.

3.3. HPgV-1 Genotypes Distribution

Out of a total of 248 HPgV-1 RNA-positive samples, 212 partial *E2* gene fragments were successfully amplified and sequenced with a success rate of 85.5% (212/248). The failure of amplification in the 36 cases was possibly due to low viral load and weak primer specificity. Phylogenetic analyses were performed based on *E2* fragments of HPgV-1. Of these, 118 (55.7%), 74 (34.9%), 10 (4.7%), and 7 (3.3%) sequences were identified as corresponding to genotypes 7, 3, 4 and 2, respectively (Figure 2A). The remaining three strains (DL27S, DL73621 and LC5718S) formed a distinct monophyletic branch with a bootstrap value of 99%, distantly related to all known HPgV-1 genotypes, indicating that these may represent a novel genotype especially when recombination was excluded by bootscanning analyses (no breakpoint was found). Further, the coordinate result obtained by a principal coordinate analysis (PCOOD) indicated that the three strains formed a single group.

Based on the transmission route of HPgV-1, the genotypic distribution among the 116 IDUs was as follows: 85 (73.3%, 85/116), genotype 7; 22 (19.0%, 22/116), genotype 3; 6 (5.2%, 6/116), genotype 2; 2 (1.7%, 2/116), genotype 4; and 1 (0.9%, 1/116), an unclassified group. Among the 96 heterosexuals, genotype 3 was the most common genotype (54.2%, 52/96), followed by genotype 7 (34.4%, 33/96), genotype 4 (5.2%, 5/96), genotype 2 (4.2%, 4/96), and the unclassified group (2.1%, 2/96). Notably, the distribution of genotype 7 among the IDUs (73.3%, 85/116) was significantly higher than among the heterosexuals (34.4%, 33/96) ($p < 0.05$). The percentage of genotype 3 prevalence among heterosexuals (54.2%, 52/96) was significantly higher than in the IDUs (19.0%, 22/116) ($p < 0.05$) (Figure 2C).

In the sampled prefectures, HPgV-1 strains circulating in Baoshan, Dehong, Lincang, Puer, Xishuangbannan and Dali exhibited extremely higher genotypic diversity compared with those in the other prefectures (Figure S1B).

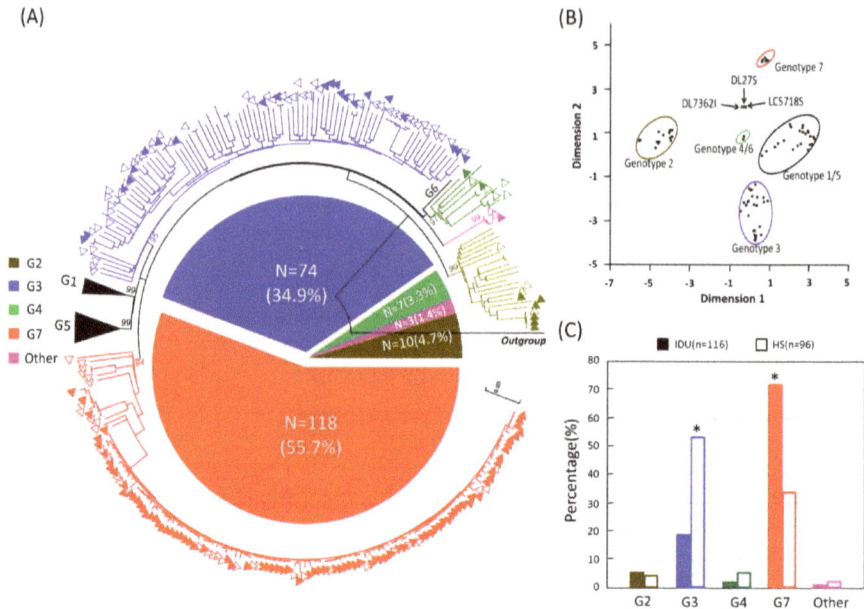

Figure 2. Genotypic analyses of human Pegivirus-1 (HPgV-1) and its distribution between injected drug users and heterosexual individuals in Yunnan, China. (**A**) A circular phylogenetic tree based on the partial *E2* sequences amplified from the 212 human immunodeficiency virus-1 (HIV-1)-infected individuals. The different genotypes are shown in different colors, as indicated on the left of the tree (other means the unclassified group) and the open and closed triangles indicate injecting drug users (IDUs) and heterosexual contact (HS), respectively. The pie chart inside the tree shows the percentages of the various HPgV-1 genotypes; (**B**) Multivariate principal coordinate analysis (PCOORD) of partial *E2* sequences of HPgV-1 (filled circles); (**C**) Comparison of HPgV-1 genotype distributions among IDUs and heterosexuals in Yunnan, China. The distribution significance of genotype 7 and 3 among the IDUs and heterosexuals were presented with asterisks ($p < 0.05$).

3.4. Co-infection of HPgV-1 with HIV/HCV and Its Clinical Effect

Out of the 1062 HIV-1-infected patients, HCV RNA was detected in plasma samples from 287 (27.02%, 287/1062) subjects using RT-nPCR based on the HCV NS5B region. In combination with the results for HPgV-1 detection, 621 (58.47%, 621/1062) subjects were found to be infected with HIV-1 alone, 193 (18.17%, 193/1062) were co-infected with HIV-1 and HCV, 154 (14.50%, 154/1062) were co-infected with HIV-1 and HPgV-1, and 94 (8.85%, 94/1062) with HIV-1, HCV, and HPgV-1. Taken together, the individuals were further grouped into two comparative groups; group 1 of HIV/HCV co-infected subjects with or without HPgV-1 infection and group 2 of HIV-infected subjects with or without HPgV-1 infection (Figure 3).

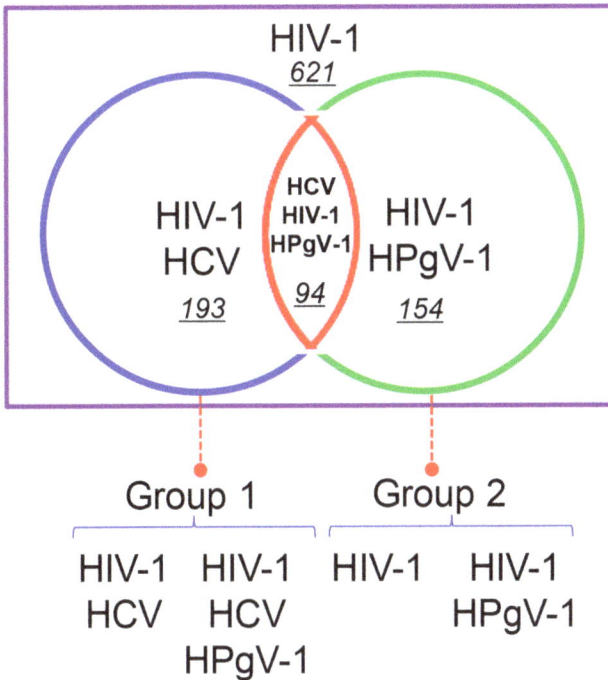

Figure 3. A graphical representation of HIV-1 mono-, HIV-1/Hepatitis C Virus (HCV) co-, and HIV-1/HCV/HPgV-1 triple-infection cases in our cohort in Yunnan, China. 621: the numbers of HIV-1 mono-infection, 193: HIV-1 and HCV co-infection, 154: HIV-1 and HPgv-1 co-infection, and 94: HIV-1, HCV and HPgv-1 triple-infection. Different groups are shown in different colors, as indicated at the bottom of the graphic.

Clinical characteristics for patients with and without HPgV-1 infection in the two groups above are summarized in Table 1. The mean (Standard deviation, SD) HIV-1 viral load of HPgV-1-infected patients was lower than patients negative for HPgV-1 (Group 1: log = 3.88 ± 0.47 vs. 4.02 ± 0.61, $p = 0.040$; Group 2: 3.93 ± 0.52 vs. 4.04 ± 0.69, $p = 0.048$, respectively) and the mean (SD) T CD4$^+$ lymphocyte cell counts was higher than patients negative for HPgV-1 (Group 1: cells/µL = 322 ± 193 vs. 280 ± 148, $p = 0.046$; Group 2: 322 ± 184 vs. 288 ± 154, $p = 0.048$, respectively). In contrast, there was no significant difference in terms of mean (SD) gender, ALT, AST, transmission route, HCV viral load, and HCV genotype ($p > 0.05$) between the HPgV-1-infected and uninfected patients.

Interestingly, in the two groups (only the most common genotype of 3 and 7 were compared and genotype 2 and 4 were not included for lacking enough number of individuals), HPgV-1 infection with genotype 7 was significantly associated with lower HIV-1 viral load (Group 1: log = 3.85 ± 0.49 vs. 4.02 ± 0.61, $p = 0.042$; Group 2: 3.83 ± 0.64 vs. 4.04 ± 0.69, $p = 0.038$, respectively) and higher CD4$^+$ cell counts (Group 1: cells/µL = 332 ± 204 vs. 280 ± 148, $p = 0.037$; Group 2: 341 ± 164 vs. 288 ± 154, $p = 0.015$, respectively) compared with the HPgV-1 negative group ($p < 0.05$) (Table 2). These findings suggest that infection with HPgV-1 genotype 7 may be associated with slower disease progression in HIV-1-positive individuals. In addition, when HPgV-1 genotype 7 compared with genotype 3, there was no significant difference showed.

Table 1. Differences between HPgV-1-infected and HPgV-1-uninfected patients among HIV-1-infected individuals with or without HCV.

Variable	Group 1: HIV-1/HCV (n = 287)			Group 2: HIV-1 (n = 775)		
	HPgV-1-Infected	HPgV-1-Uninfected	p	HPgV-1-Infected	HPgV-1-Uninfected	p
Patients, n (%)	94 (32.75)	193 (62.25)	N/A	154 (19.87)	621 (80.13)	N/A
Gender, male:female [2]	79:15	150:43	0.273	86:71	386:240	0.145
Mean (SD) age, y [1]	37.99 (5.80)	38.33 (6.05)	0.639	36.90 (11.5)	39.14 (11.34)	0.029
Mean (SD) ALT, IU/L [1]	51.45 (37.42)	55.48 (43.29)	0.227	31.15 (26.71)	33.03 (29.94)	0.834
Mean (SD) AST, IU/L [1]	55.26 (36.20)	56.22 (40.03)	0.843	34.46 (23.33)	35.06 (26.24)	0.237
Mean (SD) CD4$^+$ count, cells/Ul [1]	322 (193)	280 (148)	0.046	322 (184)	288.5 (154)	0.048
Mean (SD) HIV-1 RNA log copies/mL [1]	3.88 (0.47)	4.02 (0.61)	0.040	3.93 (0.52)	4.04 (0.69)	0.048
HIV-1 transmission route no. (%) [2]			0.530			0.278
Injection drug users	78 (82.98)	154 (79.79)		51 (33.12)	181 (28.91)	N/A
Heterosexual	16 (17.02)	39 (20.21)		103 (66.88)	445 (71.09)	N/A
Mean (SD) HCV RNA logcopies/mL [1]	4.98 (0.21)	5.12 (0.31)	0.378	N/A	N/A	N/A
HCV genotype no. (%) [2]			0.563			N/A
1	17 (18.09)	40 (20.73)		N/A	N/A	
2	0 (0.00)	2 (1.04)		N/A	N/A	
3	61 (64.89)	112 (58.03)		N/A	N/A	
6	16 (17.02)	39 (20.20)		N/A	N/A	

N/A, not applicable; *n*, number; y, years; SD, standard deviation; ALT, Alanine aminotransferase; AST, Aspartate transaminase; [1] *t* test; [2] Fisher exact test.

Table 2. Differences between HPgV-1-infected and HPgV-1-uninfected patients among HIV-1-infected individuals with or without HCV.

Variable	HPgV-1-Uninfected	HPgV-1-Infected				
		G3	P1	G7	P2	P3
Group 1: HCV+, n (%)	193 (62.25)	11 (3.83)	N/A	66 (23.00)	N/A	N/A
Gender, male:female [2]	150:43	8:3	0.714	57:9	0.156	0.363
Mean (SD) age, y [1]	38.33 (6.05)	37.91 (7.35)	0.822	38.11 (5.33)	0.783	0.915
Mean (SD) ALT, IU/L [1]	55.48 (43.29)	55.48 (43.29)	0.703	50.93 (38.8)	0.463	0.968
Mean (SD) AST, IU/L [1]	56.22 (40.03)	50.62 (22.56)	0.647	54.61 (36.25)	0.781	0.726
Mean (SD) CD4+ count, cells/uL [1]	280 (148)	352 (199)	0.144	332 (204)	0.037	0.503
Mean (SD) HIV-1 RNA log10 copies/mL [1]	4.02 (0.61)	3.97 (0.37)	0.827	3.85 (0.49)	0.042	0.481
HIV-1 transmission route IDUs:HS [2]	154:39	10:1	0.696	56:10	0.467	0.999
HCV genotype no. (%) [2]			0.757		0.535	0.753
1	40 (20.73)	2 (18.18)		10 (15.15)		
2	2 (1.04)	0 (0.00)		0 (0.00)		
3	112 (58.03)	8 (72.73)		44 (66.67)		
6	39 (20.20)	1 (9.09)		12 (18.18)		
Group 2: HCV-, n (%)	621 (84.37)	63 (8.56)	N/A	52 (7.07)	N/A	N/A
Gender, male:female [2]	386:240	33:30	0.999	31:21	0.769	0.457
Mean (SD) age, yrs [1]	39.14 (11.34)	34.73 (12.04)	0.004	38.27 (11.60)	0.596	0.113
Mean (SD) ALT [1]	33.03 (29.94)	34.5 (4.04)	0.248	33.71 (28.37)	0.879	0.324
Mean (SD) AST [1]	35.06 (26.24)	34.42 (25.90)	0.855	34.57 (19.62)	0.899	0.896
Mean (SD) CD4$^+$ count, cells/uL [1]	288.5 (154)	310 (195)	0.366	341 (164)	0.015	0.387
Mean (SD) HIV-1 RNA log10 copies/mL [1]	4.04 (0.69)	3.97 (0.57)	0.258	3.83 (0.64)	0.038	0.349
HIV-1 transmission route IDUs:HS [2]	181:445	16:47	0.661	24:28	0.012	0.030

P1, HPgV-1 G3 vs. HPgV-1-uninfected; P2, HPgV-1 G7 vs. HPgV-1-uninfected; P3, HPgV-1 G3 vs. HPgV-1 G7; [1] *t* test; [2] Fisher exact test.

4. Discussion

In the present study, we evaluated the frequency, genotypic distribution, and clinical impact of HPgV-1 infection on HIV-1 disease progression in 1062 HIV-1-infected individuals in Yunnan, China. The overall positive detection rate of HPgV-1 RNA was 23.4% in our cohort (28.5% in IDUs and 19.4% in heterosexuals), which is in accordance with other previous reports stating that HPgV-1 RNA prevalence is about 17%–41% among HIV-positive individuals [19]. Our data indicated that the frequency of HPgV-1 infection in IDUs was significantly higher than in heterosexuals, implying that intravenous injection may represent a more efficient transmission route of HPgV-1 than heterosexual contact. The observation agreed with the finding that intravenous drug use constitutes an important risk behavior for the transmission of blood-borne viruses such as HIV-1, HCV, HBV, and so on [27–29].

Moreover, we found that the rate of HPgV-1 infection was significantly higher in younger patients (age: 16-30 years) ($p = 0.029$), and that the frequency of HPgV-1 infection declines with increasing age ($p = 0.019$). Similar trends have been identified in several previous studies [30,31]. This may be attributed to younger individuals being more sexually active and exhibiting higher incidence of drug use than older patients [31–33]. Another possible explanation is that E2 antibodies against HPgV-1 reinfection are found more frequently in older people [30,31]. Unfortunately, in the present study, the HPgV-1 E2 antibody levels could not be estimated due to the lack of a commercially available serology assay [34]. In addition, with the increase of HIV-1 RNA viral load, a significant decline in the overall HPgV-1 infection rate was observed, indicating that HPgV-1 infection exhibits an inverse relationship with HIV-1 load. However, we could not draw the conclusion that a negative correlation exists between HPgV-1 and HIV-1 RNA levels. Therefore, further studies are needed to determine the HPgV-1 RNA viral load.

Our results indicate that genotypic diversity and complexity of HPgV-1 circulating in Yunnan, China is high. Multiple genotypes, namely 2, 3, 4, 7, and an unclassified group, were identified in this study; these findings differ from those of previous reports which suggest the predominance of genotype 3 in Beijing [35], and the Hubei province of China [6]. However, our results are consistent with observations in Singapore [36], and Indonesia [37] where high prevalence of HPgV-1 multiple genotypes have been described in patients with HIV-1 infection. These findings may be explained by the special geographic location of Yunnan province. Yunnan is located in southwestern China and borders the opium-producing "Golden Triangle" region composed of Myanmar, Laos, Thailand, and Vietnam [7,38,39]. The transmission of several human viruses is closely associated with illegal drug trafficking [40,41]. Additionally, large numbers of commercial sex workers from Myanmar and Vietnam are active at the Yunnan border [42,43]. These high-risk factors have facilitated the spread of numerous human viruses in Yunnan; for example, HIV-1 subtypes B and C were introduced into Yunnan from Myanmar and India by IDUs [44], and CRF01_AE was identified among commercial sex workers returning from Thailand to Yunnan. In addition [44], such factors have resulted in the transmission of HPgV-1 genotype 4 and 7 in Yunnan. In contrast, genotype 2 predominates in Europe and America [8,9]. A small number of genotype 2 infections detected in Yunnan may have been imported as a result of the frequent exchange of personnel between China and the USA, which has arisen due to the close economic ties between the two countries. Interestingly, we found that HPgV-1 genotype 7 predominated in IDUs, whereas genotype 3 was the most common genotype in heterosexual populations. This pattern and prevalence of HPgV-1 infections may be closely related to the origin of the two kinds of genotypes. HPgV-1 genotype 7 originated in IDUs from Southeast Asia described in our previous study. Genotype 3 may have been derived from China or Japan. To date, little is known about the evolution, origin, and migration of HPgV-1 genotype 3 and 7. Therefore, further molecular epidemiological surveys are needed to monitor the epidemiology and phylogenetic linkages implicated in the HPgV-1 epidemic in Asia.

Several epidemiological studies have indicated a beneficial effect of HPgV-1 viremia on disease progression of HIV-1-infected individuals. HPgV-1 infection has been associated with higher CD4$^+$ cell counts and lower HIV-1 loads [7–9]. These findings are further supported by increasing evidence for an attenuating effect of HPgV-1 on HIV-1 replication in vitro [45–47]. Furthermore, studies of the genetic diversity of HPgV-1 revealed the existence of seven distinct genotypes worldwide; these genotypes present clear differences in regional distribution [8,9]. Interestingly, several reports have suggested that genotypes 2 and 5 are associated with delayed AIDS progression to a greater extent than the other genotypes [7,14,18,19]. Another hypothesis suggests that HPgV-1 may influence the progression of HIV-1 disease; this is the subject of numerous epidemiological surveys [8,9]. However, similar results have been rarely found in cohorts of HIV-1- and/or HCV-positive patients [8,14,16]. This may be attributed to the small number of subjects investigated in these studies, which results in lower accuracy of statistical analyses. In the present study, an investigation of a large cohort of HIV-1-infected patients (1062 subjects) suggested that the recently identified genotype 7 may be more strongly associated with slow HIV-1 disease progression than negative, supporting the hypothesis outlined above.

Our study lacked a long-term follow-up epidemiological survey aimed at HPgV-1 strains with different genotypes. Another limitation of the study is that some demographic and clinical variables, such as race, HPgV-1 E2 antibody levels, HPgV-1 RNA viral load, HIV-1 subtype, and treatment regimen, were not evaluated.

5. Conclusions

In summary, our data suggest that HPgV-1 infection among HIV-1-infected populations is extremely common; the HPgV-1 epidemic in Yunnan was characterized by infection with multiple genotypes. We additionally found that HPgV-1 genotype 7 may be associated with slower disease progression in HIV-1-infected individuals. These present findings provide novel insights into the genotypic diversity of HPgV-1 in co-infected HIV-1-positive individuals and are potentially useful for the development of HPgV-1-based bio-therapies for AIDS.

Supplementary Materials: The following are available online at www.mdpi.com/1999-4915/9/2/28/s1, Table S1: Oligonucleotide primer sequences used to amplify HPgV-1, Table S2: The distribution of HPgV-1 infection rates and genotypes for each prefecture in Yunnan. Figure S1: Demographic and clinical characteristics among HIV-1-infected population.

Acknowledgments: This study was supported financially by National Natural Science Foundation of China (No. 81360247,81601767 and 81460509), Yunnan provincial innovation team project (2015HC030), New Product Development Projects of Yunnan Province (2016BC005) and Funded by Open Research Fund Program of the State Key Laboratory of Virology of China (2014KF009).

Author Contributions: Y.F., X.D. and X.X. designed the experiments. L.G., M.Z., J.L. and X.W. conducted samples collection. Z.M., L.G., Y.S., M.Y. and Y.Z. performed nRT-PCR assay. Z.M., Y.F. and X.X. performed database searches and sequence analyses. Z.M. and Y.F. wrote the main manuscript. All authors reviewed the manuscript.

Conflicts of Interest: The authors declare no conflict of interest.

References

1. Berg, M.G.; Lee, D.; Coller, K.; Frankel, M.; Aronsohn, A.; Cheng, K.; Forberg, K.; Marcinkus, M.; Naccache, S.N.; Dawson, G.; et al. Discovery of a novel human pegivirus in blood associated with hepatitis C virus co-infection. *PLoS Pathog.* **2015**, *11*, e1005325. [CrossRef] [PubMed]
2. Chivero, E.T.; Bhattarai, N.; Rydze, R.T.; Winters, M.A.; Holodniy, M.; Stapleton, J.T. Human pegivirus RNA is found in multiple blood mononuclear cells in vivo and serum-derived viral RNA-containing particles are infectious in vitro. *J. Gen. Virol.* **2014**, *95*, 1307–1319. [CrossRef] [PubMed]
3. Adams, M.J.; Lefkowitz, E.J.; King, A.M.; Carstens, E.B. Ratification vote on taxonomic proposals to the international committee on taxonomy of viruses (2014). *Arch. Virol.* **2014**, *159*, 2831–2841. [CrossRef] [PubMed]
4. Chivero, E.T.; Stapleton, J.T. Tropism of human pegivirus (formerly known as GB virus C/hepatitis G virus) and host immunomodulation: Insights into a highly successful viral infection. *J. Gen. Virol.* **2015**, *96*, 1521–1532. [CrossRef] [PubMed]
5. AbuOdeh, R.O.; Al-Absi, E.; Ali, N.H.; Khalili, M.; Al-Mawlawi, N.; Hadwan, T.A.; Althani, A.A.; Nasrallah, G.K. Detection and phylogenetic analysis of human pegivirus (GBV-C) among blood donors and patients infected with hepatitis B virus (HBV) in Qatar. *J. Med. Virol.* **2015**, *87*, 2074–2081. [CrossRef] [PubMed]
6. Wu, H.; Padhi, A.; Xu, J.; Gong, X.; Tien, P. Evidence for within-host genetic recombination among the human pegiviral strains in HIV infected subjects. *PLoS ONE* **2016**, *11*, e0161880. [CrossRef] [PubMed]
7. Feng, Y.; Zhao, W.; Feng, Y.; Dai, J.; Li, Z.; Zhang, X.; Liu, L.; Bai, J.; Zhang, H.; Lu, L.; et al. A novel genotype of GB virus C: Its identification and predominance among injecting drug users in Yunnan, China. *PLoS ONE* **2011**, *6*, e21151. [CrossRef] [PubMed]

8. Bhattarai, N.; Stapleton, J.T. GB virus C: The good boy virus? *Trends Microbiol.* **2012**, *20*, 124–130. [CrossRef] [PubMed]

9. Giret, M.T.; Kallas, E.G. GBV-C: State of the art and future prospects. *Curr. HIV/AIDS Rep.* **2012**, *9*, 26–33. [CrossRef] [PubMed]

10. Rendina, D.; Vigorita, E.; Bonavolta, R.; D'Onofrio, M.; Iura, A.; Pietronigro, M.T.; Laccetti, R.; Bonadies, G.; Liuzzi, G.; Borgia, G.; et al. HCV and GBV-C/HGV infection in HIV positive patients in southern Italy. *Eur. J. Epidemiol.* **2001**, *17*, 801–807. [CrossRef] [PubMed]

11. Tenckhoff, S.; Kaiser, T.; Bredeek, F.; Donfield, S.; Menius, E.; Lail, A.; Mossner, J.; Daar, E.S.; Tillmann, H.L. Role of GB virus C in HIV-1-infected and hepatitis C virus-infected hemophiliac children and adolescents. *J. Acquir. Immune Defic. Syndr.* **2012**, *61*, 243–248. [CrossRef] [PubMed]

12. Mohr, E.L.; Stapleton, J.T. GB virus type C interactions with HIV: The role of envelope glycoproteins. *J. Viral Hepat.* **2009**, *16*, 757–768. [CrossRef] [PubMed]

13. Berzsenyi, M.D.; Woollard, D.J.; McLean, C.A.; Preiss, S.; Perreau, V.M.; Beard, M.R.; Scott Bowden, D.; Cowie, B.C.; Li, S.; Mijch, A.M.; et al. Down-regulation of intra-hepatic T-cell signaling associated with GB virus C in a HCV/HIV co-infected group with reduced liver disease. *J. Hepatol.* **2011**, *55*, 536–544. [CrossRef] [PubMed]

14. Feng, Y.; Liu, L.; Feng, Y.M.; Zhao, W.; Li, Z.; Zhang, A.M.; Song, Y.; Xia, X. GB Virus C infection in patients with HIV/hepatitis C virus coinfection: Improvement of the liver function in chronic hepatitis C. *Hepat. Mon.* **2014**, *14*, e14169. [CrossRef] [PubMed]

15. Berzsenyi, M.D.; Bowden, D.S.; Kelly, H.A.; Watson, K.M.; Mijch, A.M.; Hammond, R.A.; Crowe, S.M.; Roberts, S.K. Reduction in hepatitis C-related liver disease associated with GB virus C in human immunodeficiency virus coinfection. *Gastroenterology* **2007**, *133*, 1821–1830. [CrossRef] [PubMed]

16. Blackard, J.T.; Ma, G.; Welge, J.A.; King, C.C.; Taylor, L.E.; Mayer, K.H.; Klein, R.S.; Celentano, D.D.; Sobel, J.D.; Jamieson, D.J.; et al. GB virus C (GBV-C) infection in hepatitis C virus (HCV) seropositive women with or at risk for HIV infection. *PLoS ONE* **2014**, *9*, e114467. [CrossRef] [PubMed]

17. Berzsenyi, M.D.; Bowden, D.S.; Roberts, S.K.; Revill, P.A. GB virus C genotype 2 predominance in a hepatitis C virus/HIV infected population associated with reduced liver disease. *J. Gastroenterol. Hepatol.* **2009**, *24*, 1407–1410. [CrossRef] [PubMed]

18. Muerhoff, A.S.; Tillmann, H.L.; Manns, M.P.; Dawson, G.J.; Desai, S.M. GB virus C genotype determination in GB virus-C/HIV co-infected individuals. *J. Med. Virol.* **2003**, *70*, 141–149. [CrossRef] [PubMed]

19. Alcalde, R.; Nishiya, A.; Casseb, J.; Inocencio, L.; Fonseca, L.A.; Duarte, A.J. Prevalence and distribution of the GBV-C/HGV among HIV-1-infected patients under anti-retroviral therapy. *Virus Res.* **2010**, *151*, 148–152. [CrossRef] [PubMed]

20. Lu, L.; Wang, M.; Xia, W.J.; Tian, L.W.; Xu, R.; Li, C.H.; Wang, J.X.; Rong, X.; Xiong, H.P.; Huang, K.; et al. Migration patterns of hepatitis C virus in China characterized for five major subtypes based on samples from 411 volunteer blood donors from 17 provinces and municipalities. *J. Virol.* **2014**, *88*, 7120–7129. [CrossRef] [PubMed]

21. China outline map. Available online: http://www.zonu.com/fullsize-en/2011-07-28-14186/China-outline-map.html (accessed on 12 October 2016).

22. Tamura, K.; Stecher, G.; Peterson, D.; Filipski, A.; Kumar, S. MEGA6: Molecular Evolutionary Genetics Analysis version 6.0. *Mol Biol Evol.* **2013**, *30*, 2725–2729. [CrossRef] [PubMed]

23. Sievers, F.; Wilm, A.; Dineen, D. Fast, scalable generation of high-quality protein multiple sequence alignments using Clustal Omega. *Mol. Syst.* **2011**, *7*, 539. [CrossRef] [PubMed]

24. Standard Nucleotide BLAST. Available online: https://blast.ncbi.nlm.nih.gov/Blast.cgi?PROGRAM=blastn&PAGE_TYPE=BlastSearch&LINK_LOC=blasthome (accessed on 12 October 2016).

25. BioEdit v7.2.5. Available online: http://www.mbio.ncsu.edu/BioEdit/page2.html (accessed on 12 October 2016).

26. The reference sequences of GBV-C E2 gene. Available online: https://www.ncbi.nlm.nih.gov/nuccore/?term=GB+Virus+C+E2 (accessed on 12 October 2016).

27. Vickerman, P.; Hickman, M.; May, M.; Kretzschmar, M.; Wiessing, L. Can hepatitis C virus prevalence be used as a measure of injection-related human immunodeficiency virus risk in populations of injecting drug users? An ecological analysis. *Addiction* **2010**, *105*, 311–318. [CrossRef] [PubMed]

28. Nelson, P.K.; Mathers, B.M.; Cowie, B.; Hagan, H.; Des Jarlais, D.; Horyniak, D.; Degenhardt, L. Global epidemiology of hepatitis B and hepatitis C in people who inject drugs: Results of systematic reviews. *Lancet* **2011**, *378*, 571–583. [CrossRef]

29. Zhu, T.F.; Wang, C.H.; Lin, P.; He, N. High risk populations and HIV-1 infection in China. *Cell Res.* **2005**, *15*, 852–857. [CrossRef] [PubMed]

30. Schwarze-Zander, C.; Blackard, J.T.; Zheng, H.; Addo, M.M.; Lin, W.; Robbins, G.K.; Sherman, K.E.; Zdunek, D.; Hess, G.; Chung, R.T. GB virus C (GBV-C) infection in hepatitis C virus (HCV)/HIV-coinfected patients receiving HCV treatment: Importance of the GBV-C genotype. *J. Infect. Dis.* **2006**, *194*, 410–419. [CrossRef] [PubMed]

31. Da Mota, L.D.; Nishiya, A.S.; Finger-Jardim, F.; Barral, M.F.; Silva, C.M.; Nader, M.M.; Goncalves, C.V.; Da Hora, V.P.; Silveira, J.; Basso, R.P.; et al. Prevalence of human pegivirus (HPgV) infection in patients carrying HIV-1C or non-C in southern Brazil. *J. Med. Virol.* **2016**, *88*, 2106–2114. [CrossRef] [PubMed]

32. Bien, C.H.; Cai, Y.; Emch, M.E.; Parish, W.; Tucker, J.D. High adult sex ratios and risky sexual behaviors: A systematic review. *PLoS ONE* **2013**, *8*, e71580. [CrossRef] [PubMed]

33. Yu, X.M.; Guo, S.J.; Sun, Y.Y. Sexual behaviours and associated risks in chinese young people: A meta-analysis. *Sex. Health* **2013**, *10*, 424–433. [CrossRef] [PubMed]

34. Ng, K.T.; Takebe, Y.; Chook, J.B.; Chow, W.Z.; Chan, K.G.; Abed Al-Darraji, H.A.; Kamarulzaman, A.; Tee, K.K. Co-infections and transmission networks of HCV, HIV-1 and HPgV among people who inject drugs. *Sci. Rep.* **2015**, *5*, 15198. [PubMed]

35. Liu, Z.; Li, L.; Chen, Z.; Xu, M.; Zhang, T.; Jiao, Y.; Sheng, B.; Chen, D.; Wu, H. Prevalence of GB virus type C viraemia in MSM with or without HIV-1 infection in Beijing, China. *Epidemiol. Infect.* **2012**, *140*, 2199–2209. [CrossRef] [PubMed]

36. Lee, C.K.; Tang, J.W.; Chiu, L.; Loh, T.P.; Olszyna, D.; Chew, N.; Archuleta, S.; Koay, E.S. Epidemiology of GB virus type C among patients infected with HIV in Singapore. *J. Med. Virol.* **2014**, *86*, 737–744. [CrossRef] [PubMed]

37. Anggorowati, N.; Yano, Y.; Subronto, Y.W.; Utsumi, T.; Heriyanto, D.S.; Mulya, D.P.; Rinonce, H.T.; Widasari, D.I.; Lusida, M.I.; Soetjipto, Y.H.; et al. GB virus C infection in indonesian HIV-positive patients. *Microbiol. Immunol.* **2013**, *57*, 298–308. [CrossRef] [PubMed]

38. Feng, Y.; Takebe, Y.; Wei, H.; He, X.; Hsi, J.H.; Li, Z.; Xing, H.; Ruan, Y.; Yang, Y.; Li, F.; et al. Geographic origin and evolutionary history of China's two predominant HIV-1 circulating recombinant forms, CRF07_BC, CRF08_BC. *Sci. Rep.* **2016**, *6*, 19279. [CrossRef] [PubMed]

39. Chen, M.; Ma, Y.; Chen, H.; Luo, H.; Dai, J.; Song, L.; Yang, C.; Mei, J.; Yang, L.; Dong, L.; et al. Multiple introduction and naturally occuring drug resistance of HCV among HIV-infected intravenous drug users in Yunnan: An origin of China's HIV/HCV epidemics. *PLoS ONE* **2015**, *10*, e0142543. [CrossRef] [PubMed]

40. Pont, J.; Neuwald, C.; Salzner, G. Antibody prevalence of parenterally transmitted viruses (HIV-1, HTLV-i, HBV, HCV) in austrian intravenous drug users. *Infection* **1991**, *19*, 427–430. [CrossRef] [PubMed]

41. Xia, X.; Lu, L.; Tee, K.K.; Zhao, W.; Wu, J.; Yu, J.; Li, X.; Lin, Y.; Mukhtar, M.M.; Hagedorn, C.H.; et al. The unique HCV genotype distribution and the discovery of a novel subtype 6u among IDUs co-infected with HIV-1 in Yunnan, China. *J. Med. Virol.* **2008**, *80*, 1142–1152. [CrossRef] [PubMed]

42. Reilly, K.H.; Wang, J.J.; Zhu, Z.B.; Li, S.H.; Yang, T.H.; Ding, G.W.; Qian, H.Z.; Kissinger, P.; Wang, N. HIV and associated risk factors among male clients of female sex workers in a chinese border region. *Sex. Transm. Dis.* **2012**, *39*, 750–755. [CrossRef] [PubMed]

43. Wei, H.M.; Xing, H.; Hsi, J.H.; Jia, M.H.; Feng, Y.; Duan, S.; He, C.; Yao, S.T.; Ruan, Y.H.; He, X.; et al. The sexually driven epidemic in youths in China's southwestern border region was caused by dynamic emerging multiple recombinant HIV-1 strains. *Sci. Rep.* **2015**, *5*, 11323. [CrossRef] [PubMed]

44. Jia, M.; Luo, H.; Ma, Y.; Wang, N.; Smith, K.; Mei, J.; Lu, R.; Lu, J.; Fu, L.; Zhang, Q.; et al. The HIV epidemic in Yunnan province, China, 1989–2007. *J. Acquir. Immune Defic. Syndr.* **2010**, *53* (Suppl. 1), S34–S40. [CrossRef] [PubMed]

45. Xiang, J.; McLinden, J.H.; Chang, Q.; Kaufman, T.M.; Stapleton, J.T. An 85-aa segment of the GB virus type C NS5A phosphoprotein inhibits HIV-1 replication in CD4+ Jurkat T cells. *Proc. Natl. Acad. Sci. USA* **2006**, *103*, 15570–15575. [CrossRef] [PubMed]

46. Herrera, E.; Tenckhoff, S.; Gomara, M.J.; Galatola, R.; Bleda, M.J.; Gil, C.; Ercilla, G.; Gatell, J.M.; Tillmann, H.L.; Haro, I. Effect of synthetic peptides belonging to E2 envelope protein of GB virus C on human immunodeficiency virus type 1 infection. *J. Med. Chem.* **2010**, *53*, 6054–6063. [CrossRef] [PubMed]
47. Jung, S.; Eichenmuller, M.; Donhauser, N.; Neipel, F.; Engel, A.M.; Hess, G.; Fleckenstein, B.; Reil, H. HIV entry inhibition by the envelope 2 glycoprotein of GB virus C. *Aids* **2007**, *21*, 645–647. [CrossRef] [PubMed]

MDPI AG

St. Alban-Anlage 66

4052 Basel, Switzerland

Tel. +41 61 683 77 34

Fax +41 61 302 89 18

http://www.mdpi.com

Viruses Editorial Office

E-mail: viruses@mdpi.com

http://www.mdpi.com/journal/viruses

www.ingramcontent.com/pod-product-compliance
Lightning Source LLC
Chambersburg PA
CBHW051852210326
41597CB00033B/5866